An Introduction to Numerical Linear Algebra

The Prindle, Weber & Schmidt Series in Developmental, Business, and Technical Mathematics

Baley and Holstege, *Algebra: A First Course, Third Edition*
Cass and O'Connor, *Beginning Algebra*
Cass and O'Connor, *Beginning Algebra with Fundamentals*
Cass and O'Connor, *Fundamentals with Elements of Algebra: A Bridge to College Mathematics, Second Edition*
Ewen and Nelson, *Elementary Technical Mathematics, Fifth Edition*
Geltner and Peterson, *Geometry for College Students, Second Edition*
Johnston, Willis, and Hughes, *Developmental Mathematics, Third Edition*
Johnston, Willis, and Lazaris, *Essential Algebra, Sixth Edition*
Johnston, Willis, and Lazaris, *Essential Arithmetic, Sixth Edition*
Johnston, Willis, and Lazaris, *Intermediate Algebra, Fifth Edition*
Jordan and Palow, *Integrated Arithmetic and Algebra, Preliminary Edition*
Kennedy and Green, *Prealgebra for College Students*
Lee, *Self-Paced Business Mathematics, Fourth Edition*
McCready, *Business Mathematics, Sixth Edition*
McKeague, *Basic Mathematics, Third Edition*
McKeague, *Introductory Mathematics*
McKeague, *Prealgebra, Second Edition*
Mangan, *Arithmetic for Self-Study, Second Edition*
Pierce and Tebeaux, *Operational Mathematics for Business, Second Edition*
Proga, *Arithmetic and Algebra, Third Edition*
Proga, *Basic Mathematics, Third Edition*
Rogers, Haney, and Laird, *Fundamentals of Business Mathematics*
Szymanski, *Algebra Facts*
Szymanski, *Math Facts: Survival Guide to Basic Mathematics*
Weltman, Perez, Byrum, and Polito, *Beginning Algebra: A Worktext*
Weltman, Perez, Byrum, and Polito, *Intermediate Algebra: A Worktext*
Wood, Capell, and Hall, *Developmental Mathematics, Fourth Edition*

The Prindle, Weber & Schmidt Series in Precalculus, Liberal Arts Mathematics, and Teacher Training

Barnett, *Analytic Trigonometry with Applications, Fifth Edition*
Bean, Sharp, and Sharp, *Precalculus*
Boye, Kavanaugh, and Williams, *Elementary Algebra*
Boye, Kavanaugh, and Williams, *Intermediate Algebra*
Davis, Murphy, and Moran, *Precalculus in Context: Functioning in the Real World*
Drooyan and Franklin, *Intermediate Algebra, Seventh Edition*
Franklin and Drooyan, *Modeling, Functions, and Graphs*
Fraser, *Elementary Algebra*
Fraser, *Intermediate Algebra: An Early Functions Approach*
Gantner and Gantner, *Trigonometry*
Gobran, *Beginning Algebra, Fifth Edition*
Grady, Drooyan, and Beckenbach, *College Algebra, Eighth Edition*
Hall, *Beginning Algebra*
Hall, *Intermediate Algebra*
Hall, *Algebra for College Students, Second Edition*
Hall, *College Algebra with Applications, Third Edition*
Holder, *A Primer for Calculus, Sixth Edition*
Huff and Peterson, *College Algebra Activities for the TI-81 Graphics Calculator*
Johnson and Mowry, *Mathematics: A Practical Odyssey*
Kaufmann, *Elementary Algebra for College Students, Fourth Edition*

KAUFMANN, *Intermediate Algebra for College Students, Fourth Edition*
KAUFMANN, *Elementary and Intermediate Algebra: A Combined Approach*
KAUFMANN, *Algebra for College Students, Fourth Edition*
KAUFMANN, *Algebra with Trigonometry for College Students, Third Edition*
KAUFMANN, *College Algebra, Third Edition*
KAUFMANN, *Trigonometry, Second Edition*
KAUFMANN, *College Algebra and Trigonometry, Third Edition*
KAUFMANN, *Precalculus, Second Edition*
LAUGHBAUM, *Intermediate Algebra, A Graphical View, Revised Preliminary Edition*
LAVOIE, *Discovering Mathematics*
MCCOWN AND SEQUEIRA, *Patterns in Mathematics: From Counting to Chaos*
RICE AND STRANGE, *Plane Trigonometry, Sixth Edition*
RIDDLE, *Analytic Geometry, Fifth Edition*
RUUD AND SHELL, *Prelude to Calculus, Second Edition*
SGROI AND SGROI, *Mathematics for Elementary School Teachers*
SWOKOWSKI AND COLE, *Fundamentals of College Algebra, Eighth Edition*
SWOKOWSKI AND COLE, *Fundamentals of Algebra and Trigonometry, Eighth Edition*
SWOKOWSKI AND COLE, *Fundamentals of Trigonometry, Eighth Edition*
SWOKOWSKI AND COLE, *Algebra and Trigonometry with Analytic Geometry, Eighth Edition*
SWOKOWSKI AND COLE, *Precalculus: Functions and Graphs, Seventh Edition*
WELTMAN AND PEREZ, *Beginning Algebra, Second Edition*
WELTMAN AND PEREZ, *Intermediate Algebra, Third Edition*

THE PRINDLE, WEBER & SCHMIDT SERIES IN CALCULUS AND UPPER-DIVISION MATHEMATICS

ALTHOEN AND BUMCROT, *Introduction to Discrete Mathematics*
ANDRILLI AND HECKER, *Linear Algebra*
BURDEN AND FAIRES, *Numerical Analysis, Fifth Edition*
CROOKE AND RATCLIFFE, *A Guidebook to Calculus with Mathematica*
CULLEN, *An Introduction to Numerical Linear Algebra*
CULLEN, *Linear Algebra and Differential Equations, Second Edition*
DENTON AND NASBY, *Finite Mathematics, Preliminary Edition*
DICK AND PATTON, *Calculus*
DICK AND PATTON, *Single Variable Calculus*
DICK AND PATTON, *Technology in Calculus: A Sourcebook of Activities*
EDGAR, *A First Course in Number Theory*
EVES, *In Mathematical Circles*
EVES, *Mathematical Circles Revisited*
EVES, *Mathematical Circles Squared*
EVES, *Return to Mathematical Circles*
FAIRES AND BURDEN, *Numerical Methods*
FINIZIO AND LADAS, *Introduction to Differential Equations*
FINIZIO AND LADAS, *Ordinary Differential Equations with Modern Applications, Third Edition*
FLETCHER, HOYLE, AND PATTY, *Foundations of Discrete Mathematics*
FLETCHER AND PATTY, *Foundations of Higher Mathematics, Second Edition*
GILBERT AND GILBERT, *Elements of Modern Algebra, Third Edition*
GORDON, *Calculus and the Computer*
HARTFIEL AND HOBBS, *Elementary Linear Algebra*
HARTIG, *Guidebook to Linear Algebra for Theorist*
HILL, ELLIS, AND LODI, *Calculus Illustrated*
HILLMAN AND ALEXANDERSON, *Abstract Algebra: A First Undergraduate Course, Fifth Edition*
HUMI AND MILLER, *Boundary-Value Problems and Partial Differential Equations*

LAUFER, *Discrete Mathematics and Applied Modern Algebra*
LEINBACH, *Calculus Laboratories Using Derive*
MARON AND LOPEZ, *Numerical Analysis, Third Edition*
MIECH, *Calculus with Mathcad*
MIZRAHI AND SULLIVAN, *Calculus with Analytic Geometry, Third Edition*
MOLLUZZO AND BUCKLEY, *A First Course in Discrete Mathematics*
NICHOLSON, *Elementary Linear Algebra with Applications, Third Edition*
NICHOLSON, *Introduction to Abstract Algebra*
O'NEIL, *Advanced Engineering Mathematics, Third Edition*
PENCE, *Calculus Activities for Graphic Calculators*
PENCE, *Calculus Activities for the TI Graphic Calculator, Second Edition*
PLYBON, *An Introduction to Applied Numerical Analysis*
POWERS, *Elementary Differential Equations with Boundary-Value Problems*
POWERS, *Elementary Differential Equations with Linear Algebra*
PRESCIENCE CORPORATION, *The Student Edition of Theorist*
RIDDLE, *Calculus and Analytic Geometry, Fourth Edition*
SCHELIN AND BANGE, *Mathematical Analysis for Business and Economics, Second Edition*
SENTILLES, *Applying Calculus in Economics and Life Science*
SWOKOWSKI, OLINICK, AND PENCE, *Calculus, Sixth Edition*
SWOKOWSKI, OLINICK, AND PENCE, *Calculus of a Single Variable, Second Edition*
SWOKOWSKI, *Calculus, Fifth Edition (Late Trigonometry Version)*
SWOKOWSKI, *Elements of Calculus with Analytic Geometry: High School Edition*
TAN, *Applied Finite Mathematics, Fourth Edition*
TAN, *Calculus for the Managerial, Life, and Social Sciences, Third Edition*
TAN, *Applied Calculus, Third Edition*
TAN, *College Mathematics, Third Edition*
TRIM, *Applied Partial Differential Equations*
VENIT AND BISHOP, *Elementary Linear Algebra, Third Edition*
VENIT AND BISHOP, *Elementary Linear Algebra, Alternate Second Edition*
WATTENBERG, *Calculus in a Real and Complex World*
WIGGINS, *Problem Solver for Finite Mathematics and Calculus*
ZILL, *Calculus, Third Edition*
ZILL, *A First Course in Differential Equations, Fifth Edition*
ZILL AND CULLEN, *Differential Equations with Boundary-Value Problems, Third Edition*
ZILL AND CULLEN, *Advanced Engineering Mathematics*

THE PRINDLE, WEBER & SCHMIDT SERIES IN ADVANCED MATHEMATICS

EHRLICH, *Fundamental Concepts of Abstract Algebra*
EVES, *Foundations and Fundamental Concepts of Mathematics, Third Edition*
JUDSON, *Abstract Algebra: Theory and Applications*
KEISLER, *Elementary Calculus: An Infinitesimal Approach, Second Edition*
KIRKWOOD, *An Introduction to Analysis*
PATTY, *Foundations of Topology*
RUCKLE, *Modern Analysis: Measure Theory and Functional Analysis with Applications*
SIERADSKI, *An Introduction to Topology and Homotopy*
STEINBERGER, *Algebra*
STRAYER, *Elementary Number Theory*
TROUTMAN, *Boundary-Value Problems of Applied Mathematics*

An Introduction to Numerical Linear Algebra

CHARLES G. CULLEN
University of Pittsburgh

PWS PUBLISHING COMPANY
BOSTON

PWS Publishing Company

Editor: Steve Quigley
Production Editor: Abigail M. Heim
Manufacturing Coordinator: Lisa M. Flanagan
Interior and Cover Design: Abigail M. Heim
Composition and Illustration: The Universities Press (Belfast) Ltd.
Cover Printer: New England Book Components, Inc.
Text Printer and Binder: Arcata Graphics Corporation/Martinsburg

ITP™

International Thomson Publishing
The trademark ITP is used under license

PWS Publishing Company is a division of Wadsworth, Inc.

Printed in the United States of America
1 2 3 4 5 6 7 8 9—98 97 96 95 94

This book is printed on acid-free, recycled paper.

Library of Congress Cataloging-in-Publication Data

Cullen, Charles G.
An introduction to numerical linear algebra / Charles G. Cullen
p. cm.
Includes bibliographical references and index.
ISBN 0-534-93690-3 (alk. paper)
1. Algebras, Linear—Data processing. 2. Numerical calculations—Data processing. I. Title
QA185.D37C85 1994 93-20737
512′.5′0285—dc20 CIP

Contents

CHAPTER 4

Transformation Methods for the Eigenvalue Problem 118

CHAPTER 5

The Linear Least Squares Problem 162

CHAPTER 6

Iterative Methods 183

Appendices

Preface

An Introduction to Numerical Linear Algebra discusses methods for using a computer to solve the standard problems of linear algebra. This text is intended for use in a one-term, upper-division course for mathematics, computer science, or engineering students. The text presupposes a one-term introductory course in linear algebra, some mathematical maturity (most theorems are proven), and some computer experience. There is a brief review of elementary linear algebra in Chapter 1; more advanced topics are reviewed as the need arises.

INTERACTIVE LINEAR ALGEBRA SOFTWARE

This book has a strong experimental flavor that is especially evident in the exercises. It is therefore important that the student have access to some interactive linear algebra software: MATLAB is the preferred choice. As an alternative to MATLAB, however, MATALG (The MATrix ALGebraist), an instructional program developed at the University of Pittsburgh by John Burkhardt and the author, is included with this text. The version provided runs on IBM PCs and compatibles; a Macintosh version is available from the author (c/o University of Pittsburgh, Dept. of Mathematics and Statistics, Pittsburgh, PA 15260). Because it was designed as a teaching aid rather than as a tool for practicing scientists, the MATALG program is less powerful than MATLAB and its matrices can be no larger than 12×12. But MATALG, like MATLAB, uses state-of-the-art techniques for doing matrix calculations. All of the calculations in the text were done using one or the other of these two programs.

Many sections end with some software notes. These notes point out features of MATLAB and MATALG that are relevant to the material discussed in the section and that will be useful in the exercises of that section.

You should be alert to the fact that the same program, running on different machines, can produce quite different results. This is due to differences in architecture and in the design of the arithmetic units of the various machines. Different programs implementing the same algorithm also can produce slightly different results, even on the same machine.

CONTENTS

Chapter 1 contains a brief review of elementary linear algebra, as well as a short description of machine arithmetic, an introduction to round-off error, and some examples of pitfalls in machine calculations.

Chapter 2 deals with direct methods for solving the linear equations problem. Several matrix factorizations are introduced, as are notions of matrix and vector norms. Some results about error analysis for Gaussian elimination with partial pivoting are included, and the special case of positive definite matrices is discussed. An opportunity to work with the LINPACK package is provided.

Chapter 3 reviews the basic theory of eigenvalues and eigenvectors and introduces the Gerschgorin circle theorem, using it to discuss some perturbation theory for the eigenvalue problem. The power method and inverse power method are presented and various special cases are described.

Chapters 4, 5, and 6 are essentially independent and can be covered in any order after the first three chapters have been completed.

Chapter 4 discusses the QR method for the eigenvalue problem. Both the Householder and the Given's methods for the QR factorization and the reduction to Hessenberg form are included. A brief introduction to the implicitly shifted double QR algorithm is included and an opportunity to explore the EISPACK package is provided.

Chapter 5 presents the linear least squares problem and some applications. This chapter also includes the singular value decomposition.

Chapter 6 discusses iterative methods for solving the linear equations problem, including the Jacobi, Gauss-Seidel, SOR, and conjugate gradient methods.

There are five appendices describing software: MATLAB, MATALG, LINPACK, EISPACK, and the subroutines DECOMP and SOLVE. There is also an appendix that contains an interactive FORTRAN driver that can be used to enter the data, call the appropriate subroutines, and output the results. This program may require minor modifications to run on a particular mainframe. If LINPACK and EISPACK are not available, a few exercises will need to be omitted.

Exercises

The exercises form an important part of the text. There are many computational exercises designed to illustrate the theory and to provide experience with the software. Using the software should help to emphasize the interpretation of the calculations rather than the calculations themselves. The answers to most computational problems are included at the end of the text. Most of the results in the answers section were obtained using MATLAB.

There are also a goodly number of theoretical exercises, designed to either complete or extend the theory. The answers section contains some solutions and many hints for these exercises.

References and Notation

Equations, definitions, theorems, figures, and tables each have a two-part number of the form $(c.\,n)$ where c is the chapter number. Examples are single-numbered consecutively within each chapter. References to exercises include the section number only if the cited exercise is not in the current section. Algorithms are numbered consecutively along with equations. (Thus Algorithm (3.5) appears in the text between Equation (3.4) and Equation (3.6).)

Algorithms are presented in the text in a pseudo-code that can be used to develop code in FORTRAN, Pascal, or some other programming language. This pseudo-code is frequently very similar to the MATLAB instructions that would implement the algorithm.

The notation used is quite standard. Matrices are denoted by capital italic letters $(A, J, I, 0)$ and matrix-valued functions have names that begin with a capital letter (Dg{. . .}). Scalar-valued functions of scalars or matrices have names that begin with lowercase letters, [det(A), tr(A)]. Vectors are denoted by matrix notation or by lowercase boldface roman letters, $(X, B, \mathbf{e}, \mathbf{x}, \mathbf{b})$. Sets normally are denoted by capital script letters, $[\mathcal{R}, \mathcal{C}, \mathcal{R}^n, \mathcal{C}^n, \mathcal{R}^{n\times n}, \mathcal{C}^{n\times n}, \mathcal{N}\mathcal{S}(A)]$. When discussing sequences of vectors or matrices, we use superscripts to designate the sequence number $[a_{ij}^{(k)}, x_i^{(k)}]$.

ACKNOWLEDGMENTS

I have many people to thank for their help and encouragement on this project. First on the list are the Math 1080 students who worked through the three preliminary drafts. Their suggestions were invaluable and their willing participation is much appreciated. Professor Werner C. Rheinboldt, Andrew Mellon Professor of Mathematics at the University of Pittsburgh, encouraged me to undertake this project and I have benefited greatly from discussing the project with him. The Mathematics and Statistics Department at the University of Pittsburgh (Stuart Hastings, chairman) was generous in both its consent and support. The semifinal version of the manuscript was prepared while I was on sabbatical leave at the University of Texas at Austin. Professor David Young and his colleagues at the Center for Numerical Analysis were most hospitable and provided me with a pleasant and stimulating environment in which to work. Professor Larry Snyder checked the answers and taught me several new MATLAB tricks. I am also indebted to the reviewers of the preliminary versions. Their suggestions were all carefully weighed and greatly influenced the final form of the text. The reviewers include Robert D. Adams, The University of Kansas; Robert D. Grone, San Diego State University; and Larry Snyder, Ohio University. The PWS editorial and production staff has been most helpful, encouraging, and cooperative. Many thanks to Steve Quigley, John Ward, and Abby Heim.

Charles G. Cullen

An Introduction to Numerical Linear Algebra

Introduction

This text discusses methods for solving the standard problems of linear algebra on a computer. It is assumed that the reader is familiar with the contents of some elementary text on linear algebra and has had some experience with a digital computer. This text has a substantial experimental component. To participate in the experiments without doing unreasonable amounts of arithmetic by hand, you will need to learn to use either MATLAB or MATALG. If you already know how to use the symbolic manipulator MATHEMATICA (Wolfram, 1988), you may wish to use it to do your experimental work.

MATLAB is a popular interactive matrix-manipulation package. It uses state-of-the-art techniques for doing all kinds of matrix calculations. The program is available for a variety of machines and is extremely versatile and powerful. If the program is available to you, it should be your choice as a computational tool while reading this text, even though we will not use all of its features. MATLAB is a powerful computational tool that will be useful to you not only in studying this text but in a wide range of other applications as well. Although the professional version of this program is quite expensive, a student version is available at moderate cost. There are versions that run on either PCs or Macs (The Math Works, 1992).

MATALG (The MATrix ALGebraist) is an instructional program developed at the University of Pittsburgh by John Burkardt and the author. It is included with this text. The version provided runs on IBM PCs and compatibles. The MATALG program is less powerful than MATLAB and is limited in scope (matrices can be no larger than 12×12) because it was designed as a teaching aid rather than as a tool for practicing scientists. MATALG, like MATLAB, uses state-of-the-art techniques for doing matrix calculations. A Mac version of this program is available from the author for $5.00 (duplication, postage, and handling).

The MATALG disk also contains the MATMAN (The MATrix MANipulator) program. This program is an interactive row-reduction program that will operate in either real or rational arithmetic. Elementary linear algebra students can use it as can the intended audience for this text. MATMAN also contains a tableau implementation of the simplex method for linear programming.

All three programs have on-line help facilities, are menu driven, and are easy to use. Most students will be able to get started with MATLAB or MATALG in a very short time. Once the initial entry is accomplished, you will learn the program as you use it.

MATHEMATICA is a combined numerical and symbolic manipulator frequently used to help teach calculus. This program contains most of the matrix functions you will encounter in this text. MATHEMATICA treats a matrix as a list of vectors and evaluates functions of matrices and vectors element by element. Unless told to do otherwise, MATHEMATICA will attempt to solve many problems exactly; it will do rational arithmetic with exact data. This can lead to very complicated algebraic constructions.

MATHEMATICA has excellent programming ability and outstanding graphics capabilities. If you are already familiar with this program, it is reasonable to use it as the computational tool of choice while working through this text. Be sure the Linear Algebra Package is loaded. If you do not already know MATHEMATICA, you will be better off with either MATLAB or MATALG.

The following table compares the commands for some familiar matrix operations using MATLAB, MATALG/MATMAN, or MATHEMATICA. See Appendices A and B or the on-line help files for more details.

COMMON COMMANDS IN MATLAB, MATALG, AND MATHEMATICA

	MATLAB	MATALG/MATMAN	MATHEMATICA
Transcript file	diary filename	D, filename	Notebooks (FED)
Case sensitive?	YES $A \neq a$ (etc.; can be reset)	NO $A = a$ (etc.)	YES
Enter the matrix $A = \begin{bmatrix} 1 & 2 & 3 & 4 & 5 \\ 6 & 7 & 8 & 9 & 10 \\ 11 & 12 & 13 & 14 & 15 \end{bmatrix}$	A=[1,2,3,4,5;6,7,8,9,10; 11,12,13,14,15] (Rows can also be separated by CR.)	E,A,3,5 1,2,3,4,5 6,7,8,9,10 11,12,13,14,15	A={{1, 2, 3, 4, 5},{6, 7, 8, 9, 10}, {11, 12, 13, 14, 15}} (The Table and Array commands can also be used to set up matrices.)
Display the matrix A	A	=A or T, A	A or MatrixForm [A]
$\det(A)$	det(A)	=det(A)	Det[A]
$B = A^T$	B=A´	B=trans(A)	B=Transpose[A]
A^{-1}	inv(A)	=inv(A)	Inverse[A]
$C = AB$	C=A*B	C=A*B	C=A.B
I	eye(n) or eye(A)	=Id(n)	IdentityMatrix[n]
$S = 2A + 3B + 5I$	S=2*A+3*B+5*eye(A)	S=2*A+3*B+5	S=2A+3B+ 5 IdentityMatrix[n]
Solve $AX = B$	X=A\B or rref([A, B])	X=inv(A)*B or use MATMAN	LinearSolve[A, B]
$p(A)$ if the coefficients of p are in v	polyvalm(v, A)	=poly(v, A)	Sum[v[[i]]MatrixPower[A,i–1], {i,Length[v]}]
$\det(\lambda I - A)$	poly(A)	P, A	Det[A– x IdentityMatrix[n]]
Eigenvalues of A	roots(poly(A)) or eig(A)	=eval(A)	Eigenvalues[N[A]]
Eigenvectors of A	[V,D]=eig(A)	V=evec(A)	V=Transpose [Eigenvectors[A]]
$\lVert x \rVert_1$	norm(x,1)	=norm1(x)	Sum[Abs[x[[i]]],{i,Length[x]}]
$\lVert x \rVert_2$	norm(x)	=norm2(x)	Sqrt[Sum[x[[i]]^2,{i,Length[x]}]]
$\lVert x \rVert_\infty$	norm(x,inf)	=norm0(x)	Max[x]
LU factorization of A	[L,U,P]=lu(A)	M,LU,A	LUFactor[A]
QR factorization of A	[Q,R]=qr(A)	M,QR,A	{Q,R}=QRDecomposition[A]
SVD factorization of A	[U S V]=svd(A)	M, SVD, A	{U, S, V}=SingularValues[N[A]]
ERO $R_2 \leftarrow k{*}R_1 + R_2$	A(2,:)=k*A(1,:)+A(2,:)	MATMAN A k 1 2	A[[2]]=kA[[1]]+A[[2]]
ERO $R_3 \leftarrow k{*}R_3$	A(3,:)=k*A(3,:)	MATMAN M 3 k	A[[3]]=kA[[3]]
ERO $R_2 \leftrightarrow R_3$	T=A(2,:) A(2,:)=A(3,:) A(3,:)=T	MATMAN I 2 3	T=A[[2]] A[[2]]=A[[3]] A[[3]]=T
Add comments to file	begin line with %	N command	Yes (FED)
On-line help	help or help "command"	H or ?	?Name or ??Name
Precision	all calculations done in double precision	single precision	16 decimal digits
Programmable	yes	no	yes
Matrix size	no limit (32×32 on Student Edition)	no bigger than 12×12	no limit
Data entry from file	load "filename"	X "filename"	<< filename.m or Readlist[filename.m]
Data storage to file	save "filename"	W "filename"	x >> filename.m or Save[filename.m,list]
Graphs	both 2-D and 3-D capability	none	both 2-D and 3-D capability

(FED) means Front End dependent.

1

Preliminaries

The basic problems of numerical linear algebra include:

1. *Linear algebraic equations*: Given the $m \times n$ matrix A and the $m \times 1$ matrix B, find all $n \times 1$ matrices X such that $AX = B$.

2. *The algebraic eigenvalue problem*: Given the $n \times n$ matrix A, find all scalars λ and all vectors $X \neq 0$ such that $AX = \lambda X$.

3. *The linear least-squares problem*: Given the $m \times n$ matrix A, with $m > n$, and the $m \times 1$ matrix B, find the $n \times 1$ matrix X that most nearly satisfies $AX = B$.

Problems 1 and 2, at least, are treated in every beginning linear algebra course. But in beginning linear algebra, the methods presented, although essential for a fundamental understanding of the problems and the nature of their solutions, are not always appropriate methods for solving problems of these types on a modern digital computer. In this course we will develop and analyze efficient techniques for solving these three problems on a computer.

We will begin with a brief review of elementary matrix theory and a discussion of computer arithmetic. More technical results from linear algebra will be introduced or reviewed as needed.

1.1 ■ REVIEW OF MATRIX THEORY

Let A be an $m \times n$ matrix and let B and C be $n \times t$ matrices. The **matrix product** $P = AB$ is the $m \times t$ matrix determined by

$$p_{ij} = \text{Row}_i(A)\,\text{Col}_j(B) = \sum_{k=1}^{n} a_{ik}b_{kj}; \quad \text{for } i = 1, 2, \ldots, m \text{ and } j = 1, 2, \ldots, t.$$

The product is defined only if the number of rows of B is equal to the number of columns of A.

The **matrix sum** $S = B + C$ is the $n \times t$ matrix defined by

$$s_{ij} = b_{ij} + c_{ij}; \quad \text{for } i = 1, 2, \ldots, n \text{ and } j = 1, 2, \ldots, t.$$

The sum is undefined if the matrices are not the same size.

The **scalar multiple** $M = kA$ is the $m \times n$ matrix defined by

$$m_{ij} = ka_{ij}; \quad \text{for } i = 1, 2, \ldots, m \text{ and } j = 1, 2, \ldots, n.$$

Matrix arithmetic has many properties in common with scalar arithmetic, but there are some important differences of which we must be constantly aware. The major differences are:

1. Matrix operations are not always defined.
2. Matrix multiplication is not commutative, that is, generally $AB \neq BA$.
3. $AB = 0$ does not imply that $A = 0$ or $B = 0$ (nonzero matrix divisors of 0 exist).

In the following table we list five familiar properties of scalar arithmetic and indicate the status of each of these properties for the matrix operations just defined. It is assumed that all the matrices are of the proper size so that the indicated operations are defined.

	Matrix Multiplication	Matrix Addition	Scalar Multiplication
Associativity	$A(BC) = (AB)C$	$A + (B + C) = (A + B) + C$	$(kA)B = A(kB) = k(AB)$
Commutativity	(Not true)	$A + B = B + A$	$kA = Ak$
Identity element	$AI_n = I_mA = A$	$A + 0 = A$	$1A = A$
Inverse elements	(Not always)	$A + (-A) = 0$	$(1/k)(kA) = A$
Distributivity	$A(B + C) = AB + AC$	$(B + C)A = BA + CA$	$k(A + B) = kA + kB$ $(h + k)A = hA + kA$

The **multiplicative identity element** I_n is the $n \times n$ diagonal matrix, each of whose diagonal entries is $+1$; the subscript is normally omitted if the size is clear from the context. The **additive identity element** 0 is the appropriate sized matrix, all of whose entries are 0; the size should always be clear from the context.

Only square matrices can have multiplicative inverses. The fact that nonzero matrix divisors of 0 exist ($AB = 0$ with $A \neq 0$ and $B \neq 0$) is logically equivalent to the fact that not all matrices have multiplicative inverses. If the $n \times n$ matrix A has a multiplicative inverse, then it is the unique matrix A^{-1} having the property that

$$AA^{-1} = A^{-1}A = I.$$

The matrix A is said to be **nonsingular** or **invertible** if A^{-1} exists, and **singular** if A^{-1} does not exist.

If the coefficient matrix A is nonsingular, then the first of our problems is theoretically easy; in this case the unique solution of $AX = B$ is $X = A^{-1}B$. If the coefficient matrix is singular, then there may be either an infinite number of solutions or no solution at all.

The basic method for solving the linear system $AX = B$ is called **Gaussian elimination.** It involves simplifying the augmented matrix $[A \mid B]$ via a finite sequence of **elementary row operations** of the following types:

Type I. $R_i \leftarrow kR_i$: Each element of row i is multiplied by $k \neq 0$.

Type II. $R_i \leftrightarrow R_j$: Rows i and j are interchanged.

Type III. $R_i \leftarrow kR_j + R_i$: Row i is replaced by itself plus k times row j.

Since each of these operations leaves the solutions of the system fixed, we can solve the system by reducing the original system to an equivalent system for which the solution is easy. The solution is easy if the augmented matrix is upper-triangular and even easier if the augmented matrix is diagonal. You normally proceed from left to right, simplifying one column at a time. The arithmetic involved in this solution process can become a burden if the system is large or if the numbers are not integers. However, MATMAN is an interactive program that can help in handling the details of the reduction in a way that should minimize the frustration associated with minor arithmetic indiscretions. You can set this program to do the row reduction automatically, using either real or rational arithmetic. You can also use the MATLAB program to do elementary row operations on matrices in an interactive mode. MATLAB also has an automatic row reducer but works only in real arithmetic. David Hill has written a MATLAB instructional m-file that allows you to use MATLAB to do row reductions in rational arithmetic (Kolman and Hill, 1993).

To find the inverse of the matrix A, you need only solve the matrix equation $AX = I$ by row reducing the augmented matrix $[A \mid I]$ to $[I \mid A^{-1}]$. We will now illustrate this procedure by finding the inverse of a 4×4 matrix.

■ EXAMPLE 1

Find the inverse of the matrix $A = \begin{bmatrix} 1 & -3 & 0 & -2 \\ 3 & -12 & -2 & -6 \\ -1 & 6 & 1 & 3 \\ -2 & 10 & 2 & 5 \end{bmatrix}$.

SOLUTION We first simplify the first column of the augmented matrix $[A \mid I]$ using the row operations $(R_2 \leftarrow -3R_1 + R_2)(R_3 \leftarrow 1R_1 + R_3)$, $(R_4 \leftarrow 2R_1 + R_4)$. The corresponding MATMAN commands are

(A,−3,1,2), (A,1,1,3), (A,2,1,4).

This step of the reduction could be accomplished in MATLAB by setting $A = [A \quad I]$ and then using the commands

```
A(2,:) = - 3*A(1,:)+A(2,:),    A(3,:) = A(1,:)+A(3,:),    A(4,:) = 2*A(1,:)+A(4,:).
```

These operations yield:

$$[A \mid I] = \left[\begin{array}{rrrr|rrrr} 1 & -3 & 0 & -2 & 1 & 0 & 0 & 0 \\ 3 & -12 & -2 & -6 & 0 & 1 & 0 & 0 \\ -1 & 6 & 1 & 3 & 0 & 0 & 1 & 0 \\ -2 & 10 & 2 & 5 & 0 & 0 & 0 & 1 \end{array}\right]$$

$$\rightarrow \left[\begin{array}{rrrr|rrrr} 1 & -3 & 0 & -2 & 1 & 0 & 0 & 0 \\ 0 & -3 & -2 & 0 & -3 & 1 & 0 & 0 \\ 0 & 3 & 1 & 1 & 1 & 0 & 1 & 0 \\ 0 & 4 & 2 & 1 & 2 & 0 & 0 & 1 \end{array}\right].$$

The second column is now reduced using the operations $(R_2 \leftarrow (-1/3)R_2)$, $(R_1 \leftarrow 3R_2 + R_1)$, $(R_3 \leftarrow -3R_2 + R_3)$, and $(R_4 \leftarrow -4R_2 + R_4)$; the corresponding MATMAN commands are **(M,2,-1/3), (A,3,2,1), (A,-3,2,3), (A,-4,2,4)**. These operations yield

$$\left[\begin{array}{rrrr|rrrr} 1 & 0 & 2 & -2 & 4 & -1 & 0 & 0 \\ 0 & 1 & \frac{2}{3} & 0 & 1 & -\frac{1}{3} & 0 & 0 \\ 0 & 0 & -1 & 1 & -2 & 1 & 1 & 0 \\ 0 & 0 & -\frac{2}{3} & 1 & -2 & \frac{4}{3} & 0 & 1 \end{array}\right] \rightarrow \left[\begin{array}{rrrr|rrrr} 1 & 0 & 0 & 0 & 0 & 1 & 2 & 0 \\ 0 & 1 & 0 & \frac{2}{3} & -\frac{1}{3} & \frac{1}{3} & \frac{2}{3} & 0 \\ 0 & 0 & 1 & -1 & 2 & -1 & -1 & 0 \\ 0 & 0 & 0 & \frac{1}{3} & -\frac{2}{3} & \frac{2}{3} & -\frac{2}{3} & 1 \end{array}\right]$$

where the second matrix in the previous line is obtained from the first, using the operations $(R_3 \leftarrow -R_3)$, $(R_2 \leftarrow (-2/3)R_3 + R_2)$, $(R_1 \leftarrow -2R_3 + R_1)$, and $(R_4 \leftarrow (2/3)R_3 + R_4)$. The operations $(R_4 \leftarrow 3R_4)$, $(R_3 \leftarrow R_4 + R_3)$, and $(R_2 \leftarrow (-2/3)R_4 + R_2)$ then yield

$$\left[\begin{array}{rrrr|rrrr} 1 & 0 & 0 & 0 & 0 & 1 & 2 & 0 \\ 0 & 1 & 0 & 0 & 1 & -1 & 2 & -2 \\ 0 & 0 & 1 & 0 & 0 & 1 & -3 & 3 \\ 0 & 0 & 0 & 1 & -2 & 2 & -2 & 3 \end{array}\right] = [I \mid A^{-1}].$$

■

The relation between matrix multiplication and elementary row operations will be important in what follows. Recall that if $\mathscr{E}$ is an elementary row operation, then

$$\mathscr{E}(A) = \mathscr{E}(I)A.$$

The matrices $\mathscr{E}(I)$ are called **elementary matrices**. For example, it is easy to verify the following special cases of the above result:

$$\begin{bmatrix} 1 & 0 & 0 \\ 0 & 0 & 1 \\ 0 & 1 & 0 \end{bmatrix} \begin{bmatrix} a & b & c \\ d & e & f \\ g & h & i \end{bmatrix} = \begin{bmatrix} a & b & c \\ g & h & i \\ d & e & f \end{bmatrix}$$

and

$$\begin{bmatrix} 1 & 0 & 0 \\ 0 & 1 & 0 \\ k & 0 & 1 \end{bmatrix} \begin{bmatrix} a & b & c \\ d & e & f \\ g & h & i \end{bmatrix} = \begin{bmatrix} a & b & c \\ d & e & f \\ ka + g & kb + h & kc + i \end{bmatrix}.$$

Elementary row operations are also the most efficient way of evaluating the determinant function on a square matrix. The **determinant function** is defined inductively as follows: If A is 1×1, then $\det(A) = a_{11}$. Otherwise,

$$\det(A) \equiv \sum_{j=1}^{n} a_{1j}(-1)^{1+j} \det(Min_{1j}(A))$$

where $Min_{ij}(A)$ is the $n-1 \times n-1$ submatrix of A obtained by deleting row i and column j of A.

It is shown in every elementary linear algebra text that

1. $\det(A^T) = \det(A)$.
2. $\det(T) = t_{11}t_{22} \cdots t_{nn}$ if T is a triangular matrix.
3. If B is obtained from A by the operation $(R_i \leftarrow (1/k)R_i)$, then $\det(A) = k \det(B)$.
4. If B is obtained from A by the operation $(R_i \leftrightarrow R_j)$, then $\det(A) = -\det(B)$.
5. If B is obtained from A by the operation $(R_i \leftarrow kR_j + R_i)$, then $\det(A) = \det(B)$.
6. $\det(AB) = \det(A)\det(B)$.

The most efficient way to evaluate $\det(A)$ is to use elementary row operations to reduce A to a triangular matrix T and then evaluate $\det(T)$. For the matrix of Example 1 it follows that

$$\det(A) = \det \begin{bmatrix} 1 & -3 & 0 & -2 \\ 0 & -3 & -2 & 0 \\ 0 & 3 & 1 & 1 \\ 0 & 4 & 2 & 1 \end{bmatrix} = -3 \det \begin{bmatrix} 1 & -3 & 0 & -2 \\ 0 & 1 & \frac{2}{3} & 0 \\ 0 & 0 & -1 & 1 \\ 0 & 0 & -\frac{2}{3} & 1 \end{bmatrix}$$

$$= -3 \det \begin{bmatrix} 1 & -3 & 0 & -2 \\ 0 & 1 & \frac{2}{3} & 0 \\ 0 & 0 & -1 & 1 \\ 0 & 0 & 0 & \frac{1}{3} \end{bmatrix} = (-3)(-\tfrac{1}{3}) = 1\,.$$

The basic object of study in a theoretical linear algebra course is a vector space $\mathscr{V}$ over a field $\mathscr{F}$. We will restrict our attention here to the vector spaces $\mathscr{R}^{m\times n} = \{m \times n \text{ real matrices}\}$ and $\mathscr{C}^{m\times n} = \{m \times n \text{ complex matrices}\}$. The special cases $\mathscr{R}^n = \mathscr{R}^{n\times 1}$ and $\mathscr{C}^n = \mathscr{C}^{n\times 1}$ will occur frequently. Recall that a subset $\mathscr{W} \subseteq \mathscr{V}$ is itself a vector space if and only if it is nonempty and is closed under arbitrary linear combinations. In this case $\mathscr{W}$ is called a **subspace** of $\mathscr{V}$.

A set of vectors $\{\mathbf{v}_1, \mathbf{v}_2, \ldots, \mathbf{v}_k\}$ from a subspace $\mathscr{W}$ is said to **span** $\mathscr{W}$ if every vector $\mathbf{w}$ in $\mathscr{W}$ can be written as a linear combination of the $\mathbf{v}_i$; that is, there exist scalars x_i such that $\mathbf{w} = x_1\mathbf{v}_1 + x_2\mathbf{v}_2 + \cdots + x_k\mathbf{v}_k$. The set of vectors $\{\mathbf{v}_1, \mathbf{v}_2, \ldots, \mathbf{v}_k\}$ is said to be **linearly independent** if

$$x_1\mathbf{v}_1 + x_2\mathbf{v}_2 + \cdots + x_k\mathbf{v}_k = 0 \quad \text{only if all } x_i = 0$$

and is said to be **linearly dependent** if there exist scalars x_i, not all zero, such that

$$x_1\mathbf{v}_1 + x_2\mathbf{v}_2 + \cdots + x_k\mathbf{v}_k = 0.$$

A **basis** for a vector space $\mathscr{V}$ is a linearly independent spanning set. If $\{\mathbf{v}_1, \mathbf{v}_2, \ldots, \mathbf{v}_n\}$ is a basis for $\mathscr{V}$, then it follows that every vector $\mathbf{w}$ in $\mathscr{V}$ is uniquely expressible as a linear combination of the basis vectors: $\mathbf{w} = x_1\mathbf{v}_1 + x_2\mathbf{v}_2 + \cdots + x_n\mathbf{v}_n$. The **dimension** of a vector space $\mathscr{V}$ is the number of vectors in any basis for $\mathscr{V}$.

It is frequently useful to think of an $m \times n$ matrix A as defining a linear transformation from the vector space $\mathscr{R}^n = \mathscr{R}^{n\times 1}$ to the vector space $\mathscr{R}^m = \mathscr{R}^{m\times 1}$. There are several subspaces associated with the matrix A that will be of interest to us. The **null space of A** is defined by $\mathscr{N}\mathscr{S}(A) = \{X \text{ in } \mathscr{R}^n : AX = 0\}$. This subspace of $\mathscr{R}^n$ consists of all the solutions of the homogeneous system $AX = 0$. The **column space of A** is the subspace of $\mathscr{R}^m$ spanned by the columns of A; $\mathscr{C}\mathscr{S}(A) = \{AX : X \text{ in } \mathscr{R}^n\}$. The column space of A is also called the **range of A** and its dimension is called the **rank of A**. The **row space of A** is the subspace of $\mathscr{R}^{1\times n}$ spanned by the rows of A; $\mathscr{R}\mathscr{S}(A) = \{YA : Y \text{ in } \mathscr{R}^{1\times m}\}$. The dimension of the row space of A is also equal to the rank of A. A very important result from elementary linear algebra is that, for any $m \times n$ matrix A,

$$\text{rank}(A) + \dim(\mathscr{N}\mathscr{S}(A)) = n. \tag{1.1}$$

This result is known as Sylvester's Law of Nullity. All of the above results are still true if the real field $\mathscr{R}$ is replaced by the complex field $\mathscr{C}$. An interesting and useful corollary of this result is contained in Theorem 1.1. The matrix $A^H = \bar{A}^T$ is called the **Hermitian conjugate** of A.

THEOREM 1.1

For any $m \times n$ complex matrix A, A and A^HA have the same rank.

PROOF If X is in $\mathscr{N}\mathscr{S}(A)$, then $AX = 0$ and $A^HAX = 0$, so that X is in $\mathscr{N}\mathscr{S}(A^HA)$. This shows that $\mathscr{N}\mathscr{S}(A) \subseteq \mathscr{N}\mathscr{S}(A^HA)$. If Y is in $\mathscr{N}\mathscr{S}(A^HA)$, then $A^HAY = 0$ and $Y^HA^HAY = (AY)^H(AY) = 0$. It follows that $AY = 0$ and Y is in $\mathscr{N}\mathscr{S}(A)$. Therefore $\mathscr{N}\mathscr{S}(A^HA) \subseteq \mathscr{N}\mathscr{S}(A)$, and it follows that $\mathscr{N}\mathscr{S}(A) = \mathscr{N}\mathscr{S}(A^HA)$. Since A and A^HA have the same number of columns, it follows from (1.1) that $\text{rank}(A^HA) = n - \dim(\mathscr{N}\mathscr{S}(A^HA)) = n - \dim(\mathscr{N}\mathscr{S}(A)) = \text{rank}(A)$. ■

The following theorem from elementary linear algebra lists several conditions equivalent to the invertibility of a square matrix A.

THEOREM 1.2

For any $n \times n$ matrix A the following are equivalent:

1. A^{-1} exists.
2. A has either a right or a left inverse.
3. $AX = 0$ implies that $X = 0$.
4. A is row equivalent to I.
5. A is a product of elementary matrices.
6. $\det(A) \neq 0$.
7. $\text{rank}(A) = n$.
8. The rows of A are linearly independent.
9. The columns of A are linearly independent.
10. $AX = B$ has a unique solution for any $n \times s$ matrix B. ■

Matrix operations are not difficult to program for a digital computer and in fact they are often used as exercises in beginning programming courses. The most efficient way to proceed is to construct subroutines for each of the matrix operations. These subroutines should work for different-sized matrices and should be free of input and output statements. Standard programming practice is to define each array in the calling program and then pass the relevant dimension information to the subroutine as arguments.

If the matrices you are considering are all smaller than 50×100, then the calling program dimensions the arrays as 50×100. If a particular problem involves an $m \times n$ matrix A, then it is useful to think of the matrix A as being stored in the upper-left (northwest) corner of the 50×100 array A, as pictured in the following diagram:

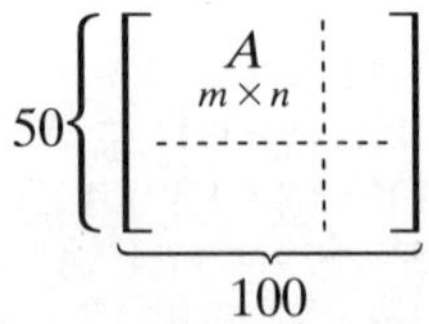

In this case, a subroutine would need both the size of A and the size of the array in which A is stored. See Exercise 10.

The following FORTRAN subroutine would compute the sum of two

$m \times n$ matrices A and B and store the sum in the array C. The matrices are all stored in arrays that are $mdim = 50$ by $ndim = 100$.

```
      Subroutine MTXADD(m, n, A, B, mdim, ndim, C)
      Integer m, n, mdim, ndim
      Real A(mdim, ndim), B(mdim, ndim), C(mdim, ndim)
C
C     Compute C=A+B where A and B are m  n matrices
C
      Do 20 i=1, m
         Do 10 j=1, n
            C(i,j)=A(i,j)+B(i,j)
10       Continue
20    Continue
      Return
      End
```

You could structure a calling program for the subroutine MTXADD as follows:

```
      Integer m, n, mdim, ndim
      Parameter (mdim=50, ndim=100)
      Real A(mdim, ndim), B(mdim, ndim), C(mdim, ndim)
                 .
                 .
      (Input m, n, A, B)
                 .
C
C     Compute C=A+B
C
      Call MTXADD(m, n, A, B, mdim, ndim, C)
                 .
      (Output C if desired)
                 .
      Stop
      End
```

You could then use this program, without change, to compute the sum of two 3×5 matrices or the sum of two 42×84 matrices. If you needed to add larger matrices, you would only need to change the parameter statement in the calling program.

Finely tuned subroutines for matrix operations are contained in the BLAS (Basic Linear Algebra Subroutines), which are part of the LINPACK and LAPACK packages. An excellent guide to the construction of efficient matrix programs is (Coleman and van Loan, 1988).

1.1 ■ Exercise Set

1. Find all X such that $AX = B$ if

$$A = \begin{bmatrix} 1 & 0 & 1 \\ 2 & 1 & 0 \\ 0 & 0 & 1 \\ 1 & -2 & 0 \end{bmatrix} \quad \text{and} \quad B = \begin{bmatrix} 0 & 0 & 2 \\ 1 & 0 & 3 \\ -1 & 1 & 1 \\ 3 & -5 & -1 \end{bmatrix}.$$

2. Find the inverse of $A = \begin{bmatrix} -3 & 5 & 6 \\ -1 & 2 & 2 \\ 1 & -1 & -1 \end{bmatrix}$.

3. Use MATMAN, with rational arithmetic, to find all solutions of

$$\begin{bmatrix} 2 & 4 & 3 & 2 & 5 \\ 3 & 6 & 5 & 2 & -7 \\ 2 & 5 & 2 & -3 & 4 \\ 4 & 5 & 14 & 14 & 2 \\ 3 & 10 & -4 & 13 & 0 \end{bmatrix} X = \begin{bmatrix} 5 \\ -16 \\ 18 \\ -16 \\ -71 \end{bmatrix}.$$

If you get an overflow message try changing to real arithmetic.

4. Show that the rows of the matrix product AB are linear combinations of the rows of B and that the columns of AB are linear combinations of the columns of A.

5. Let A be an $m \times n$ matrix A with $m > n$. Show that A^TA has an inverse if and only if the columns of A are linearly independent.

6. Let $A = \begin{bmatrix} 2 & 4 & 3 & 2 \\ 3 & 6 & 5 & 2 \\ 2 & 5 & 2 & -3 \\ 4 & 5 & 14 & 14 \end{bmatrix}$, $B = \begin{bmatrix} 1 & 0 & 1 \\ 2 & 1 & 0 \\ 0 & 0 & 1 \\ 1 & -2 & 0 \end{bmatrix}$,

$$C = \begin{bmatrix} 3 & -2 & 1 & 6 & 0 \\ 5 & -4 & 2 & 11 & 1 \\ -2 & 2 & -1 & -5 & 3 \\ 1 & -6 & 3 & 10 & 4 \\ -7 & 7 & -3 & -16 & 2 \end{bmatrix}.$$

Use MATLAB or MATALG to compute the following:

a. AB and BA

b. $\det(C)$, $\det(C^{-1})$, and C^{-1}

c. B^{-1}

d. $(B^TB)^{-1}$

e. $A^3 - 5A^2 + 11A + I$

f. $p(A)$ where $p(x) = 5x^2 - 72x + 11$ (Use the **poly** command in MATALG or **polyvalm** in MATLAB.)

g. X^TAX for $X = [1\ 1\ 1\ 1]^T$

7. Use MATMAN, with rational arithmetic, to find the exact inverse of

$$H = \begin{bmatrix} 1 & \frac{1}{2} & \frac{1}{3} & \frac{1}{4} & \frac{1}{5} \\ \frac{1}{2} & \frac{1}{3} & \frac{1}{4} & \frac{1}{5} & \frac{1}{6} \\ \frac{1}{3} & \frac{1}{4} & \frac{1}{5} & \frac{1}{6} & \frac{1}{7} \\ \frac{1}{4} & \frac{1}{5} & \frac{1}{6} & \frac{1}{7} & \frac{1}{8} \\ \frac{1}{5} & \frac{1}{6} & \frac{1}{7} & \frac{1}{8} & \frac{1}{9} \end{bmatrix}.$$

Now use MATLAB or MATALG to compute an approximation to the inverse of H. Compare with the MATMAN result and then compute the products HH^{-1} and $H^{-1}H$. Note that this matrix can be generated in MATLAB by the command **H=hilb(5)**.

8. Write a FORTRAN subroutine MTXMULT that will multiply an $m \times n$ matrix A and an $r \times s$ matrix B for $m, n, r, s \leq 50$. Include a test for compatibility with an appropriate error flag if the product is not defined. Write a driver program to call your subroutine, and test it for the matrices of Exercise 6.

9. Write a matrix multiplication subroutine that will overwrite B with the product AB. You should be able to do this using only two matrices and one vector.

10. a. Assume that the elements of the array A are stored, by columns, in consecutive memory locations. If $A(1, 1)$ is stored in memory location addr$[A(1, 1)]$ and A is dimensioned to be *mdim* by *ndim*, then show that

$A(i, j)$ is stored in memory location

$$\text{addr}[A(i, j)] = \text{addr}[A(1, 1)] + (j - 1)*mdim + (i - 1).$$

b. Use the result of Part **a** to explain why it is not really necessary to pass *ndim* to the subroutine MTXADD. In this case one can replace *ndim* in the subroutine dimension statements by *.

11. Let A, B, and C be $n \times n$ matrices such that $AB = 0$ and $CA = I$. Show that $B = 0$.

12. Let A be an $m \times n$ matrix. Show that $AY = 0$ for all vectors Y in $\mathscr{R}^n$ only if $A = 0$.

1.2 ■ MACHINE ARITHMETIC

In ordinary hand calculations and in most of your previous work in mathematics you have used the real number system. This system contains an infinite number of elements of the form

$$x = \pm d_1 d_2 d_3 \cdots \times 10^e = \pm\left(\frac{d_1}{10} + \frac{d_2}{10^2} + \frac{d_3}{10^3} + \cdots\right) \times 10^e$$
$$= \pm d_1 d_2 \cdots d_e d_{e+1} d_{e+2} \cdots$$

where the d_i are decimal digits $(0, 1, 2, \ldots, 9)$ with $d_1 \neq 0$.

A digital computer cannot represent all of these numbers exactly. In a computer, numbers are represented in **normalized floating-point form.** To describe this representation we introduce the following parameters:

β, the base (usually 2, 8, 10, or 16),

t, the precision,

U, the upper exponent range, and

L, the lower exponent range.

A typical nonzero floating-point number on a machine with these parameters is then

$$x = \pm\left(\frac{d_1}{\beta} + \frac{d_2}{\beta^2} + \frac{d_3}{\beta^3} + \cdots + \frac{d_t}{\beta^t}\right) \times \beta^e$$

where e is an integer, $L \leq e \leq U$, and $d_1 \neq 0$.

These floating-point parameters vary from machine to machine; for the VAX family of computers, $\beta = 2$, $t = 24$, $L = -127$, and $U = 127$. A typical number in this machine thus requires 32 bits and can be represented as follows:

±	$d_1\ d_2\ d_3 \cdots d_{24}$	±	$e_1\ e_2 \cdots e_6$

where $d_i = 1$ or 0 and $d_1 \neq 0$.

Note that there are only a finite number of such numbers; $2(\beta - 1)\beta^{t-1}(U - L + 1) + 1$ of them, to be precise. For the VAX, there are $2^{24}(255) + 1 = 4{,}278{,}190{,}081$ machine numbers. This means that the VAX cannot do federal budget calculations exactly in single precision. There are a lot of machine numbers, all of which are real numbers, but not all real numbers are machine numbers.

When you enter a real number x into the machine, the machine replaces the number you enter with the normalized floating-point number, $\mathrm{fl}(x)$, nearest to x. Thus the machine rounds the number x to t significant figures.

When we are comparing the values of a number x and an approximation x^* for x, we will call $|x^* - x|$ the **absolute error** in the approximation and we will call $|(x^* - x)/x|$ the **relative error** in the approximation. In general the relative error is a more useful measure of the error made in the approximation.

■ EXAMPLE 2

Let $x = 103$ and $y = 0.03127$, and let $x^* = 103.2$ and $y^* = 0.03125$ be approximations for x and y.

SOLUTION The absolute errors are

$$|x^* - x| = |103.2 - 103| = 0.2 \quad \text{and} \quad |y^* - y| = 0.00002.$$

Note that the absolute error in x^* is much greater than the absolute error in y^*. The relative errors are

$$\left|\frac{x^* - x}{x}\right| = \frac{0.2}{103} = 1.9 \times 10^{-3} \quad \text{and} \quad \left|\frac{y^* - y}{y}\right| = \frac{0.00002}{0.03127} = 6.39 \times 10^{-4}.$$

Note that the relative errors are about the same size, reflecting the fact that both approximations are correct to three significant figures. ■

■ DEFINITION 1.1 ■

The number x^* is said to approximate x, **correct to t significant digits**, if t is the largest nonnegative integer for which

$$\left|\frac{x^* - x}{x}\right| < 5 \times 10^{-t}.$$

In most cases this definition agrees with our intuitive notion of what it means for two numbers to agree to t significant figures. See, for example, the following table of approximations to $x = 3.437326$:

x^*	Relative Error	Digits of Accuracy
3.00000	1.3×10^{-1}	1
3.40000	1.1×10^{-2}	2
3.43000	2.1×10^{-3}	3
3.43700	0.95×10^{-4}	4
3.43730	0.76×10^{-5}	5
3.43732	1.7×10^{-6}	6

Consider again the normalized floating-point number $\text{fl}(x) = 0.d_1d_2 \cdots d_t \times \beta^e$. Since $|\text{fl}(x) - x| \leq (\beta/2) \times \beta^{-(t+1)} \times \beta^e$ and $|x| \geq (1/\beta) \times \beta^e$, it follows that the relative error made in the replacement of x by $\text{fl}(x)$ is

$$\left|\frac{\text{fl}(x) - x}{x}\right| \leq \frac{1}{2}\beta^{1-t} \ (5.96 \times 10^{-8} \text{ for the VAX}).$$

The number $\mu = \beta^{1-t}/2$ is called **unit round-off** for the machine in question. It follows that $\text{fl}(x) = x(1 + \rho)$ where $|\rho| \leq \mu = 1/2\beta^{1-t}$. Note that μ is the smallest floating-point number such that $\text{fl}(1 + \mu) > 1$.

The error (unavoidable) made in entering data into the machine is one example of **round-off error**. Note that the real numbers 0.1, 1/3, and 1/5 cannot be represented exactly on a binary machine like the VAX, while 1/2 and $49/16 = 3 + 2^{-4}$ can be represented exactly.

The machine approximates the real product xy by $\text{fl}(\text{fl}(x)*\text{fl}(y))$; that is, the machine computes the exact product of the machine approximations to x and y, and then rounds the product to obtain a machine number. Thus, **floating-point multiplication**, $\otimes$, is defined by

$$\text{fl}(x) \otimes \text{fl}(y) = \text{fl}(\text{fl}(x)*\text{fl}(y)).$$

Similarly, **floating-point addition**, $\oplus$, is defined by

$$\text{fl}(x) \oplus \text{fl}(y) = \text{fl}(\text{fl}(x) + \text{fl}(y)).$$

The other floating-point operations are defined similarly.

■ EXAMPLE 3

Consider a machine with $\beta = 10$, $t = 5$, and $-L = U = 99$. Compare $\text{fl}(xy)$ with $\text{fl}(x) \otimes \text{fl}(y)$, $\text{fl}(x + y)$ with $\text{fl}(x) \oplus \text{fl}(y)$, and $\text{fl}(y + z)$ with $\text{fl}(y) \oplus \text{fl}(z)$ if $x = 3256.78$, $y = -0.0432697$, and $z = 0.0432735$.

SOLUTION The machine approximations to x, y, and z are $\text{fl}(x) = 0.32568 \times 10^4$, $\text{fl}(y) = -0.43270 \times 10^{-1} = -0.0000043270 \times 10^4$, and $\text{fl}(z) = 0.43274 \times$

10^{-1}. By direct calculation we obtain

$$xy = -140.9198936 \quad \text{and} \quad \text{fl}(x)*\text{fl}(y) = -0.140921736 \times 10^3;$$

thus

$$\text{fl}(x) \otimes \text{fl}(y) = \text{fl}(\text{fl}(x)*\text{fl}(y)) = -0.14092 \times 10^3,$$

a very satisfactory result. Similarly we have

$$x + y = 3256.73673 \quad \text{and} \quad \text{fl}(x) + \text{fl}(y) = 0.325675673 \times 10^4,$$

so

$$\text{fl}(x) \oplus \text{fl}(y) = \text{fl}(\text{fl}(x) + \text{fl}(y)) = 0.32568 \times 10^4.$$

This is again a quite acceptable result, but we should note, perhaps with a bit of alarm, that

$$\text{fl}(x) \oplus \text{fl}(y) = \text{fl}(x) \quad \text{does not imply that } \text{fl}(y) = 0.$$

Finally we have

$$y + z = 0.0000038, \quad \text{fl}(y) + \text{fl}(z) = 0.00004 \times 10^{-1},$$

and

$$\text{fl}(y) \oplus \text{fl}(z) = \text{fl}(\text{fl}(y) + \text{fl}(z)) = 0.40000 \times 10^{-5}.$$

This result is not very good at all, since we have agreement in only one place. The relative error in the last approximation is

$$\left|\frac{(\text{fl}(y) \oplus \text{fl}(z)) - (y + z)}{y + z}\right| = \left|\frac{0.02 \times 10^{-5}}{0.38 \times 10^{-5}}\right| = 0.0526 = 0.526 \times 10^{-1},$$

which is much larger than the relative error in the first two calculations where we have

$$\left|\frac{(\text{fl}(x) \otimes \text{fl}(y)) - (xy)}{xy}\right| = 0.755 \times 10^{-6}$$

and

$$\left|\frac{(\text{fl}(x) \oplus \text{fl}(y)) - (x + y)}{x + y}\right| = 0.19427 \times 10^{-4}.$$

This example illustrates one of the greatest causes of inaccuracies in machine calculations. The error comes from the catastrophic cancellation that occurs when you subtract two numbers that have very nearly the same value.

In constructing computer algorithms for solving mathematical problems you should take care to avoid this kind of catastrophic cancellation.

The next example illustrates how round-off error can affect familiar calculations.

EXAMPLE 4

Find the roots of $x^2 - 75.023x + 1 = 0$. Assume that all calculations are to be done on a five-digit decimal machine ($t = 5$).

SOLUTION The quadratic formula tells us that the roots of the equation $ax^2 + bx + c = 0$ are

$$x_1 = \frac{-b + \sqrt{b^2 - 4ac}}{2a} \quad \text{and} \quad x_2 = \frac{-b - \sqrt{b^2 - 4ac}}{2a}.$$

Since b is negative, the smallest root of the given equation is

$$x_1 = \frac{75.023 - \sqrt{75.023^2 - 4}}{2}.$$

Using this formula our five-digit machine will compute approximations to x_1 and x_2 as follows:

$$75.023^2 = 5628.5$$
$$5628.5 - 4 = 5624.5$$
$$\sqrt{5624.5} = 74.997$$
$$\bar{x}_1 = (75.023 - 74.997)/2 = 0.026000/2 = 0.013000$$
$$\bar{x}_2 = (75.023 + 74.997)/2 = 150.02/2 = 75.010.$$

Very accurate values for the roots of this equation are $x_1 = 0.013331614$ and $x_2 = 75.00966839$. The relative errors in our five-digit calculations are thus

$$\left|\frac{\bar{x}_1 - x_1}{x_1}\right| = \left|\frac{0.013 - 0.013331614}{0.013331614}\right| = 0.248 \times 10^{-1} \text{ for } x_1,$$

and

$$\left|\frac{\bar{x}_2 - x_2}{x_2}\right| = \left|\frac{75.010 - 75.00966839}{75.00966839}\right| = 0.442 \times 10^{-5} \text{ for } x_2.$$

Clearly the approximation for x_1 is much less accurate than the approximation for x_2. The computed value of x_2 is correct to five significant figures, which is all you can expect on a five-digit machine, while the computed value of x_1 is correct to only two significant figures, a very unsatisfactory result.

The loss of accuracy in the calculation of x_1 comes at the last step where we have subtracted two values that are nearly equal. No such catastrophic cancellation occurs in the calculation of x_2. Note that you could use the accurate value for x_2, and the elementary fact that $x_1x_2 = c = 1$, to find a much better approximation for x_1; that is, $x_1 = 1/x_2 = 0.013332$. This new approximation for x_1 is correct to five significant figures, which is the most we could hope for on a five-digit machine.

This simple example demonstrates the need for being very careful when developing algorithms for use on finite precision machines. A more satisfactory algorithm for solving the quadratic equation $x^2 + bx + c = 0$ is obtained if one first tests for the algebraic sign of b: if $b > 0$, compute x_2 from the above formula and then compute $x_1 = c/x_2$; if $b < 0$, compute x_1 from the formula and then compute $x_2 = c/x_1$. ■

We have already observed that, on a t-digit base β machine,

$$\text{fl}(x) = x(1 + \rho) \quad \text{where } |\rho| \le \tfrac{1}{2}\beta^{1-t}.$$

It follows (see Exercise 10) that the errors in the floating-point approximations to the ordinary operations of arithmetic, on a t-digit decimal machine, satisfy

$$\begin{aligned}
&\text{fl}(x) = x(1 + \rho) \quad \text{where } |\rho| < v \times 10^{-t} \\
&x \otimes y = \text{fl}(\text{fl}(x)*\text{fl}(y)) = xy(1 + \rho) \quad \text{where } |\rho| < v \times 10^{-t} \\
&\text{fl}(\text{fl}(x)/\text{fl}(y)) = (x/y)(1 + \rho) \quad \text{where } |\rho| < v \times 10^{-t} (y \neq 0) \\
&x \oplus y = \text{fl}(\text{fl}(x) + \text{fl}(y)) = x(1 + \rho) + y(1 + \sigma) \\
&\qquad\qquad\qquad\qquad \text{where } |\rho|, |\sigma| < v \times 10^{-t}.
\end{aligned}$$

Here v is a number of order unity that depends on the machine.

It follows from the above bounds that rounding, floating-point multiplication, and floating-point division all have small relative errors. It also follows that the relative error in floating-point addition, which is

$$\left|\frac{x \oplus y - (x + y)}{x + y}\right| = \left|\frac{x\rho - y\sigma}{x + y}\right|,$$

can be quite large if x and y have opposite signs and nearly equal magnitudes.

Another item of concern in floating-point calculations is overflow and underflow. If the machine computes a floating-point number for which the exponent $e > U$, then **overflow** has occurred, and the machine will normally stop and give an error message. If the machine computes a floating-point number for which $e < L$, then **underflow** has occurred. Underflow is normally not a fatal error; the machine will treat this small number as 0 and proceed. Careful analysis as well as careful programming can help to reduce problems associated with overflow. For example, the quotient

$$72!/70! = 72 \times 71 = 5112$$

could easily overflow on a machine where $U = 99$ (most pocket calculators) if one attempted the calculation without the indicated simplification. This simple example indicates that both analysis and programming skills are needed to get good numerical results from a computer.

As the final example of this section, we consider the calculation of a function that is defined recursively. Again, careful analysis is necessary to ensure good results.

EXAMPLE 5

Compute $E_0, E_1, \ldots, E_{10}$, where $E_n \equiv \int_0^1 x^n e^{x-1}\,dx$. Assume that the calculations are done on a six-digit decimal machine.

SOLUTION It follows from a simple integration by parts that

$$E_n \equiv \int_0^1 x^n e^{x-1}\,dx = x^n e^{x-1}\Big|_0^1 - n\int_0^1 x^{n-1}e^{x-1}\,dx = 1 - nE_{n-1}.$$

Since $E_0 = \int_0^1 e^{x-1}\,dx = e^{x-1}\big|_0^1 = 1 - (1/e)$, we can use the simple recursion $E_n = 1 - nE_{n-1}$ to compute the other E_i. Note also that $0 < E_n < \int_0^1 x^n\,dx = 1/(n+1)$. The value of E_0, correct to six places, is $E_0 = 0.632120$. We use this value and the recursion relation $E_n = 1 - nE_{n-1}$ to compute $E_1, E_2, \ldots, E_{10}$. The following table compares these values with those computed on a 12-digit decimal machine:

	$t = 6$	$t = 12$	Absolute Error
E_0	0.632120	0.6321205588	$\varepsilon_0 = 5.5883 \times 10^{-7}$
$E_1 = 1 - E_0$	0.367880	0.367879441	$\varepsilon_1 = -5.5883 \times 10^{-7} = -\varepsilon_0$
$E_2 = 1 - 2E_1$	0.264240	0.264241117	$\varepsilon_2 = 1.11766 \times 10^{-6} = 2\varepsilon_0$
$E_3 = 1 - 3E_2$	0.207280	0.207276647	$\varepsilon_3 = -3.35298 \times 10^{-6} = -6\varepsilon_0$
$E_4 = 1 - 4E_3$	0.170880	0.170893411	$\varepsilon_4 = 1.341192 \times 10^{-5} = 24\varepsilon_0$
$E_5 = 1 - 5E_4$	0.145600	0.145532940	$\varepsilon_5 = -6.70596 \times 10^{-5} = -5!\varepsilon_0$
$E_6 = 1 - 6E_5$	0.126400	0.126802357	$\varepsilon_6 = 4.023576 \times 10^{-4} = 6!\varepsilon_0$
$E_7 = 1 - 7E_6$	0.115200	0.112383496	$\varepsilon_7 = -2.81650 \times 10^{-3} = -7!\varepsilon_0$
$E_8 = 1 - 8E_7$	0.0784000	0.100932025	$\varepsilon_8 = 0.022532 = 8!\varepsilon_0$
$E_9 = 1 - 9E_8$	0.294400	0.091611769	$\varepsilon_9 = -0.202788 = -9!\varepsilon_0$
$E_{10} = 1 - 10E_9$	−1.94400	0.083882304	$\varepsilon_{10} = 2.027882 = 10!\varepsilon_0$

Clearly our six-digit value of E_{10} is very bad because, as noted earlier, $0 < E_n < 1/(n+1)$. Note that all the error in the six-digit calculation is due to the unavoidable round-off error in entering E_0. What is happening here is that the error in E_1 is multiplied by -2 when E_2 is computed. Then the error in E_2 is multiplied by -3 when E_3 is computed, and so on. It follows easily that the error in E_n is $\varepsilon_n = (-1)^n n!\varepsilon_0$. This is an example of an **unstable algorithm**; small changes in the data (E_0) generate large changes in the answers.

In this case we can improve our results dramatically by rewriting the recursion relation as

$$E_{n-1} = \frac{1 - E_n}{n}.$$

Using this version of the recursion relation we see that the error in E_n would be divided by n in the computation of E_{n-1}. Since $0 < E_{20} < 1/21$, let us take 0 as an approximation for E_{20} and then use the modified recursion relation to compute E_{10}. In this case the error in E_{10} will be

$$\frac{1}{11} \times \frac{1}{12} \times \cdots \times \frac{1}{20}\varepsilon_{20} = \frac{10!}{20!}\varepsilon_0 \leq 1.49 \times 10^{-12} \times \frac{1}{21}.$$

The results of the calculation follow; note that the values of E_9 and E_{10} agree quite well with the 12-digit results given previously.

	$t = 6$
E_{20}	0
$E_{19} = (1 - E_{20})/20$	0.05
$E_{18} = (1 - E_{19})/19$	0.05
$E_{17} = (1 - E_{18})/18$	0.0527778
$E_{16} = (1 - E_{17})/17$	0.0557190
$E_{15} = (1 - E_{16})/16$	0.0590176
$E_{14} = (1 - E_{15})/15$	0.0627322
$E_{13} = (1 - E_{14})/14$	0.0669477
$E_{12} = (1 - E_{13})/13$	0.0717733
$E_{11} = (1 - E_{12})/12$	0.0773522
$E_{10} = (1 - E_{11})/11$	0.0838771
$E_9 = (1 - E_{10})/10$	0.0916123

■

1.2 ■ Exercise Set

1. Find the normalized floating-point representatives, on a machine with $\beta = 10$ and $t = 4$, of the real numbers $\sqrt{3}$, π, e, $1/(57!)$, $68!$.

2. Two common rational approximations for π are 22/7 and 355/113. Find the absolute error and the relative error for each of these approximations.

3. Find examples to show that floating-point multiplication is not associative and that floating-point multiplication does not distribute over floating-point addition. That is, find x, y, and z such that

$$x \otimes (y \otimes z) \neq (x \otimes y) \otimes z \quad \text{and} \quad x \otimes (y \oplus z) \neq x \otimes y \oplus x \otimes z.$$

4. The nth degree Taylor approximation for e^x is the polynomial

$$e^x \approx \sum_{k=0}^{n} \frac{x^k}{k!} = 1 + x + \frac{x^2}{2} + \frac{x^3}{3!} + \cdots + \frac{x^n}{n!}.$$

a. Use this formula to approximate e^{-6} and e^6 on a machine with $t = 4$. Use $n = 20$ and be sure to round properly at each step. Note that $x^k/k! = (x^{k-1}/(k-1)!)(x/k)$.

b. Find the relative error in each calculation. Why is one result so much better than the other?

5. Describe all the real numbers that are represented, on a four-digit decimal machine, by

a. 0.1234×10^{-2}

b. 0.3567×10^3.

6. Suppose the polynomial $p(x) = x^3 - 6x^2 + 3x - 0.649$ is to be evaluated on a four-digit decimal machine.

a. What would be the computed value of $p(4.72)$?

b. What value would be obtained for $p(4.72)$ if $p(x)$ is rewritten as $p(x) = ((x - 6)x + 3)x - 0.649$?

c. Find the relative error in each computed value. Which is better?

d. Count the number of arithmetic operations required for the two evaluations.

7. What advantages are there in computing

$$\sqrt{x^2 + y^2 + z^2}$$

using the equivalent expression

$$|x|\sqrt{1 + \left(\frac{y}{x}\right)^2 + \left(\frac{z}{x}\right)^2}\ ?$$

8. How could one modify the algorithm for solving the quadratic equation $ax^2 + bx + c = 0$ to handle the case where b is so large that b^2 would cause fatal overflow?

9. Consider the following algorithm:

```
E = 1.0, EP1 = 2
While EP1 > 1
    E = 0.5*E
    EP1 = E + 1
PRINT E
```

Describe the number produced by this algorithm on a finite precision machine.

10. **a.** Show that if $0 \le \mu < 1$, then $1 - n\mu < (1 - \mu)^n$. *Hint*: Consider the Taylor Series for $f(x) = (1 - x)^n$.

b. Show that if $0 \le n\mu \le 0.01$, then $(1 + \mu)^n \le 1 + 1.01n\mu$. *Hint*: Show that $e^x < 1 + 1.01x$ for $0 \le x \le 0.01$ and hence that $(1 + \mu)^n \le e^{n\mu} \le 1 + 1.01n\mu$.

c. Show that if $|\rho_i| < \mu$ for $i = 1, 2, \ldots, n$, and if $n\mu \le 0.01$, then

$$1 - n\mu \le \prod_{i=1}^{n} (1 + \rho_i) \le 1 + 1.01n\mu.$$

11. Suppose that the roots of the quadratic equation $x^2 + bx + c = 0$ are x_1 and x_2.

a. Show that $x_1 + x_2 = -b$ and that $x_1 x_2 = c$.

b. By differentiating $x_1^2 + bx_1 + c = 0$ with respect to c or b, show that $\dfrac{\partial x_1}{\partial c} = -\dfrac{1}{x_1 - x_2}$ and that $\dfrac{\partial x_1}{\partial b} = -\dfrac{x_1}{x_1 - x_2}$.

c. What can you conclude from the results in Part b if x_1 and x_2 are of the same sign and approximately the same magnitude?

2 Direct Methods for Linear Systems

2.1 ■ GAUSSIAN ELIMINATION

The basic method for solving the linear equations problem $AX = B$ is a suitably modified version of the elimination method described in Section 1.1. We would like to present a method that is reasonably **robust** (it works well for a large class of problems), is **stable** (small changes in the data generate only small changes in the computed solution), and is **efficient** (it does not require an unreasonable amount of computer time).

The Gaussian reduction of an $m \times n$ matrix A to upper triangular form, as described in Section 1.1, can be written in algorithmic language as follows:

GAUSSIAN REDUCTION TO TRIANGULAR FORM (Preliminary)

1. INPUT: $A(m, n)$
2. For $i = 1, 2, \ldots, m - 1$
3.   For $k = i + 1, \ldots, m$
4. $T = \dfrac{-a_{ki}}{a_{ii}}$
5.   For $j = i, \ldots, n$
6. $a_{kj} = T^*a_{ij} + a_{kj}$
7. OUTPUT: Upper triangular matrix A
8. Stop

(2.1)

This simple form of the algorithm assumes that $m \le n$ and that no $a_{ii} = 0$ at the time that it appears in the denominator of the fraction defining the multiplier T. If a zero appears on the diagonal at an inopportune time, a row interchange is required to bring a nonzero entry to the diagonal position. In this case Algorithm (2.1) would fail. The diagonal entries of the final matrix are called the **pivots**. Thus Algorithm (2.1) will not fail if all the pivots are nonzero.

As a measure of the amount of work required to implement an algorithm on a computer, we will count the number of floating-point multiplications required. For Algorithm (2.1) there is one multiplication each time line six is executed and one multiplication each time line four is executed. Thus the total number of multiplications involved in the implementation of Algorithm (2.1) is

$$\begin{aligned}\sum_{i=1}^{m-1}\sum_{k=i+1}^{m}\left(1+\sum_{j=i}^{n}1\right) &= \sum_{i=1}^{m-1}\sum_{k=i+1}^{m}(1+n-i+1)\\ &= \sum_{i=1}^{m-1}(m-i)(n+2-i)\\ &= \sum_{i=1}^{m-1}(m(n+2)-i(m+n+2)+i^2)\\ &= (m-1)m(n+2)-\frac{(m+n+2)m(m-1)}{2}\\ &\quad+\frac{(m-1)m(2m-1)}{6}\end{aligned}\tag{2.2}$$

where in the evaluation of the sum we have used the familiar formulae

$$\sum_{i=1}^{n} i = \frac{n(n+1)}{2} \quad \text{and} \quad \sum_{i=1}^{n} i^2 = \frac{n(n+1)(2n+1)}{6}.$$

If $m = n$, (2.2) reduces to $(n^3/3) + (n^2/2) - (5n/6)$. If $m = n - 1$, (2.2) reduces to $(n^3/3) - (7n/3) + 2$. In most cases we will use only the highest-order term of such operation counts, because as n becomes large the lower-order terms are insignificant. Thus we say that Algorithm (2.1), when applied to an $n \times n$ matrix A, requires about $(n^3/3)$ multiplications.

Our first example illustrates how Algorithm (2.1) works if no zero diagonal entries are encountered.

■ EXAMPLE 1

Find the solution of the system

$$\begin{aligned} x_1 - 3x_2 &= 0\\ 2x_1 + x_2 + x_3 &= 0\\ 3x_1 - x_2 + 2x_3 &= 1.\end{aligned}$$

SOLUTION The augmented matrix and the reduction to upper triangular form are given below.

$$\left[\begin{array}{rrr|r} 1 & -3 & 0 & 0\\ 2 & 1 & 1 & 0\\ 3 & -1 & 2 & 1\end{array}\right] \xrightarrow[R_3\leftarrow -3R_1+R_3]{R_2\leftarrow -2R_1+R_2} \left[\begin{array}{rrr|r} 1 & -3 & 0 & 0\\ 0 & 7 & 1 & 0\\ 0 & 8 & 2 & 1\end{array}\right]$$

$$\xrightarrow[R_3\leftarrow -(8/7)R_2+R_3]{} \left[\begin{array}{rrr|r} 1 & -3 & 0 & 0\\ 0 & 7 & 1 & 0\\ 0 & 0 & \frac{6}{7} & 1\end{array}\right]$$

The routine solution of the resulting triangular system (back substitution) now yields

$$\frac{6}{7}x_3 = 1 \Rightarrow x_3 = \frac{7}{6}$$

$$7x_2 + x_3 = 0 \Rightarrow x_2 = -\frac{x_3}{7} = -\frac{1}{6}$$

$$x_1 - 3x_2 = 0 \Rightarrow x_1 = -\frac{3}{6} = -\frac{1}{2}.$$

An illustration of how round-off error can affect the machine implementation of the Gaussian elimination algorithm is contained in the next example.

EXAMPLE 2

Use Gaussian elimination to solve the following system on a machine with $\beta = 10$ and $t = 5$:

$$\begin{cases} 10x_1 - 7x_2 = 7 \\ -3x_1 + 2.099x_2 + 6x_3 = 3.901 \\ 5x_1 - x_2 + 5x_3 = 6 \end{cases}$$

SOLUTION The exact solution of this system is $x_1 = 0$, $x_2 = -1$, and $x_3 = 1$, as can be verified by direct substitution. The augmented matrix of the system is

$$\begin{bmatrix} 10 & -7 & 0 & 7 \\ -3 & 2.099 & 6 & 3.901 \\ 5 & -1 & 5 & 6 \end{bmatrix} \rightarrow \begin{bmatrix} 10 & -7 & 0 & 7 \\ 0 & -1.0 \times 10^{-3} & 6 & 6.001 \\ 0 & 2.5 & 5 & 2.5 \end{bmatrix}$$

where the second matrix is obtained from the first by the row operations $R_2 \leftarrow (3/10)R_1 + R_2$ and $R_3 \leftarrow (-5/10)R_1 + R_3$ on our five-digit machine. If we further reduce this matrix using the operation $R_3 \leftarrow (2.5 \times 10^3)R_2 + R_3$, we obtain the upper triangular matrix

$$\begin{bmatrix} 10 & -7 & 0 & 7 \\ 0 & -1.0 \times 10^{-3} & 6 & 6.001 \\ 0 & 0 & 1.5005 \times 10^4 & 1.5006 \times 10^4 \end{bmatrix}.$$

If we now solve the upper triangular system (the process is called **back**

substitution), we obtain

$$x_3 = \frac{1.5006}{1.5005} = 1.0001$$

$$x_2 = \frac{6.0010 - 6(1.0001)}{-1.0 \times 10^{-3}} = \frac{0.00040}{-1.0 \times 10^{-3}} = -0.4$$

$$x_1 = \frac{7 + 7(-0.4)}{10} = \frac{9.8}{10} = 0.98\,.$$

This result is clearly very bad; for example, the relative error in x_2 is $\left|\frac{-0.4 - [-1]}{-1}\right| = 0.6\,.$

The loss of accuracy in this calculation is due to the fact that the $(2, 2)$ entry of the second matrix (the pivot) was very small. If that entry had been zero we would have needed to interchange the last two rows. Because of the small pivot, we needed to use a large multiplier when reducing the third row. This resulted in a loss of accuracy in computing the third row of the final matrix. This example suggests that very small pivots should be avoided whenever possible.

Suppose we replace the last step in the above reduction with the operation $R_2 \leftrightarrow R_3$ followed by $R_3 \leftarrow (4 \times 10^{-4})R_2 + R_3$. This yields

$$\begin{bmatrix} 10 & -7 & 0 & 7 \\ 0 & 2.5 & 5 & 2.5 \\ 0 & -1.0 \times 10^{-3} & 6 & 6.001 \end{bmatrix} \to \begin{bmatrix} 10 & -7 & 0 & 7 \\ 0 & 2.5 & 5 & 2.5 \\ 0 & 0 & 6.0020 & 6.0020 \end{bmatrix}.$$

The back substitution now yields the very accurate results

$$x_3 = \frac{6.0020}{6.0020} = 1.0000$$

$$x_2 = \frac{2.5000 - 5(1.0000)}{2.5000} = -1.0000$$

$$x_1 = \frac{7 + 7(-1.000)}{10} = 0.$$

■

Example 2 illustrates that very small diagonal elements are as dangerous as zero diagonal elements because they lead to very large row multipliers. For this reason it is essential to include in the algorithm a way of preventing the occurrence of large row multipliers. The most convenient way of doing this is called **partial pivoting**. Partial pivoting involves moving the largest element below the diagonal to the diagonal position of the column being reduced. Partial pivoting guarantees that all the multipliers will have absolute values less than or equal to one. The inclusion of partial pivoting in Algorithm (2.1) can be shown to significantly reduce the growth of round-off error (Golub and van Loan, 1989).

If we include the partial pivoting strategy in Algorithm (2.1), we obtain Algorithm (2.3). If $m = n$ or $n - 1$, the number of multiplications required is still about $(n^3/3)$.

GAUSSIAN REDUCTION WITH PARTIAL PIVOTING

1. INPUT: $A(m, n)$
2. $m1 = \min\{n, m - 1\}$
3. For $i = 1, 2, \ldots, m1$
4.   Choose j so that $|a_{ji}| \geq |a_{ki}|$ for $k = i, \ldots, m$.
5.   Interchange rows i and j.
6.   If $a_{ii} = 0$, increment i.
7.   For $k = i + 1, \ldots, m$
8. $T = \dfrac{-a_{ki}}{a_{ii}}$ (2.3)
9. $a_{ki} = 0$
10.   For $j = i + 1, \ldots, n$
11. $a_{kj} = T^*a_{ij} + a_{kj}$
12. OUTPUT: Upper triangular matrix $A(m, n)$
13. Stop

The test for a zero pivot (line six) is included to prevent the algorithm from crashing if a zero column is encountered. In the presence of rounding error, exact zeros are very unlikely, so it would be reasonable to replace the test for "$a_{ii} = 0$" by a test for "$a_{ii} <$ tol" where tol is some small number of the order of machine round off.

If the input to Algorithm (2.3) is the augmented matrix $[A \mid B]$ of a nonsingular linear system $AX = B$, then the output from (2.3) will be an upper triangular matrix R and a column vector C such that $RX = C$. This triangular matrix will have nonzero diagonal entries and the unique solution of the system $RX = C$ will be the unique solution of the original system. In practice R will overwrite A, and C will overwrite B. The triangular system can be solved using the **back substitution** Algorithm (2.4).

BACK SUBSTITUTION

INPUT: $A(n, n)$, upper triangular and nonsingular; $B(n, 1)$

$x_n = \dfrac{b_n}{a_{nn}}$

For $i = n - 1, n - 2, \ldots, 2, 1$

  $x_i = b_i$

  For $j = i + 1, i + 2, \ldots, n$

    $x_i = x_i - a_{ij} * x_j$

  $x_i = \dfrac{x_i}{a_{ii}}$

OUTPUT: $X(n, 1)$ such that $AX = B$

(2.4)

In Exercise 10b you are asked to show that Algorithm (2.4) takes about $(n^2/2)$ multiplications.

■ SOFTWARE NOTES ■

MATLAB can be used to count the number of floating-point operations (flops) it uses in the calculations done by the program. In general the flop count will be about double the multiplication count. Type **help flops** for a description of this command.

MATMAN and MATLAB can both be used to do row reductions in an interactive way. The automatic reductions in either program use partial pivoting. The password for the **Z** command in MATMAN is **31984**. The MATLAB command is **rref(A)**.

The MATLAB command **X=A\B** will solve the linear system $AX = B$. If A is singular, **rref([A,B])** will yield better results. The MATALG command **=inv(A)*B** will produce the solution if A is nonsingular. MATALG does not include a full-purpose linear equations solver; use MATMAN for the singular case.

2.1 ■ Exercise Set

1. How would the Gaussian elimination Algorithm (2.1) need to be modified

a. if $m > n$?

b. if you wanted the first nonzero entry of each row of the reduced matrix to be 1?

c. if you wanted the final matrix to be in row-reduced-echelon form?

d. if the exact zeros in the pivot column were to be set rather than computed?

2. Apply Algorithm (2.1) to each of the matrices below. Use MATMAN or MATLAB if you like.

$$A = \begin{bmatrix} 5 & 7 & 6 & 5 \\ 7 & 10 & 8 & 7 \\ 6 & 8 & 10 & 9 \\ 5 & 7 & 9 & 10 \end{bmatrix}, \quad B = \begin{bmatrix} 4 & -1 & -1 & 0 & 1 \\ -1 & 4 & -1 & -1 & 2 \\ -1 & -1 & 4 & -1 & 3 \\ 0 & -1 & -1 & 4 & 4 \end{bmatrix}$$

3. Apply Algorithm (2.3) to each of the matrices below. Use MATMAN or MATLAB if you like.

$$A = \begin{bmatrix} 5 & 7 & 6 & 5 \\ 7 & 10 & 8 & 7 \\ 6 & 8 & 10 & 9 \\ 5 & 7 & 9 & 10 \end{bmatrix}, \quad B = \begin{bmatrix} 4 & -1 & -1 & 0 & 1 \\ -1 & 4 & -1 & -1 & 2 \\ -1 & -1 & 4 & -1 & 3 \\ 0 & -1 & -1 & 4 & 4 \end{bmatrix}$$

4. Write a FORTRAN subroutine REDUCE$(A, m, n, m1, n1)$ to implement Gaussian elimination with partial pivoting for any $m \times n$ matrix A with $m \le m1$ and $n \le n1$. Write a calling program that will use REDUCE to row-reduce any matrix smaller than 50×51.

5. Assume that you are solving the following system on a three-digit decimal machine:

$$\begin{aligned} 0.000100x_1 + 1.00x_2 &= 1.00 \\ 1.00x_1 + 1.00x_2 &= 2.00 \end{aligned}$$

a. Find the solution using Gaussian elimination without partial pivoting.

b. Find the solution using partial pivoting.

c. Find the exact solution using real arithmetic.

6. Consider the linear system

$$\begin{aligned} x_1 + 0.99x_2 &= 1.99 \\ 0.99x_1 + 0.98x_2 &= 1.97 \end{aligned}$$

for which the true solution is $x_1 = x_2 = 1$.

a. Sketch the two straight lines represented by the two linear equations.

b. Solve this system on a three-digit decimal machine.

c. Does partial pivoting help here?

7. Two polynomials are of the **same order** if their highest-degree terms are the same. Verify that $\sum_{i=1}^{n} i$ and $\int_1^n i\, di$ are of the same order and that $\sum_{i=1}^{n} i^2$ and $\int_1^n i^2\, di$ are of the same order. Use these observations to find a triple integral of the same order as the triple sum in Equation (2.2).

8. **a.** Verify that $(i + 1)^3 - i^3 = 3i^2 + 3i + 1$. Now, sum both sides of this identity from $i = 1$ to n and thus prove that

$$\sum_{i=1}^{n} i = \frac{n(n+1)}{2} \Rightarrow \sum_{i=1}^{n} i^2 = \frac{n(n+1)(2n+1)}{6}.$$

b. Find a similar identity for $(i+1)^4 - i^4$ and develop a formula for $\sum_{i=1}^{n} i^3$.

c. Verify that $\sum_{i=1}^{n} i^3$ and $\int_1^n i^3\, di$ are of the same order.

9. Consider the simple suspension bridge shown here. Given the values h_0 and h_6 and the loads $L_1, L_2, \ldots, L_5$, we wish to determine the lengths $h_1, h_2, \ldots, h_5$ so that the system will be in equilibrium.

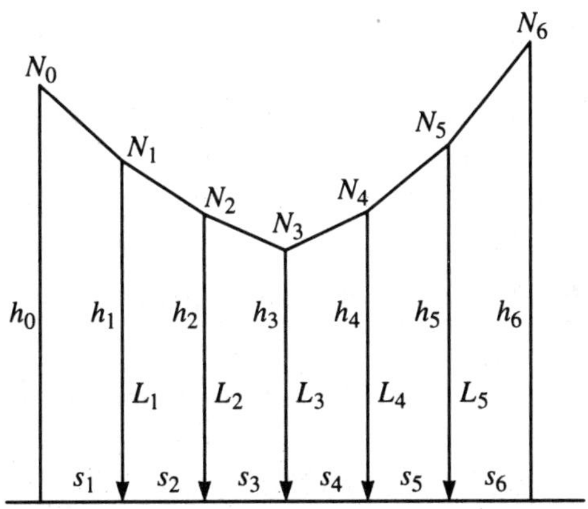

Consider the enlarged view of node 2.

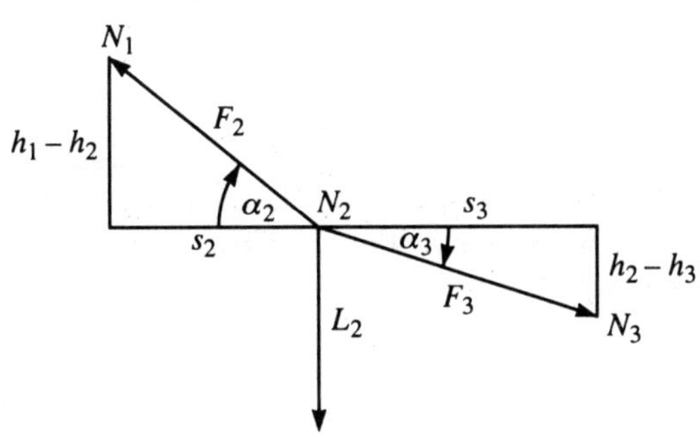

a. Show that balancing the horizontal forces at node 2 requires that

$$F_3 \cos \alpha_3 = F_2 \cos \alpha_2 = T \tag{1}$$

where T is the constant horizontal component of the tension. Show also that balancing the vertical forces at node 2 leads to the equation

$$F_2 \sin \alpha_2 = F_3 \sin \alpha_3 + L_2. \tag{2}$$

b. Use Equation (1) to eliminate F_2 and F_3 from Equation (2) to obtain

$$T \tan \alpha_2 - T \tan \alpha_3 = L_2. \tag{3}$$

Note that $\tan \alpha_2 = \dfrac{h_1 - h_2}{s_2}$ and $\tan \alpha_3 = \dfrac{h_2 - h_3}{s_3}$ so that Equation (3) is equivalent to

$$\frac{h_1 - h_2}{s_2} - \left(\frac{h_2 - h_3}{s_3}\right) = \frac{L_2}{T}$$

or

$$\left(\frac{1}{s_2}\right)h_1 - \left(\frac{1}{s_2} - \frac{1}{s_3}\right)h_2 + \left(\frac{1}{s_3}\right)h_3 = \frac{L_2}{T}. \tag{4}$$

c. Carry out similar analyses for the other nodes to obtain a system of five linear equations in the five unknowns $h_1, h_2, \ldots, h_5$.

10. a. Count the number of multiplications required to reduce an $n \times (n + 1)$ matrix A to upper triangular form using Algorithm (2.3).

b. Count the number of multiplications required to solve an $n \times n$ system by Algorithm (2.4) if the coefficient matrix is upper triangular. (Assume that divisions take as much time as multiplications.)

11. Count the total number of floating-point arithmetic operations (flops) in Algorithm (2.1).

12. What happens if you apply Gaussian elimination to a singular matrix?

2.2 ■ MATRIX FACTORIZATIONS

In this section we will develop some matrix factorizations and show how they are related to the Gaussian elimination algorithm for solving the linear system $AX = B$.

■ **DEFINITION 2.1** ■

A square matrix C is said to be **lower triangular** if $c_{ij} = 0$ for $i < j$, and **unit lower triangular** if it is lower triangular and each $c_{ii} = 1$. The matrix U is **upper triangular** if its transpose U^T is lower triangular. For example:

$$L = \begin{bmatrix} 1 & 0 & 0 \\ -5 & 1 & 0 \\ 3 & 0 & 1 \end{bmatrix} \text{ is unit lower triangular}$$

while

$$U = \begin{bmatrix} 2 & 5 & -3 \\ 0 & 0 & 1 \\ 0 & 0 & 6 \end{bmatrix} \text{ is upper triangular.}$$

Note that the only $n \times n$ matrix that is both unit lower triangular and upper triangular is the $n \times n$ identity matrix I.

Recall that if $\mathscr{E}$ is an elementary row operation, then the result of performing $\mathscr{E}$ on A can be obtained by premultiplying A by an identity matrix on which $\mathscr{E}$ has been performed, that is, $\mathscr{E}(A) = \mathscr{E}(I)A$. If $\mathscr{E}$ is the Type III elementary row operation $R_i \leftarrow kR_j + R_i$, with $i > j$, then the elementary matrix $\mathscr{E}(I)$ is a unit lower triangular matrix whose only nonzero subdiagonal entry is a k in position (i, j).

Suppose for the moment that no row interchanges are needed in the reduction of the $n \times n$ matrix A to upper triangular form. In that case we would have

$$E_r E_{r-1} \cdots E_3 E_2 E_1 A = U$$

where each E_i is an elementary matrix of Type III ($R_i \leftarrow kR_j + R_i$ with $i > j$), which is also a unit lower triangular matrix. Now it is easy to see that the set of $n \times n$ unit lower triangular matrices is closed under matrix multiplication (Exercise 1), so the matrix $L_1 = E_r E_{r-1} \cdots E_3 E_2 E_1$ is also a unit lower triangular matrix. Because the inverse of a unit lower triangular matrix is also a unit lower triangular matrix (Exercise 2), it follows that

$$A = (L_1)^{-1} U = LU$$

is a factorization of A as a product of a unit lower triangular matrix L and an upper triangular matrix U. A factorization of this type will be called an **LU factorization** of the matrix A. It follows easily, from the above discussion, that such a factorization is possible if no row interchanges are required in the reduction of A to upper triangular form, that is, if no zero appears in the pivot position during the reduction. Conversely, if $A = LU$, where L is unit lower triangular, then $L^{-1}A = U$, where the unit lower triangular matrix L^{-1} is a product of Type III elementary matrices. Thus A can be reduced to U without row interchanges.

THEOREM 2.1

The $n \times n$ matrix A has an LU factorization if and only if no row interchanges are required in the Gaussian reduction of A to upper triangular form. If A is nonsingular, the factorization is unique.

PROOF The first assertion has already been proven. To establish the uniqueness we assume that $A = L_1U_1 = L_2U_2$, where L_1 and L_2 are unit lower triangular matrices and U_1 and U_2 are upper triangular. Because A is nonsingular, so are L_2 and U_1. Now

$$L_1U_1 = L_2U_2 \Rightarrow L_2^{-1}L_1 = U_2U_1^{-1}$$

where the matrix on the left is a unit lower triangular matrix and the matrix on the right is an upper triangular matrix. Because I is the only unit lower triangular matrix that is also upper triangular, it follows that

$$L_2^{-1}L_1 = U_2U_1^{-1} = I \Rightarrow L_1 = L_2 \quad \text{and} \quad U_1 = U_2.$$

■

If A is 4×4, then L_1, and hence $L = L_1^{-1}$, is the product of six elementary matrices of Type III: $L_1 = E_6(E_5E_4)(E_3E_2E_1)$. Note that all the operations used to simplify a given column can be lumped together, for example:

$$E_3E_2E_1 = \begin{bmatrix} 1 & 0 & 0 & 0 \\ 0 & 1 & 0 & 0 \\ 0 & 0 & 1 & 0 \\ k_3 & 0 & 0 & 1 \end{bmatrix} \begin{bmatrix} 1 & 0 & 0 & 0 \\ 0 & 1 & 0 & 0 \\ k_2 & 0 & 1 & 0 \\ 0 & 0 & 0 & 1 \end{bmatrix} \begin{bmatrix} 1 & 0 & 0 & 0 \\ k_1 & 1 & 0 & 0 \\ 0 & 0 & 1 & 0 \\ 0 & 0 & 0 & 1 \end{bmatrix}$$

$$= \begin{bmatrix} 1 & 0 & 0 & 0 \\ k_1 & 1 & 0 & 0 \\ k_2 & 0 & 1 & 0 \\ k_3 & 0 & 0 & 1 \end{bmatrix}.$$

It follows that $L = L_1^{-1} = (E_3E_2E_1)^{-1}(E_5E_4)^{-1}E_6^{-1}$. Thus,

$$L = L_1^{-1} = \begin{bmatrix} 1 & 0 & 0 & 0 \\ k_1 & 1 & 0 & 0 \\ k_2 & 0 & 1 & 0 \\ k_3 & 0 & 0 & 1 \end{bmatrix}^{-1} \begin{bmatrix} 1 & 0 & 0 & 0 \\ 0 & 1 & 0 & 0 \\ 0 & k_4 & 1 & 0 \\ 0 & k_5 & 0 & 1 \end{bmatrix}^{-1} \begin{bmatrix} 1 & 0 & 0 & 0 \\ 0 & 1 & 0 & 0 \\ 0 & 0 & 1 & 0 \\ 0 & 0 & k_6 & 1 \end{bmatrix}^{-1}$$

$$= \begin{bmatrix} 1 & 0 & 0 & 0 \\ -k_1 & 1 & 0 & 0 \\ -k_2 & 0 & 1 & 0 \\ -k_3 & 0 & 0 & 1 \end{bmatrix} \begin{bmatrix} 1 & 0 & 0 & 0 \\ 0 & 1 & 0 & 0 \\ 0 & -k_4 & 1 & 0 \\ 0 & -k_5 & 0 & 1 \end{bmatrix} \begin{bmatrix} 1 & 0 & 0 & 0 \\ 0 & 1 & 0 & 0 \\ 0 & 0 & 1 & 0 \\ 0 & 0 & -k_6 & 1 \end{bmatrix}$$

$$= \begin{bmatrix} 1 & 0 & 0 & 0 \\ -k_1 & 1 & 0 & 0 \\ -k_2 & -k_4 & 1 & 0 \\ -k_3 & -k_5 & -k_6 & 1 \end{bmatrix}.$$

Note that the element in the (i, j) position of L is the negative of the multiplier used to eliminate the entry in the (i, j) position of A. This observation makes it possible to construct L as the reduction proceeds. A particularly efficient way to handle this on a computer is to store the elements of L in the subdiagonal positions as they are zeroed. Note that if you use this compact storage method, you are restricted to using the pivot row as the operator row for all Type III operations. For an $n \times n$ matrix A the algorithm would then take the form

LU FACTORIZATION ALGORITHM (Preliminary)

(2.5)

1. INPUT: $A(n, n)$
2. For $i = 1, 2, \ldots, n - 1$
3.   For $k = i + 1, \ldots, n$
4. $T = \dfrac{-a_{ki}}{a_{ii}}$
5.   For $j = i + 1, \ldots, n$
6. $a_{kj} = T^*a_{ij} + a_{kj}$
7. $a_{ki} = -T$
8. OUTPUT: LU factorization in compact form
9. Stop

Algorithm (2.5), like Algorithm (2.1), requires about $(n^3/3)$ multiplications (Exercise 17). Note that Algorithm (2.5), like Algorithm (2.1), will fail if a zero pivot is encountered, that is, if any $a_{ii} = 0$ when it is used in the calculation of T.

■ EXAMPLE 3

Find the LU factorization of $A = \begin{bmatrix} 2 & -4 & 4 \\ -2 & 2 & 0 \\ 4 & 1 & -8 \end{bmatrix}$.

SOLUTION The operations used to simplify the first column of A are $R_2 \leftarrow R_1 + R_2$ and $R_3 \leftarrow -2R_1 + R_3$, so the (2, 1) entry of L is -1 and the (3, 1) entry of L is 2. The reduced matrix and the compact storage array, with the entries of L in bold type, are

$$A \to \begin{bmatrix} 2 & -4 & 4 \\ 0 & -2 & 4 \\ 0 & 9 & -16 \end{bmatrix}, \quad \text{compact storage matrix} = \begin{bmatrix} 2 & -4 & 4 \\ \mathbf{-1} & -2 & 4 \\ \mathbf{2} & 9 & -16 \end{bmatrix}.$$

The final operation needed is $R_3 \leftarrow (9/2)R_2 + R_3$ so the $(3, 2)$ entry of L is $-9/2$:

$$A \rightarrow \begin{bmatrix} 2 & -4 & 4 \\ 0 & -2 & 4 \\ 0 & 0 & 2 \end{bmatrix}, \quad \text{compact storage matrix} = \begin{bmatrix} 2 & -4 & 4 \\ \mathbf{-1} & -2 & 4 \\ \mathbf{2} & \mathbf{-\frac{9}{2}} & 2 \end{bmatrix}.$$

Both L and U are available from the compact storage matrix since we know that the diagonal entries of L are all 1:

$$L = \begin{bmatrix} 1 & 0 & 0 \\ -1 & 1 & 0 \\ 2 & -9/2 & 1 \end{bmatrix} \quad \text{and} \quad U = \begin{bmatrix} 2 & -4 & 4 \\ 0 & -2 & 4 \\ 0 & 0 & 2 \end{bmatrix}.$$

$$LU = \begin{bmatrix} 2 & -4 & 4 \\ -2 & 2 & 0 \\ 4 & 1 & -8 \end{bmatrix} = A.$$

You should verify by direct multiplication that $A = LU$. ■

Note that if we have the LU factorization of the coefficient matrix A, then the solution of the linear system $AX = B$ can be reduced to the solution of two triangular systems. If we set $UX = Y$, then $AX = L(UX) = LY = B$, so we can first solve $LY = B$ for Y (the **forward substitution**) and then solve $UX = Y$ for X (the **back substitution**). The solution of the lower triangular system $LY = B$ is precisely equivalent to the modification of the second block in the naive implementation of Gaussian elimination described in Section 2.1. In the notation of this section the reduction part of the naive method amounts to

$$[A \,\vdots\, B] \rightarrow [U \,\vdots\, Y] = [U \,\vdots\, L^{-1}B].$$

The gain in efficiency comes if one wishes to solve other systems $AX = C$ where C is not known at the time of the factorization. The number of multiplications required for the factorization is about $(n^3/3)$, while the solution of a triangular system requires only about $(n^2/2)$ multiplications. Thus it is much cheaper to solve two triangular systems than to solve the system from scratch, which requires about $(n^3/3)$ multiplications. Because the case of multiple right-hand sides occurs very frequently, it is best to include the LU factorization as the first step in an algorithm for solving the linear system $AX = B$.

■ EXAMPLE 4

Let A be the matrix of Example 3. Use the LU factorization to solve the two systems $AX = [48, -16, -28]^T$ and $AZ = X$.

SOLUTION For the first system we solve the lower triangular system

$$LY = \begin{bmatrix} 1 & 0 & 0 \\ -1 & 1 & 0 \\ 2 & -\frac{9}{2} & 1 \end{bmatrix} Y = \begin{bmatrix} 48 \\ -16 \\ -28 \end{bmatrix}$$

for which the solution is easily seen to be

$$\begin{aligned} y_1 &= 48 \\ y_2 &= -16 + y_1 = 32 \\ y_3 &= -28 - 2(48) + \frac{9}{2}(32) = 20 \end{aligned} \qquad \text{or} \quad Y = \begin{bmatrix} 48 \\ 32 \\ 20 \end{bmatrix}.$$

We next need to solve the upper triangular system

$$UX = \begin{bmatrix} 2 & -4 & 4 \\ 0 & -2 & 4 \\ 0 & 0 & 2 \end{bmatrix} X = Y = \begin{bmatrix} 48 \\ 32 \\ 20 \end{bmatrix}.$$

The solution obtained by back substitution is

$$\begin{aligned} x_3 &= 10 \\ x_2 &= -\frac{1}{2}(32 - 4(10)) = 4 \\ x_1 &= \frac{1}{2}(48 - 4(10) + 4(4)) = 12 \end{aligned}$$

or

$$X = \begin{bmatrix} 12 \\ 4 \\ 10 \end{bmatrix}.$$

We have now solved the first system and found the right-hand side for the second system. To solve the second system we first solve the lower triangular system

$$LW = \begin{bmatrix} 1 & 0 & 0 \\ -1 & 1 & 0 \\ 2 & -\frac{9}{2} & 1 \end{bmatrix} W = X = \begin{bmatrix} 12 \\ 4 \\ 10 \end{bmatrix}$$

$$\Rightarrow \begin{aligned} w_1 &= 12 \\ w_2 &= 4 + 12 = 16 \\ w_3 &= 10 - 2(12) + \frac{9}{2}(16) = 58 \end{aligned} \quad \Rightarrow W = \begin{bmatrix} 12 \\ 16 \\ 58 \end{bmatrix}.$$

Finally, we solve the upper triangular system

$$UZ = \begin{bmatrix} 2 & -4 & 4 \\ 0 & -2 & 4 \\ 0 & 0 & 2 \end{bmatrix} Z = W = \begin{bmatrix} 12 \\ 16 \\ 58 \end{bmatrix}$$

$$\Rightarrow \begin{aligned} z_3 &= 29 \\ z_2 &= -\frac{1}{2}(16 - 4(29)) = 50 \\ z_1 &= \frac{1}{2}(12 - 4(29) + 4(50)) = 48 \end{aligned} \quad \Rightarrow Z = \begin{bmatrix} 48 \\ 50 \\ 29 \end{bmatrix}.$$

■

The same considerations that led us to include the partial pivoting strategy in the basic Gaussian elimination algorithm indicate that we should also include the partial pivoting strategy in the algorithm for computing the LU factorization. Some modifications of Algorithm (2.5) are required in this case.

For a 4×4 matrix A we might have

$$L_3P_3L_2P_2L_1P_1A = U \tag{2.6}$$

where P_1 corresponds to $R_1 \leftrightarrow R_4$, P_2 corresponds to $R_2 \leftrightarrow R_4$, P_3 corresponds to $R_3 \leftrightarrow R_4$ and where

$$L_1 = \begin{bmatrix} 1 & 0 & 0 & 0 \\ k_1 & 1 & 0 & 0 \\ k_2 & 0 & 1 & 0 \\ k_3 & 0 & 0 & 1 \end{bmatrix}, \quad L_2 = \begin{bmatrix} 1 & 0 & 0 & 0 \\ 0 & 1 & 0 & 0 \\ 0 & k_4 & 1 & 0 \\ 0 & k_5 & 0 & 1 \end{bmatrix}, \quad \text{and} \quad L_3 = \begin{bmatrix} 1 & 0 & 0 & 0 \\ 0 & 1 & 0 & 0 \\ 0 & 0 & 1 & 0 \\ 0 & 0 & k_6 & 1 \end{bmatrix}.$$

If we rewrite Equation (2.6) as

$$U = L_3P_3L_2P_2L_1(P_2^{-1}P_2)P_1A = L_3P_3L_2(P_2L_1P_2^{-1})P_2P_1A,$$

we have

$$P_2L_1P_2^{-1} = P_2L_1P_2 = \begin{bmatrix} 1 & 0 & 0 & 0 \\ 0 & 0 & 0 & 1 \\ 0 & 0 & 1 & 0 \\ 0 & 1 & 0 & 0 \end{bmatrix} \begin{bmatrix} 1 & 0 & 0 & 0 \\ k_1 & 1 & 0 & 0 \\ k_2 & 0 & 1 & 0 \\ k_3 & 0 & 0 & 1 \end{bmatrix} \begin{bmatrix} 1 & 0 & 0 & 0 \\ 0 & 0 & 0 & 1 \\ 0 & 0 & 1 & 0 \\ 0 & 1 & 0 & 0 \end{bmatrix}$$

$$= \begin{bmatrix} 1 & 0 & 0 & 0 \\ k_3 & 1 & 0 & 0 \\ k_2 & 0 & 1 & 0 \\ k_1 & 0 & 0 & 1 \end{bmatrix} = L_1^{(1)}.$$

Note that $L_1^{(1)}$ is just L_1 with the second and fourth entries of column 1 interchanged. Similarly, since $P_3^{-1} = P_3$, Equation (2.6) is equivalent to

$$U = L_3P_3L_2L_1^{(1)}P_2P_1A = L_3(P_3L_2P_3)(P_3L_1^{(1)}P_3)P_3P_2P_1A = L_3L_2^{(1)}L_1^{(2)}P_3P_2P_1A$$

where, as above, $L_2^{(1)}$ and $L_1^{(2)}$ come from L_2 and $L_1^{(1)}$ by interchanging the subdiagonal entries in rows 3 and 4. Thus,

$$L_2^{(1)} = \begin{bmatrix} 1 & 0 & 0 & 0 \\ 0 & 1 & 0 & 0 \\ 0 & k_5 & 1 & 0 \\ 0 & k_4 & 0 & 1 \end{bmatrix} \quad \text{and} \quad L_1^{(2)} = \begin{bmatrix} 1 & 0 & 0 & 0 \\ k_3 & 1 & 0 & 0 \\ k_1 & 0 & 1 & 0 \\ k_2 & 0 & 0 & 1 \end{bmatrix}.$$

It follows, as in the proof of Theorem 2.1, that

$$PA = (P_3P_2P_1)A = (L_3L_2^{(1)}L_1^{(2)})^{-1}U = (L_1^{(2)})^{-1}(L_2^{(1)})^{-1}L_3^{-1}U = LU$$

where L is the unit lower triangular matrix

$$L = \begin{bmatrix} 1 & 0 & 0 & 0 \\ -k_3 & 1 & 0 & 0 \\ -k_1 & -k_5 & 1 & 0 \\ -k_2 & -k_4 & -k_6 & 1 \end{bmatrix}.$$

The matrix

$$P = P_3P_2P_1 = \begin{bmatrix} 0 & 0 & 0 & 1 \\ 1 & 0 & 0 & 0 \\ 0 & 1 & 0 & 0 \\ 0 & 0 & 1 & 0 \end{bmatrix}$$

is an example of a **permutation matrix**; it is the identity matrix with its rows rearranged. Thus PA is just the original coefficient matrix A with its rows rearranged.

The above discussion generalizes easily to any $n \times n$ matrix A so that we have a way of computing the factorization described in the following theorem.

THEOREM 2.2

For any $n \times n$ matrix A there exists a permutation matrix P so that PA has an LU factorization.

PROOF (Induction on n). The theorem is certainly obvious if $n = 1$, so assume that it is true for $n - 1$ and consider the $n \times n$ matrix

$$A = \left[\begin{array}{c|c} A_1 & \mathbf{c} \\ \hline \mathbf{r} & a_{nn} \end{array}\right]$$ where A_1 is $n - 1 \times n - 1$, $\mathbf{r}$ is a $1 \times n - 1$ row matrix, and $\mathbf{c}$ is an $n - 1 \times 1$ column matrix.

The rows of the $n \times n - 1$ matrix $\begin{bmatrix} A_1 \\ \mathbf{r} \end{bmatrix}$ must be linearly dependent, so their exists a permutation matrix P_1 such that

$$P_1A = \left[\begin{array}{c|c} A_2 & \mathbf{c}' \\ \hline \mathbf{r}' & a'_{nn} \end{array}\right] \quad \text{with } \mathbf{r}' \text{ in } \mathcal{RS}(A_2), \text{ the rowspace of } A_2.$$

By inductive hypothesis, there exists an $n - 1 \times n - 1$ permutation matrix P_2 so that P_2A_2 has an LU factorization, say $P_2A_2 = L_2U_2$. Now we have that

$$PA = \begin{bmatrix} P_2 & 0 \\ 0 & 1 \end{bmatrix} P_1A = \left[\begin{array}{c|c} P_2A_2 & P_2\mathbf{c}' \\ \hline \mathbf{r}' & a'_{nn} \end{array}\right] = \begin{bmatrix} L_2 & 0 \\ \mathbf{x} & 1 \end{bmatrix} \begin{bmatrix} U_2 & \mathbf{y} \\ 0 & u_{nn} \end{bmatrix}$$

$$= \left[\begin{array}{c|c} L_2U_2 & L_2\mathbf{y} \\ \hline \mathbf{x}U_2 & \mathbf{xy} + u_{nn} \end{array}\right]$$

is a valid factorization of PA if and only if

$$L_2\mathbf{y} = P_2c', \quad \mathbf{x}U_2 = \mathbf{r}', \quad \text{and} \quad \mathbf{xy} + u_n = a'_{nn}.$$

The first of these equations has a unique solution for $\mathbf{y}$, since L_2 is a unit lower triangular matrix and hence is nonsingular. The second equation has a solution for $\mathbf{x}$ because $\mathbf{r}'$ is in $\mathcal{RS}(U_2) = \mathcal{RS}(A_2)$. Once the first two equations are solved, the third equation determines u_{nn}. This completes the proof. Note that neither P_1 nor $\mathbf{x}$ is certain to be unique, so the factorization is not unique. ■

Theorem 2.2 says that there is some rearrangement of the rows of A (equivalently the equations of the system) for which the resulting matrix has an LU factorization. We could do the reordering first, but it is not clear how; that is discovered as the algorithm proceeds.

We can use the compact storage scheme introduced earlier to compute this new factorization. We must keep track of P separately and be careful to interchange the rows of the full array when we interchange rows.

EXAMPLE 5

Find an LU factorization for $A = \begin{bmatrix} 2 & 4 & 3 & 2 \\ 3 & 6 & 5 & 2 \\ 2 & 5 & 2 & -3 \\ 4 & 5 & 14 & 14 \end{bmatrix}$ using partial pivoting.

SOLUTION We will list the compact storage array and the matrix P at each step. The elements of L are in bold type. The first operation required is the row interchange $R_4 \leftrightarrow R_1$, which leads to

$$\begin{bmatrix} 4 & 5 & 14 & 14 \\ 3 & 6 & 5 & 2 \\ 2 & 5 & 2 & -3 \\ 2 & 4 & 3 & 2 \end{bmatrix}, \quad P = \begin{bmatrix} 0 & 0 & 0 & 1 \\ 0 & 1 & 0 & 0 \\ 0 & 0 & 1 & 0 \\ 1 & 0 & 0 & 0 \end{bmatrix}.$$

We now reduce the first column using the three Type III operations $R_2 \leftarrow (-3/4)R_1 + R_2$, $R_3 \leftarrow (-2/4)R_1 + R_3$, and $R_4 \leftarrow (-2/4)R_1 + R_4$. This yields

$$\begin{bmatrix} 4 & 5 & 14 & 14 \\ \mathbf{0.75} & 2.25 & -5.5 & -8.5 \\ \mathbf{0.5} & 2.5 & -5 & -10 \\ \mathbf{0.5} & 1.5 & -4 & -5 \end{bmatrix}, \quad P = \begin{bmatrix} 0 & 0 & 0 & 1 \\ 0 & 1 & 0 & 0 \\ 0 & 0 & 1 & 0 \\ 1 & 0 & 0 & 0 \end{bmatrix}.$$

The pivoting strategy now requires that we interchange rows 2 and 3. This leads to

$$\begin{bmatrix} 4 & 5 & 14 & 14 \\ \mathbf{0.5} & 2.5 & -5 & -10 \\ \mathbf{0.75} & 2.25 & -5.5 & -8.5 \\ \mathbf{0.5} & 1.5 & -4 & -5 \end{bmatrix}, \quad P = \begin{bmatrix} 0 & 0 & 0 & 1 \\ 0 & 0 & 1 & 0 \\ 0 & 1 & 0 & 0 \\ 1 & 0 & 0 & 0 \end{bmatrix}.$$

Notice that we interchanged the *entire* second and third rows of the compact storage matrix. We now reduce the second column using the operations $R_3 \leftarrow (-2.25/2.5)R_2 + R_3$ and $R_4 \leftarrow (-1.5/2.5)R_2 + R_4$. The resulting matrices are

$$\begin{bmatrix} 4 & 5 & 14 & 14 \\ \mathbf{0.5} & 2.5 & -5 & -10 \\ \mathbf{0.75} & \mathbf{0.9} & -1 & 0.5 \\ \mathbf{0.5} & \mathbf{0.6} & -1 & 1 \end{bmatrix}, \quad P = \begin{bmatrix} 0 & 0 & 0 & 1 \\ 0 & 0 & 1 & 0 \\ 0 & 1 & 0 & 0 \\ 1 & 0 & 0 & 0 \end{bmatrix}.$$

No row interchanges are required at the next step. The single operation $R_4 \leftarrow -R_3 + R_4$ will complete the reduction. The result is

$$\begin{bmatrix} 4 & 5 & 14 & 14 \\ \mathbf{0.5} & 2.5 & -5 & -10 \\ \mathbf{0.75} & \mathbf{0.9} & -1 & 0.5 \\ \mathbf{0.5} & \mathbf{0.6} & \mathbf{1} & 0.5 \end{bmatrix}, \quad P = \begin{bmatrix} 0 & 0 & 0 & 1 \\ 0 & 0 & 1 & 0 \\ 0 & 1 & 0 & 0 \\ 1 & 0 & 0 & 0 \end{bmatrix}.$$

From the last compact storage array we recover the factorization

$$PA = \begin{bmatrix} 4 & 5 & 14 & 14 \\ 2 & 5 & 2 & -3 \\ 3 & 6 & 5 & 2 \\ 2 & 4 & 3 & 2 \end{bmatrix} = LU = \begin{bmatrix} 1 & 0 & 0 & 0 \\ 0.5 & 1 & 0 & 0 \\ 0.75 & 0.9 & 1 & 0 \\ 0.5 & 0.6 & 1 & 1 \end{bmatrix} \begin{bmatrix} 4 & 5 & 14 & 14 \\ 0 & 2.5 & -5 & -10 \\ 0 & 0 & -1 & 0.5 \\ 0 & 0 & 0 & 0.5 \end{bmatrix},$$

which you should check by direct multiplication. ■

The factorization of Theorem 2.2 is easy to incorporate into the solution of the linear system $AX = B$. The only change is that we must reorder the equations of the original system using the permutation matrix P. Thus we have $AX = B \Rightarrow PAX = PB \Rightarrow L(UX) = PB$. Thus if we set $UX = Y$, we first need to solve the lower triangular system $LY = PB$ for Y (the forward substitution) and then solve the upper triangular system $UX = Y$ for X (the back substitution).

It is really not necessary to compute the matrix P as we did in Example 5. All that is needed is a way of keeping track of the reordering of the rows. An efficient way to do this is to replace P with an $n \times 1$ pointer vector IP. Initially $IP = [1\ 2\ 3 \cdots n]^T$. In Example 5, the successive values of IP would be

$$IP = \begin{bmatrix}1\\2\\3\\4\end{bmatrix} \rightarrow \begin{bmatrix}4\\2\\3\\1\end{bmatrix} \rightarrow \begin{bmatrix}4\\3\\2\\1\end{bmatrix} \Rightarrow P = \begin{bmatrix}\text{Row}_4(I)\\\text{Row}_3(I)\\\text{Row}_2(I)\\\text{Row}_1(I)\end{bmatrix} = \begin{bmatrix}0&0&0&1\\0&0&1&0\\0&1&0&0\\1&0&0&0\end{bmatrix}.$$

The permutation matrix P can be easily constructed from the final value of the pointer vector IP.

If we include these modifications in Algorithm (2.5) we obtain the following algorithm for computing the factorization of Theorem 2.2. We have included the possibility that the algorithm may encounter a zero pivot; this will not happen if the matrix A is nonsingular (see Exercise 16). Algorithm (2.7), like Algorithms (2.3) and (2.5), takes about $(n^3/3)$ multiplications to factor an $n \times n$ matrix A.

LU FACTORIZATION WITH PARTIAL PIVOTING

1. INPUT: $A(n, n)$
2. $IP = [1, 2, 3, \ldots, n]$
3. For $i = 1, 2, \ldots, n - 1$
4. Choose j so that $|a_{ji}| \geq |a_{ki}|$ for $k = i, \ldots, n$.
5. Interchange rows i and j of A and of IP.
6. If $a_{ii} = 0$, next i
7. For $k = i + 1, \ldots, n$
8. $T = \dfrac{-a_{kj}}{a_{ii}}$
9. For $j = i + 1, \ldots, n$
10. $a_{kj} = T^*a_{ij} + a_{kj}$
11. $a_{ki} = -T$
12. OUTPUT: LU factorization of PA in compact form and IP
13. Stop

(2.7)

■ SOFTWARE NOTES ■

The factorization of Theorem 2.2 is computed by

the MATLAB command **[L,U,P]=lu(A)**

or

the MATALG command **M, LU, A**.

The automatic row reduction in MATMAN (the **Z** command) uses partial pivoting.

2.2 ■ Exercise Set

1. Let A and B be $n \times n$ unit lower triangular matrices. Show that AB is also a unit lower triangular matrix.

2. Show that every $n \times n$ unit lower triangular matrix is nonsingular and that its inverse is also a unit lower triangular matrix.

3. Find the LU factorizations of

$$A = \begin{bmatrix} 1 & 1 & 1 \\ 3 & -2 & -17 \\ 3 & 2 & -1 \end{bmatrix}, \quad B = \begin{bmatrix} 2 & 3 & -1 & 2 \\ 2 & 4 & 4 & -1 \\ 4 & 8 & 2 & 7 \\ -4 & -6 & 2 & 1 \end{bmatrix},$$

$$C = \begin{bmatrix} 1 & -1 & 2 & 3 \\ 2 & -1 & 0 & 2 \\ 4 & 1 & -11 & -1 \\ 1 & 2 & 33 & 84 \end{bmatrix}.$$

4. Show that the matrix $A = \begin{bmatrix} 2 & 4 & 3 & 2 \\ 3 & 6 & 5 & 2 \\ 2 & 5 & 2 & -3 \\ 4 & 5 & 14 & 14 \end{bmatrix}$ does not have an LU factorization, but that the matrix B, obtained from A by interchanging rows 2 and 3, does have an LU factorization.

5. Describe how MATMAN can be used to find the LU factorization.

6. A matrix Q is said to be **orthogonal** if $Q^TQ = I$. Show how a factorization $A = QR$, where Q is orthogonal and R is upper triangular, could be used to solve the linear system $AX = B$. Compare the work involved in this method with the work involved using the LU factorization.

7. If A is $n \times n$, the **leading principal minor** $A^{(k)}$ is the $k \times k$ submatrix appearing in the upper-left-hand corner of A. Show, by induction, that A has

an LU factorization if the leading principal minors $A^{(1)}, A^{(2)}, \ldots, A^{(n-1)}$ are all nonsingular.

8. For the following matrices, use partial pivoting to find the decomposition of Theorem 2.2.

$$A = \begin{bmatrix} 1 & 1 & 1 \\ 3 & -2 & -17 \\ 3 & 2 & -1 \end{bmatrix}, \quad B = \begin{bmatrix} 1 & -1 & 2 & 3 \\ 2 & -1 & 0 & 2 \\ 4 & 1 & -11 & -1 \\ 1 & 2 & 33 & 83 \end{bmatrix}, \quad C = \begin{bmatrix} 1 & \frac{1}{2} & \frac{1}{3} & \frac{1}{4} \\ \frac{1}{2} & \frac{1}{3} & \frac{1}{4} & \frac{1}{5} \\ \frac{1}{3} & \frac{1}{4} & \frac{1}{5} & \frac{1}{6} \\ \frac{1}{4} & \frac{1}{5} & \frac{1}{6} & \frac{1}{7} \end{bmatrix}$$

9. Given $A = \begin{bmatrix} 1 & 1 & 3 \\ 1 & -2 & 1 \\ 3 & 1 & 3 \end{bmatrix}$ and $B = \begin{bmatrix} 1 \\ 1 \\ 1 \end{bmatrix}$, find the LU decomposition of A and use it to find X_1 such that $AX_1 = B$, X_2 such that $AX_2 = X_1$, and X_3 such that $AX_3 = X_2$.

10. Use either MATLAB or MATALG to find the decomposition of Theorem 2.2 for the matrices of Exercise 3.

11. Show that if A has an LU decomposition, then it also has a decomposition of the form $A = LDR$, where L is unit lower triangular, R is unit upper triangular, and D is diagonal. Show that if A is nonsingular then all three factors are unique.

12. a. Show that if A is nonsingular and symmetric ($A^T = A$), then in the LDR decomposition of Exercise 11, $R^T = L$. If the diagonal elements of D are positive, show that $A = BB^T$ where B is lower triangular. Such a factorization is called a **Cholesky factorization**.

b. Show that if A is symmetric and positive definite ($X^TAX > 0$ for every $X \neq 0$ in $\mathscr{R}^n$), then it has a Cholesky factorization.

13. Write a subroutine DECOMP($A, n, n1$) that uses Algorithm (2.7) to compute the LU decomposition for an $n \times n$ matrix A with $n \leq n1$.

14. a. Show that an LU factorization of A, with L lower triangular and U unit upper triangular, can be computed by the following algorithm:

CROUT DECOMPOSITION

For $k = 1, 2, \ldots, n$
  For $i = k, k + 1, \ldots, n$

$$l_{ik} = a_{ik} - \sum_{r=1}^{k-1} l_{ir}u_{rk}$$

  For $j = k + 1, k + 2, \ldots, n$

$$u_{kj} = l_{kk}^{-1}\left(a_{kj} - \sum_{r=1}^{k-1} l_{kr}u_{rj}\right)$$

b. Modify this algorithm so that it uses only one $n \times n$ storage array.

15. Show that Algorithm (2.5) fails for the matrix

$$A = \begin{bmatrix} 2 & 3 & -1 & 2 \\ -4 & -6 & 7 & -3 \\ 6 & 9 & 22 & 14 \\ 2 & 3 & 9 & -2 \end{bmatrix}.$$

Use Algorithm (2.7) to find the LU factorization of A.

16. Consider the following modified version of Algorithm 2.7:

```
1.  INPUT: A(n, n)
2.  IP = [1, 2, 3, ..., n], nzc = 0
3.  For i = 1, 2, ..., n - 1
4.      l = i + nzc  (nzc = number of zero columns)
5.      If l > n, Stop
6.      Choose j so that |a_jl| >= |a_kl| for k = i, ..., n.
7.      Interchange rows i and j of A and of IP.
8.      If a_il = 0, set nzc = nzc + 1 and go to line four.
9.      For k = i + 1, ..., n
10.         T = -a_kl / a_il
11.         For j = l + 1, ..., n
12.             a_kj = T*a_ij + a_kj
13.         a_ki = -T
14. OUTPUT: LU factorization of PA in compact form and IP
15. Stop
```

a. What will this program yield if A is nonsingular?

b. What will this program yield for the matrix of Exercise 15?

17. Verify that Algorithm (2.5) takes about $(n^3/3)$ multiplications.

18. Consider the following algorithm for Gaussian elimination without pivoting:

```
For k = 1, 2, ..., n − 1
    For i = k + 1, ..., n
        a_ik = a_ik / a_kk
    For j = k + 1, ..., n
        For i = k + 1, ..., n
            a_ij = a_ij − a_ik a_kj.
```

How does this implementation differ from Algorithm (2.5)? Explain why the algorithm in this exercise would be more efficient than would Algorithm (2.5) if the compiler stored matrices by columns as does FORTRAN.

2.3 ■ MATRIX AND VECTOR NORMS

To analyze the error associated with numerical methods for solving linear systems we will need to have a way of measuring what it means for a vector to be "small," or equivalently, what it means for two vectors to be "close" to each other.

■ DEFINITION 2.2 ■

A **vector norm** is a function v from $\mathcal{R}^n$ to $\mathcal{R}$ that is

1. positive definite $(X \neq 0 \Rightarrow v(X) > 0)$,
2. homogeneous $(v(kX) = |k|\, v(X))$, and
3. satisfies the triangle inequality $(v(X + Y) \leq v(X) + v(Y))$.

There are many ways of defining a norm on $\mathcal{R}^n$. We will consider three different ways:

■ DEFINITION 2.3 ■

For any vector X in $\mathcal{R}^n$,

1. $\|X\|_1 = \sum_{i=1}^n |x_i|$ is called the **1-norm** of X.
2. $\|X\|_2 = \sqrt{X^T X} = \sqrt{\sum_{i=1}^n x_i^2}$ is called the **Euclidean norm** or the **2-norm** of X.

3. $\|X\|_\infty = \max\{|x_i| : i = 1, \ldots, n\}$ is called the **max-norm** or the **∞-norm** of X.

Note that for $n = 1$, all three of these norms reduce to $\nu(X) = |X|$ for which the three properties of the definition are certainly clear. We, of course, have in mind that a vector will be considered "small" if its norm is small and that two vectors will be considered "close" if the norm of their difference is small.

For the vectors $X = \begin{bmatrix} 5 \\ 2 \\ -1 \end{bmatrix}$ and $Y = \begin{bmatrix} -1 \\ 3 \\ 1 \end{bmatrix}$, for which $X + Y = \begin{bmatrix} 4 \\ 5 \\ 0 \end{bmatrix}$, we have

$$\begin{array}{lll} \|X\|_1 = 8, & \|Y\|_1 = 5, & \|X + Y\|_1 = 9 \\ \|X\|_2 = \sqrt{30} & \|Y\|_2 = \sqrt{11} & \|X + Y\|_2 = \sqrt{41} \\ \|X\|_\infty = 5, & \|Y\|_\infty = 3, & \|X + Y\|_\infty = 5 \end{array}$$

The details of the verification that $\|X\|_1$ and $\|X\|_\infty$ satisfy the three properties of Definition 2.2 are straightforward; they are left to the exercises. To verify the triangle inequality for the Euclidean norm, we will first establish a useful result known as the **Cauchy-Schwarz inequality**. For $n = 2$ or 3, this inequality is equivalent to the fact that $|\cos\theta| \le 1$.

THEOREM 2.3

For X, Y in $\mathcal{R}^n$, we have $|X^TY| \le \|X\|_2 \, \|Y\|_2$.

PROOF Consider first the case where $U = [u_1, u_2, \ldots, u_n]^T$ and $V = [v_1, v_2, \ldots, v_n]^T$ are unit vectors in $\mathcal{R}^n$. In this case,

$$\|U\|_2^2 = \sum_{i=1}^n u_i^2 = 1 \quad \text{and} \quad \|V\|_2^2 = \sum_{i=1}^n v_i^2 = 1.$$

It follows that

$$\begin{aligned} 0 \le \|U - V\|_2^2 &= \sum_{i=1}^n (u_i - v_i)^2 = \sum_{i=1}^n u_i^2 - 2\sum_{i=1}^n u_iv_i + \sum_{i=1}^n v_i^2 \\ &= 2 - 2U^TV \Rightarrow U^TV \le 1, \end{aligned}$$

and

$$\begin{aligned} 0 \le \|U + V\|_2^2 &= \sum_{i=1}^n (u_i + v_i)^2 = \sum_{i=1}^n u_i^2 + 2\sum_{i=1}^n u_iv_i + \sum_{i=1}^n v_i^2 \\ &= 2 + 2U^TV \Rightarrow -U^TV \le 1. \end{aligned}$$

From these two equations it follows that $|U^TV| \le 1$.

The theorem is obvious if either $X = 0$ or $Y = 0$. Now, for any nonzero vectors X and Y, $X/\|X\|_2$ and $Y/\|Y\|_2$ are unit vectors for which we have just shown that

$$\left|\left(\frac{X}{\|X\|_2}\right)^T\left(\frac{Y}{\|Y\|_2}\right)\right| = \frac{|X^TY|}{\|X\|_2\,\|Y\|_2} \leq 1.$$

It follows at once that

$$|X^TY| \leq \|X\|_2\,\|Y\|_2.$$

■

The triangle inequality for the 2-norm follows directly from Theorem 2.3.

THEOREM 2.4

For any vectors X and Y in $\mathscr{R}^n$, $\|X + Y\|_2 \leq \|X\|_2 + \|Y\|_2$.

PROOF Since

$$\begin{aligned}\|X + Y\|_2^2 &= (X + Y)^T(X + Y) = X^TX + 2X^TY + Y^TY \\ &= \|X\|_2^2 + 2X^TY + \|Y\|_2^2,\end{aligned}$$

we have, from the Cauchy-Schwarz inequality,

$$\|X + Y\|_2^2 \leq \|X\|_2^2 + 2\,\|X\|_2\,\|Y\|_2 + \|Y\|_2^2 = (\|X\|_2 + \|Y\|_2)^2.$$

The desired result follows by taking square roots. ■

There are some simple and useful relationships among the three vector norms of Definition 2.3. For example, for any X in $\mathscr{R}^m$ we have

$$\|X\|_\infty = \max_i\{|x_i|\} \leq \|X\|_1 = \sum_{i=1}^{m} |x_j| \leq \sum_{j=1}^{m} \|X\|_\infty = m\,\|X\|_\infty.$$

So, for any X in $\mathscr{R}^m$, $\|X\|_\infty \leq \|X\|_1 \leq m\,\|X\|_\infty$. From these last inequalities we also obtain $(1/m)\,\|X\|_1 \leq \|X\|_\infty \leq \|X\|_1$. We state these useful inequalities as our next theorem. Some similar results involving the 2-norm are included in the exercises.

THEOREM 2.5

For any vector X in $\mathscr{R}^m$, we have

$$\|X\|_\infty \leq \|X\|_1 \leq m\,\|X\|_\infty \quad \text{and} \quad \left(\frac{1}{m}\right)\|X\|_1 \leq \|X\|_\infty \leq \|X\|_1.$$

■

If $\{X^{(n)}\}$ is a sequence of vectors in $\mathscr{R}^m$, then we say that the sequence

$\{X^{(n)}\}$ **converges** to X, and write $\lim_{n\to\infty}\{X^{(n)}\} = X$, if each of the m component sequences $\{x_i^{(n)}\}$ converges to x_i, $i = 1, 2, \ldots, n$, or equivalently, if $\lim_{n\to\infty} |x_i^{(n)} - x_i| = 0$ for each i. Clearly $\lim_{n\to\infty}\{X^{(n)}\} = X$ if and only if $\lim_{n\to\infty}\|X^{(n)} - X\|_\infty = \lim_{n\to\infty}(\max\{|x_i^{(n)} - x_i|\}) = 0$. This is equivalent to saying that $\lim_{n\to\infty} \|X^{(n)} - X\|_\infty = 0$. From the inequalities in Theorem 2.5 it follows that $\lim_{n\to\infty}\{X^{(n)}\} = X$ if and only if $\lim_{n\to\infty} \|X^{(n)} - X\|_1 = 0$. This result holds for any vector norm, although the proof (Halmos, 1974) is beyond the scope of this text. The general result is contained in Theorem 2.6.

THEOREM 2.6

Let $\{X^{(n)}\}$ be a sequence of vectors in $\mathscr{R}^m$ and let ν be any vector norm. Then, $\lim_{n\to\infty}\{X^{(n)}\} = X$ if and only if $\lim_{n\to\infty}(\nu(X^{(n)} - X)) = 0$. An important special case is $\lim_{n\to\infty} X^{(n)} = 0 \Leftrightarrow \lim_{n\to\infty} \nu(X^{(n)}) = 0$. ■

Since the vector space $\mathscr{R}^{n\times n}$ is isomorphic to the vector space $\mathscr{R}^{n^2}$, it is possible to use the above definitions to define norms on $\mathscr{R}^{n\times n}$. This approach fails to take into account the important interpretation of $n \times n$ matrices as representatives of linear operators on $\mathscr{R}^n$ (Cullen, 1972). The standard way of defining norms for operators yields more satisfactory results.

■ DEFINITION 2.4 ■

If ν is a vector norm on $\mathscr{R}^n$, then the **induced matrix norm** on $\mathscr{R}^{n\times n}$ is the function N from $\mathscr{R}^{n\times n}$ to $\mathscr{R}$ defined by

$$\mathrm{N}(A) = \max\left\{\frac{\nu(AX)}{\nu(X)} : X \neq 0\right\} = \max\{\nu(AX) : \nu(X) = 1\}.$$

We will use the same notation for a vector norm and the induced matrix norm. Thus the matrix norm induced by the vector 1-norm is denoted $\|A\|_1$, the matrix norm induced by the Euclidean norm is $\|A\|_2$, and the matrix norm induced by the ∞-norm is $\|A\|_\infty$.

THEOREM 2.7

If ν is a vector norm on $\mathscr{R}^n$ and N is the induced matrix norm, then N is a norm on the vector space $\mathscr{R}^{n\times n}$, and the matrix and vector norms are **compatible** in the sense that

$$\nu(AX) \leq \mathrm{N}(A)\nu(X) \quad \text{for every vector } X \text{ in } \mathscr{R}^n$$

and **consistent** in the sense that $\mathrm{N}(AB) \leq \mathrm{N}(A)\mathrm{N}(B)$.

PROOF If $A \neq 0$, then there exists a vector $X \neq 0$ in $\mathscr{R}^n$ such that $AX \neq 0$.

Therefore

$$\mathrm{N}(A) = \max\left\{\frac{\nu(AX)}{\nu(X)} : X \neq 0\right\} > 0$$

and the induced norm N is positive definite. To see that N is also homogeneous, observe that

$$\mathrm{N}(kA) = \max\left\{\frac{\nu(kAX)}{\nu(X)} : X \neq 0\right\} = \max\left\{\frac{|k|\,\nu(AX)}{\nu(X)} : X \neq 0\right\}$$
$$= |k| \max\left\{\frac{\nu(AX)}{\nu(X)} : X \neq 0\right\} = |k|\mathrm{N}(A).$$

Since $\nu((A + B)X) = \nu(AX + BX) \leq \nu(AX) + \nu(BX)$, by the triangle inequality for ν, it follows that

$$\mathrm{N}(A + B) = \max\left\{\frac{\nu((A + B)X)}{\nu(X)} : X \neq 0\right\}$$
$$\leq \max\left\{\frac{(\nu(AX) + \nu(BX))}{\nu(X)} : X \neq 0\right\}$$
$$\leq \max\left\{\frac{\nu(AX)}{\nu(X)} : X \neq 0\right\} + \max\left\{\frac{\nu(BX)}{\nu(X)} : X \neq 0\right\}$$
$$= \mathrm{N}(A) + \mathrm{N}(B).$$

Thus the induced norm N satisfies the triangle inequality.

Now, since $\mathrm{N}(A) = \max\{\nu(AX)/\nu(X) : X \neq 0\}$, it follows that $\nu(AX)/\nu(X) \leq \mathrm{N}(A)$ for any X in $\mathcal{R}^n$. Thus $\nu(AX) \leq \mathrm{N}(A)\nu(X)$ for any X in $\mathcal{R}^n$, and the vector and matrix norms are compatible as claimed.

Finally, since

$$\nu(A(BX)) \leq \mathrm{N}(A)\nu(BX) \leq \mathrm{N}(A)\mathrm{N}(B)\nu(X),$$

it follows that

$$\frac{\nu(ABX)}{\nu(X)} \leq \mathrm{N}(A)\mathrm{N}(B) \quad \text{for all } X \neq 0.$$

Thus

$$\mathrm{N}(AB) = \max\left\{\frac{\nu(ABX)}{\nu(X)} : X \neq 0\right\} \leq \mathrm{N}(A)\mathrm{N}(B)$$

as claimed. ■

The definition of the induced matrix norm does not provide a reasonable way of computing the norm of a given matrix A, since the definition involves finding the maximum of an infinite set of numbers. It is not difficult to find computationally convenient expressions for the induced matrix norms $\|A\|_1$ and $\|A\|_\infty$; $\|A\|_2$ is much more difficult.

THEOREM 2.8

For any $n \times n$ matrix A, $\|A\|_1$ is the maximum of the absolute column sums and $\|A\|_\infty$ is the maximum of the absolute row sums. That is,

$$\|A\|_1 = \max\left\{\sum_{i=1}^{n} |a_{ij}| : j = 1, 2, \ldots, n\right\} = \max\{\|\mathrm{Col}_j(A)\|_1 : j = 1, 2, \ldots, n\}$$

$$\|A\|_\infty = \max\left\{\sum_{j=1}^{n} |a_{ij}| : i = 1, 2, \ldots, n\right\} = \max\{\|\mathrm{Row}_i(A)\|_1 : i = 1, 2, \ldots, n\}.$$

PROOF Let $M_1 = \max\{\|\mathrm{Col}_i(A)\|_1 : i = 1, 2, \ldots, n\}$. For any $X \neq 0$ in $\mathscr{R}^n$,

$$\begin{aligned}\|AX\|_1 &= \|x_1 \mathrm{Col}_1(A) + x_2 \mathrm{Col}_2(A) + \cdots + x_n \mathrm{Col}_n(A)\|_1 \\ &\leq |x_1|\,\|\mathrm{Col}_1(A)\|_1 + |x_2|\,\|\mathrm{Col}_2(A)\|_1 + \cdots + |x_n|\,\|\mathrm{Col}_n(A)\|_1 \\ &\leq M_1(|x_1| + |x_2| + \cdots + |x_n|) = M_1\,\|X\|_1.\end{aligned}$$

Thus

$$\|AX\|_1/\|X\|_1 \leq M_1$$

for all X. If we can find just one vector X for which this last inequality is an equality, it will follow that $\|A\|_1 = M_1$. If $M_1 = \|\mathrm{Col}_j(A)\|_1$, then for $X = \mathbf{e_j} = \mathrm{Col}_j(I)$ we have $\|A\mathbf{e_j}\|_1/\|\mathbf{e_j}\|_1 = \|\mathrm{Col}_j(A)\|_1/1 = M_1$. Thus $\|A\|_1 = \max\{\|\mathrm{Col}_i(A)\|_1 : i = 1, \ldots, n\}$ is the maximum absolute column sum as claimed.

For the ∞-norm we let X be any nonzero vector in $\mathscr{R}^n$ and observe that

$$\left|\sum_{j=1}^{n} a_{ij}x_j\right| \leq \sum_{j=1}^{n} |a_{ij}|\,|x_j| \leq \sum_{j=1}^{n} |a_{ij}|\,\|X\|_\infty = \|X\|_\infty \sum_{j=1}^{n} |a_{ij}| = \|X\|_\infty\,\|\mathrm{Row}_i(A)\|_1.$$

Therefore

$$\begin{aligned}\|AX\|_\infty &= \max\left\{\left|\sum_{j=1}^{n} a_{ij}x_j\right| : i = 1, 2, \ldots, n\right\} \\ &\leq \|X\|_\infty \max\left\{\sum_{j=1}^{n} |a_{ij}| : i = 1, 2, \ldots, n\right\}.\end{aligned}$$

It follows that

$$\begin{aligned}\frac{\|AX\|_\infty}{\|X\|_\infty} \leq M_2 &= \max\left\{\sum_{j=1}^{n} |a_{ij}| : i = 1, 2, \ldots, n\right\} \\ &= \max\{\|\mathrm{Row}_i(A)\|_1 : i = 1, 2, \ldots, n\}.\end{aligned}$$

If $M_2 = \|\mathrm{Row}_k(A)\|_1$, construct X such that $x_j = 1$ if $a_{kj} \geq 0$ and $x_j = -1$ if $a_{kj} < 0$. Then $\|X\|_\infty = 1$, $\|AX\|_\infty = M_2$, and it follows that $\|A\|_\infty = M_2 = \max\{\|\mathrm{Row}_i(A)\|_1 : i = 1, 2, \ldots, n\}$ is the maximum absolute row sum. ■

No comparable formula exists for the 2-norm of the matrix A. The best we could do is show that $\|A\|_2$ is the square root of the largest eigenvalue of A^TA; this is neither easy nor inexpensive to compute, so we will favor the other two norms when we need to use both vector and matrix norms.

■ EXAMPLE 6

Find $\|A\|_1$ and $\|A\|_\infty$ for the matrix $A = \begin{bmatrix} 3 & -5 & 2 & 1 \\ 2 & 7 & -2 & 0 \\ 4 & 3 & -5 & 2 \\ 1 & 0 & 2 & -4 \end{bmatrix}$.

SOLUTION The absolute column sums are 10, 15, 11, and 7; so $\|A\|_1 = 15$. The absolute row sums are 11, 11, 14, and 7; so $\|A\|_\infty = 14$. ■

■ SOFTWARE NOTES ■

MATLAB will compute both vector and matrix norms. The command is **norm(x,p)** where $p = 1, 2$, or inf.

The MATALG command **normp(x)** will compute vector norms but not matrix norms; here $p = 1, 2$ or 0 (for the ∞-norm). See Exercise 13 for advice on computing matrix norms.

2.3 ■ Exercise Set

1. For any norm v on $\mathcal{R}^n$, show that $|v(X) - v(Y)| \le v(X - Y)$. Interpret this result geometrically in the case $n = 3$.

2. Show that $\|X\|_1$ is a norm on $\mathcal{R}^n$.

3. Show that $\|X\|_\infty$ is a norm on $\mathcal{R}^n$.

4. Let v be any norm on $\mathcal{R}^n$ and let A be an $n \times n$ matrix. What condition on A will ensure that $m(X) \equiv v(AX)$ is a norm on $\mathcal{R}^n$?

5. The **unit ball** in the norm v is the set $\{X \text{ in } \mathcal{R}^n : v(X) \le 1\}$. Sketch the unit ball in $\mathcal{R}^2$ for each of the norms of Definition 2.3.

6. Let A be a symmetric **positive definite** matrix ($X^TAX > 0$ if $X \ne 0$ in $\mathcal{R}^n$). Show, using Exercise 12.b. in Section 2.2, that $v(X) = \sqrt{X^TAX}$ defines a vector norm.

7. Express $(X + tY)^T(X + tY) = \|X + tY\|_2^2 \ge 0$, as a quadratic polynomial in t. Show that the roots of this quadratic must be equal or nonreal and hence the discriminant of the quadratic must be ≤ 0. Use this fact to provide an alternate proof of the Cauchy-Schwarz inequality.

8. For any vector X in $\mathcal{R}^n$, show that
   a. $\|X\|_\infty \le \|X\|_2 \le \sqrt{n}\,\|X\|_\infty$

b. $\|X\|_2 \le \|X\|_1 \le \sqrt{n}\,\|X\|_2$

c. $\dfrac{1}{\sqrt{n}}\|X\|_1 \le \|X\|_2 \le \|X\|_1.$

9. For $A = \begin{bmatrix} 1 & \frac{1}{2} & \frac{1}{3} & \frac{1}{4} & \frac{1}{5} \\ \frac{1}{2} & \frac{1}{3} & \frac{1}{4} & \frac{1}{5} & \frac{1}{6} \\ \frac{1}{3} & \frac{1}{4} & \frac{1}{5} & \frac{1}{6} & \frac{1}{7} \\ \frac{1}{4} & \frac{1}{5} & \frac{1}{6} & \frac{1}{7} & \frac{1}{8} \\ \frac{1}{5} & \frac{1}{6} & \frac{1}{7} & \frac{1}{8} & \frac{1}{9} \end{bmatrix}$, select ten vectors from $\mathscr{R}^5$ and estimate $M = \|A\|_1 = \max\{\|AX\|_1/\|X\|_1 : X \neq 0\}$ and $m = \min\{\|AX\|_1/\|X\|_1 : X \neq 0\}$. (Use the software.)

10. Repeat Exercise 9 for the 2-norm.

11. Show for the ∞-norm, from Exercise 8 and Theorem 2.6, that the vector sequence $X^{(n)}$ converges to X if and only if the scalar sequence $\|X^{(n)} - X\|_2$ converges to 0.

12. Use Theorem 2.8 and Exercise 7 in Section 1.1 to find $\|A\|_1$ and $\|A^{-1}\|_1$ for the matrix A of Exercise 9.

13. Let U be an $n \times 1$ matrix with each entry 1. Show that for any $n \times n$ matrix A, with all $a_{ij} \ge 0$, that

a. $\|A\|_1 = \|U^TA\|_\infty$

b. $\|A\|_\infty = \|AU\|_\infty$.

Use these formulas and MATALG to check Exercises 9 and 12.

14. Show for any nonsingular matrix A, and show for any norm, that $\min\{\|AX\| : \|X\| = 1\} = 1/\|A^{-1}\|$.

15. Use the Cauchy-Schwarz inequality to show that $|X^TAY| \le \|A\|_2$ if X and Y are unit vectors in the 2-norm. Now show that $\|A\|_2 = \max\{|X^TAY| : \|X\|_2 = \|Y\|_2 = 1\}$.

16. Show that $\|A^T\|_2 = \|A\|_2$. Show also that $\|A^TA\|_2 = \|A\|_2^2$.

17. For any $n \times n$ matrix A, show that

a. $\|A\|_1 \le n\|A\|_\infty$

b. $\|A\|_\infty \le \sqrt{n}\,\|A\|_2$

c. $\|A\|_1 \le \sqrt{n}\,\|A\|_2$

d. $\|A\|_2 \leq \sqrt{n}\,\|A\|_\infty$

e. $\|A\|_2 \leq \sqrt{n}\,\|A\|_1$

f. $\|A\|_\infty \leq n\,\|A\|_1$.

18. Show by example that $\|X^{(n)}\|$ converging to $\|X\|$ does not imply that $X^{(n)}$ converges to X.

2.4 ■ ERROR ANALYSIS FOR GAUSSIAN ELIMINATION

We now consider the linear system $AX = B$, where the coefficient matrix A is $n \times n$ and nonsingular. Suppose that the unknown true solution is X_t and that X_c is an approximate solution computed on some floating-point machine. We would like the **error vector** $\mathbf{e} = X_t - X_c$ to be small, that is, to have a small norm, but in general we cannot compute $\mathbf{e}$ because we do not know X_t. What we can compute is the **residual vector** $\mathbf{r} = B - AX_c$. Certainly, it is clear that $\mathbf{e} = 0$ if and only if $\mathbf{r} = 0$, but as the following example shows, a small residual does not always mean that the error vector is small.

■ EXAMPLE 7

Consider the linear system $\begin{bmatrix} 0.757 & 0.546 \\ 0.913 & 0.659 \end{bmatrix}\begin{bmatrix} x \\ y \end{bmatrix} = \begin{bmatrix} 0.211 \\ 0.254 \end{bmatrix}$.

SOLUTION On a 3-digit decimal machine, the computed solution for this linear system (without pivoting or the benefit of double precison accumulation of inner products) would be $y_c = 0.500$ and $x_c = -0.0819$. For this computed solution the residual vector is

$$\mathbf{r} = \begin{bmatrix} 0.211 \\ 0.254 \end{bmatrix} - \begin{bmatrix} 0.757 & 0.546 \\ 0.913 & 0.659 \end{bmatrix}\begin{bmatrix} -0.819 \\ 0.500 \end{bmatrix}$$

$$= \begin{bmatrix} -1.7 \times 10^{-6} \\ -7.25 \times 10^{-4} \end{bmatrix}, \text{ which is certainly small.}$$

It is easy to check that the true solution of this system is $x_t = 1$ and $y_t = -1$. Thus the error vector is

$$\mathbf{e} = X_t - X_c = \begin{bmatrix} 1 \\ -1 \end{bmatrix} - \begin{bmatrix} -0.0819 \\ 0.500 \end{bmatrix}$$

$$= \begin{bmatrix} 1.0819 \\ -1.500 \end{bmatrix}, \text{ which is certainly not small.}$$

In this example the two equations represent nearly coincident lines and the residual is a measure of how far the point, determined by the computed solution, is off the lines (Exercise 1). Thus it is not surprising that a point might be close to both lines without being close to the intersection of the two lines.

EXAMPLE 8

Consider the linear systems $AX = B$ and $AX = (B + \Delta B)$ where

$$A = \begin{bmatrix} 4 & -1 & -1 & 0 & 0 \\ -1 & 4 & -1 & -1 & 0 \\ -1 & -1 & 4 & -1 & -1 \\ 0 & -1 & -1 & 4 & -1 \\ 0 & 0 & -1 & -1 & 4 \end{bmatrix}, \quad B = \begin{bmatrix} 1 \\ 2 \\ 3 \\ 4 \\ 5 \end{bmatrix}, \quad \text{and} \quad \Delta B = \begin{bmatrix} 0 \\ 0 \\ 0 \\ 0.1 \\ 0 \end{bmatrix}.$$

SOLUTION Using either MATLAB or MATALG we find that the solution of $AX = B$ is $X = \begin{bmatrix} 1.7287 \\ 2.5688 \\ 3.3462 \\ 3.2004 \\ 2.8866 \end{bmatrix}$ while the solution of $AX = (B + \Delta B)$ is $X_1 = \begin{bmatrix} 1.7377 \\ 2.5852 \\ 3.3654 \\ 3.2379 \\ 2.9008 \end{bmatrix}$. In this case the small change in the input vector B has generated a small change ΔX in the solution. If we measure the relative change using the 1-norm we obtain

$$\frac{\|X_1 - X\|_1}{\|X\|_1} = \frac{\|\Delta X\|_1}{\|X\|_1} = \frac{0.0963}{13.73} = 0.0070 \quad \text{while} \quad \frac{\|\Delta B\|_1}{\|B\|_1} = \frac{0.1}{15} = 0.0067.$$

Thus, in this case, the relative change in the solution is about the same size as the relative error in the data. We could not reasonably expect better results.

Consider now another input vector $B_1 = \begin{bmatrix} -0.618031 \\ 1 \\ -0.494327 \times 10^{-5} \\ -0.999997 \\ 0.618036 \end{bmatrix}$ and perturbation

$\Delta B_1 = \begin{bmatrix} 0.00627051 \\ 0.00840665 \\ 0.01 \\ 0.00840666 \\ 0.00627051 \end{bmatrix}$. For these vectors we have

$$X_2 = A^{-1}B_1 = \begin{bmatrix} -0.110008 \\ 0.177998 \\ -0.841916 \times 10^{-6} \\ -0.177998 \\ 0.110009 \end{bmatrix} \quad \text{and} \quad X_3 = A^{-1}(B_1 + \Delta B_1)$$

$$= \begin{bmatrix} -0.104118 \\ 0.185895 \\ 0.00939265 \\ -0.170101 \\ 0.115899 \end{bmatrix}.$$

If we again measure the relative changes using the 1-norm, we obtain

$$\frac{\|X_3 - X_2\|_1}{\|X_2\|_1} = \frac{\|\Delta X_2\|_1}{\|X_2\|_1} = 0.0641781 \quad \text{while} \quad \frac{\|\Delta B_1\|_1}{\|B_1\|_1} = 0.0121612.$$

Thus we see that in this case the solution process has magnified the relative change in B by a factor of about 5.278. This is not as good as the first case, but it is still quite acceptable. It can be shown, using results from the next chapter, that this is the worst possible case for this matrix.

Let us now repeat the above calculations using the 5×5 Hilbert matrix:

$$H = \begin{bmatrix} 1 & \frac{1}{2} & \frac{1}{3} & \frac{1}{4} & \frac{1}{5} \\ \frac{1}{2} & \frac{1}{3} & \frac{1}{4} & \frac{1}{5} & \frac{1}{6} \\ \frac{1}{3} & \frac{1}{4} & \frac{1}{5} & \frac{1}{6} & \frac{1}{7} \\ \frac{1}{4} & \frac{1}{5} & \frac{1}{6} & \frac{1}{7} & \frac{1}{8} \\ \frac{1}{5} & \frac{1}{6} & \frac{1}{7} & \frac{1}{8} & \frac{1}{9} \end{bmatrix}.$$

Now the solution of $HX = B$ is $X = \begin{bmatrix} 125 \\ 2880 \\ 14490 \\ -24640 \\ 13230 \end{bmatrix}$ and the solution of $HX =$ $(B + \Delta B)$ is $X_1 = \begin{bmatrix} -15 \\ -192 \\ 2730 \\ -6720 \\ 4410 \end{bmatrix}$. In this case the small change in the input vector

B has generated a large change in the solution. Measuring the relative changes using the 1-norm yields

$$\frac{\|X_1 - X\|_1}{\|X\|_1} = \frac{\|\Delta X\|_1}{\|X\|_1} = \frac{4.1328 \times 10^4}{5.5365 \times 10^4} = 0.7465 \quad \text{while} \quad \frac{\|\Delta B\|_1}{\|B\|_1} = 0.0067.$$

Thus we see that the relative change in the data has been magnified more than 100 times by the solution process. For this matrix this is far from the worst possible case. ■

Clearly the second problem considered in Example 8 is more sensitive to changes in the data than is the first problem. We will now examine this situation in general.

Let us now consider, for the linear system $AX = B$, the effect on the solution vector X of a change ΔB in the input vector B; this change could come from the round-off error associated with data entry or by an actual perturbation of the vector B. The resulting system $AX = B + \Delta B$, will have a solution $X + \Delta X$ for which $A(\Delta X) = \Delta B$. Now $\Delta X = A^{-1}(\Delta B)$ and, using the results of the last section, we have

$$\|\Delta X\| \le \|A^{-1}\,\Delta B\| \le \|A^{-1}\|\;\|\Delta B\|$$

$$\text{and} \quad \|B\| = \|AX\| \le \|A\|\;\|X\| \Rightarrow \frac{1}{\|X\|} \le \frac{\|A\|}{\|B\|}.$$

It follows, from the last two inequalities, that the relative error in X satisfies

$$\frac{\|\Delta X\|}{\|X\|} \le \|A^{-1}\|\;\|\Delta B\|\,\frac{\|A\|}{\|B\|} = \|A\|\;\|A^{-1}\|\,\frac{\|\Delta B\|}{\|B\|}.$$

Thus the solution process cannot magnify the relative error in the data by more than the factor $\|A\|\;\|A^{-1}\|$.

As a measure of the difficulty in solving linear systems with the coefficient matrix A we consider the condition number, which is defined next.

■ **DEFINITION** *2.5* ■

The **condition number** of the matrix A, with respect to the norm $\|A\|$, is defined by $\text{cond}(A) \equiv \|A\|\;\|A^{-1}\|$. The problem of solving the linear system $AX = B$ is said to be **well-conditioned** if the condition number is small, and **ill-conditioned** if the condition number is large.

Note that the condition number depends on the norm as well as on the matrix A. If we wish to specify the norm that we are using, we will write $\text{cond}_1(A)$, $\text{cond}_\infty(A)$, or $\text{cond}_2(A)$. It is easy to see that for any consistent matrix norm, $I = AA^{-1} \Rightarrow 1 = \|I\| \leq \|A\| \|A^{-1}\| = \text{cond}(A)$ and that $\text{cond}(kA) = \text{cond}(A)$. For a diagonal matrix $D = Dg\{d_1, d_2, \ldots, d_n\}$ and any of our norms, $\text{cond}(D) = \max\{d_i\}/\min\{d_j\}$. The first coefficient matrix in Example 8 is well-conditioned; its condition number is $\text{cond}_1(A) = 5.27729$. The second coefficient matrix in Example 8 is ill-conditioned; its condition number is $\text{cond}_1(A) = 9.44 \times 10^5$.

The discussion preceding Definition 2.5 is summarized in our next theorem.

THEOREM 2.9

If X_t is the true solution of the system $AX = B$, then the solution of the perturbed system $AX = B + \Delta B$ is $X_t + \Delta X$, where the relative change in the solution ΔX satisfies

$$\frac{\|\Delta X\|}{\|X_t\|} \leq \|A\| \|A^{-1}\| \frac{\|\Delta B\|}{\|B\|} = \text{cond}(A)\frac{\|\Delta B\|}{\|B\|}. \tag{2.8}$$

Thus we see that the condition number gives an upper bound to how much a relative error in the input vector B can be magnified by the solution process. If the condition number is large, small changes in B can generate large changes in the solution X. Note that small changes in B could result from the rounding that takes place when the data is entered into the machine. If the condition number is small, we can be confident that small changes in B will not drastically change the solution. The second part of Example 8, where the coefficient matrix was the 5×5 Hilbert matrix, shows that the actual change may be substantially less than the upper bound given in Theorem 2.9.

For the matrix of Example 7, $\text{cond}_1(A) = 7192.4$, which explains why the computed solution was so inaccurate. In general we will say that a problem is ill-conditioned if small changes in the data can produce large changes in the answer. For such a problem, no algorithm, on a finite word-length machine, can be expected to give accurate results. The condition number of the coefficient matrix is a reasonable way to measure the condition of a linear equations problem. Ideally we would like to be able to estimate the condition number of A without the expense of computing the inverse of A.

Theorem 2.9 describes how a perturbation in the input vector B may change the solution of the linear system $AX = B$. The next theorem describes how a change in the coefficient matrix might affect the solution.

THEOREM 2.10

Let A and $A + \Delta A$ be $n \times n$ and be nonsingular. If $AX = B$ and $(A + \Delta A)(X + \Delta X) = B$, then

$$\frac{\|\Delta X\|}{\|X + \Delta X\|} \leq \|A\| \, \|A^{-1}\| \frac{\|\Delta A\|}{\|A\|} = \operatorname{cond}(A) \frac{\|\Delta A\|}{\|A\|}.$$

PROOF Since $X + \Delta X = (A + \Delta A)^{-1}B$,

$$\Delta X = (A + \Delta A)^{-1}B - A^{-1}B = ((A + \Delta A)^{-1} - A^{-1})B.$$

We now use the matrix identity $M^{-1} - N^{-1} = N^{-1}(N - M)M^{-1}$ with $M = A + \Delta A$ and $N = A$ to obtain

$$\Delta X = A^{-1}(-\Delta A)(A + \Delta A)^{-1}B = A^{-1}(-\Delta A)(X + \Delta X).$$

Taking norms in the last equation yields

$$\|\Delta X\| \leq \|A^{-1}\| \, \|\Delta A\| \, \|X + \Delta X\|$$

from which it follows that

$$\frac{\|\Delta X\|}{\|X + \Delta X\|} \leq \|A\| \, \|A^{-1}\| \frac{\|\Delta A\|}{\|A\|} = \operatorname{cond}(A) \frac{\|\Delta A\|}{\|A\|}.$$

■

This last result is a bit disappointing since it gives the error relative to the solution of the perturbed system rather than relative to the solution of the original system. It does however indicate that, as in Theorem 2.9, the condition number is an upper bound on the ratio of the relative error in the solution to the relative error in the data.

A somewhat better result is contained in the next theorem.

THEOREM 2.11

Let A be $n \times n$ and nonsingular with $AX = B$ and let ΔA be a perturbation in A with $(A + \Delta A)(X + \Delta X) = B$. If ΔA is small enough that $\|A^{-1}\| \, \|\Delta A\| < 1$, then

$$\frac{\|\Delta X\|}{\|X\|} \leq \frac{\operatorname{cond}(A) \dfrac{\|\Delta A\|}{\|A\|}}{1 - \operatorname{cond}(A) \dfrac{\|\Delta A\|}{\|A\|}}.$$

PROOF As in the proof of Theorem 2.10, we find that

$$\|\Delta X\| \leq \|A^{-1}\| \, \|\Delta A\| \, \|X + \Delta X\|.$$

Applying the triangle inequality to the last factor on the right yields

$$\|\Delta X\| \leq \|A^{-1}\| \, \|\Delta A\| \, \|X + \Delta X\| \leq \|A^{-1}\| \, \|\Delta A\|(\|X\| + \|\Delta X\|)$$

from which we obtain

$$\begin{aligned} \|\Delta X\|\{1 - \|A^{-1}\| \, \|\Delta A\|\} &\leq \|A^{-1}\| \, \|\Delta A\| \, \|X\| \\ &= \|A^{-1}\| \, \|A\| \left(\frac{\|\Delta A\|}{\|A\|}\right) \|X\|. \end{aligned}$$

If $\|A^{-1}\| \, \|\Delta A\| < 1$, then the bracketed quantity on the left is positive, and it follows that

$$\frac{\|\Delta X\|}{\|X\|} \leq \frac{\|A\| \, \|\Delta A\|}{1 - \|A^{-1}\| \, \|\Delta A\|} = \frac{\|A^{-1}\| \, \|A\| \dfrac{\|\Delta A\|}{\|A\|}}{1 - \|A^{-1}\| \, \|A\| \dfrac{\|\Delta A\|}{\|A\|}}$$

$$= \frac{\text{cond}(A) \dfrac{\|\Delta A\|}{\|A\|}}{1 - \text{cond}(A) \dfrac{\|\Delta A\|}{\|A\|}}.$$

■

If the relative error in A is small and cond(A) is not too large, then the denominator of the above fraction is close to 1; Theorem 2.11 says that the ratio of the relative error in the solution vector X, to the relative error in A, is bounded by γ cond(A) where γ is a constant of the order of unity.

The errors discussed so far can all be attributed to inaccuracies in the data. Another source of error is the round-off error in the machine calculations. For Gaussian elimination with partial pivoting it can be shown that the computed solution X_c is the exact solution of a system

$$(A + \Delta A)X_c = B \quad \text{where } |\Delta a_{ij}| \leq 10n^2 \mu \rho$$

and (2.9)

$$\rho = \max\{|a_{ij}^{(k)}| : 1 \leq i, j, k \leq n\}.$$

This result, from James Wilkinson (Wilkinson, 1965), is the basic result of the "inverse error analysis" for Gaussian elimination. It says that unless ρ is large, the Gaussian elimination algorithm, with partial pivoting, is **stable** in the sense that the computed solution is the exact solution of a problem that is close to the original problem; it would be naive to expect more. Although there are examples for which ρ is large, in most practical problems partial pivoting is enough to keep ρ reasonably small. Thus for all practical purposes, Gaussian elimination with partial pivoting is a stable algorithm for moderate-sized problems ($n\mu < 1/3$). An excellent proof of Equation 2.9 is given in (Kincaid and Cheney, 1991). We will not pursue the analysis of round-off error any further in this text.

Note, from Equation (2.9), that the residual $\mathbf{r} = B - AX_c = (\Delta A)X_c$, so that

$$\|\mathbf{r}\| = \|B - AX_c\| = \|(\Delta A)X_c\| \le \|\Delta A\|\,\|X_c\|,$$

and hence

$$\frac{\|\mathbf{r}\|}{\|A\|\,\|X_c\|} = \frac{\|B - AX_c\|}{\|A\|\,\|X_c\|} \le \frac{\|\Delta A\|}{\|A\|}.$$

From the bounds given in Equation (2.9), we can expect that $\|\Delta A\|/\|A\|$ will be small. Thus this last equation indicates that the residual for the computed solution is always small, even when the computed solution is very inaccurate, as it was in Example 7. Therefore, simply computing the residual, as a check of accuracy, is really a waste of time and effort; it will always be small.

The final result of this section shows how you can use the residual and the condition number to estimate the relative error in an approximate solution.

THEOREM 2.12

Let X_c be an approximate solution of $AX = B$ for which the residual vector is $\mathbf{r} = B - AX_c$. Then

$$\frac{\|X_c - X\|}{\|X\|} = \frac{\|\Delta X\|}{\|X\|} \le \operatorname{cond}(A)\frac{\|\mathbf{r}\|}{\|B\|}.$$

PROOF Since $\mathbf{r} = B - AX_c$, $AX_c = B - \mathbf{r}$ and $A(X + X_c - X) = B - \mathbf{r}$. This is equivalent to $A(X + \Delta X) = B + \Delta B$ where $\Delta B = -\mathbf{r}$ and $\Delta X = X_c - X$. From Theorem 2.9 we have

$$\frac{\|X_c - X\|}{\|X\|} = \frac{\|\Delta X\|}{\|X\|} \le \operatorname{cond}(A)\frac{\|\Delta B\|}{\|B\|} = \operatorname{cond}(A)\frac{\|\mathbf{r}\|}{\|B\|}.$$

■

The last inequality clearly indicates that knowing that the condition number of the coefficient matrix A is not too large would substantially increase faith in the computed solutions of the linear system $AX = B$. It would be very desirable to have a Gaussian elimination program able to estimate the condition number of the coefficient matrix without having to compute the inverse. This is actually done in the LINPACK subroutine SGECO and also in the subroutine DECOMP from Forsythe, Malcolm, and Moler (1977). The source code for these subroutines is given in Appendices C and D. DECOMP is written in standard FORTRAN and is quite easy to follow. The LINPACK subroutine is very similar but is a bit harder to follow because of the use of the BLAS (Basic Linear Algebra Subroutines). Note however that it is the use of the BLAS that makes the LINPACK and LAPACK subroutines extremely efficient.

The subroutines DECOMP and SGECO estimate $\text{cond}_1(A)$ by computing $M = \|A\|$ and then estimating

$$\frac{1}{m} = \|A^{-1}\| = \frac{\|A^{-1}Y\|}{\|Y\|} = \frac{\|Z\|}{\|Y\|} \quad \text{where } AZ = Y$$

and $A^TY = \mathbf{v}$ where the entries of $\mathbf{v}$ are ± 1 and are chosen to maximize $\|Y\|$ during the back substitution.

■ SOFTWARE NOTES ■

The MATLAB command **cond(A)** will compute $\text{cond}_2(A)$.

The MATLAB command **norm(A,1)*norm(inv(A),1)** will compute $\text{cond}_1(A)$.

The MATLAB command **norm(A,inf)*norm(inv(A),inf)** will compute $\text{cond}_\infty(A)$.

The subroutine naming convention for LINPACK, is described in Appendix D. A complete description of LINPACK is contained in the user's manual (Dongarra et al., 1979).

Appendix E contains an interactive FORTRAN driver, which can be used to call the subroutines from LINPACK, LAPACK, or the subroutines DECOMP and SOLVE.

2.4 ■ Exercise Set

1. Recall, from calculus, that the distance from the point (x_1, y_1) to the line $ax + by - c = 0$ is $d = \dfrac{ax_1 + by_1 - c}{\sqrt{a^2 + b^2}}$. Use this fact to measure the sum of the distances of the point (x_1, y_1) from the two lines

$$a_{11}x + a_{12}y = c_1 \quad \text{and} \quad a_{21}x + a_{22}y = c_2.$$

Compare with the residual at the approximate solution (x_1, y_1).

2. Show that if A is an **orthogonal matrix** ($A^TA = I$), then $\text{cond}_2(A) = 1$. Show that $\text{cond}_1(A) = \text{cond}_\infty(A)$ for an orthogonal matrix A.

3. Find $\text{cond}_1(A)$ for each of the following matrices:

$$A = \begin{bmatrix} 100 & 99 \\ 99 & 98 \end{bmatrix}, \quad A = \begin{bmatrix} 1 & 2 & 3 \\ 4 & 5 & 6 \\ 7 & 8 & 0 \end{bmatrix}, \quad A = \begin{bmatrix} 6 & 13 & -17 \\ 13 & 29 & -38 \\ -17 & -38 & 50 \end{bmatrix}.$$

4. Find $\text{cond}_\infty(H)$ for the matrix H of Exercise 7 in Section 1.1.

5. For a diagonal matrix $D = Dg\{d_1, d_2, \ldots, d_n\}$, verify that, for any of the three norms of Definition 2.3, $\text{cond}(D) = \max\{d_i\}/\min\{d_i\}$.

6. Let P be a matrix for which $\|P\| < 1$.

   a. Show that if $X \neq 0$, then $\|(I - P)X\| \neq 0$. *Hint*: Use Exercise 1 in Section 2.3.

   b. Show that $(I - P)$ is nonsingular and $\|(I - P)^{-1}\| \leq 1/(1 - \|P\|)$.

   c. Show that $(I - P)^{-1} = I + P + P^2 + P^3 + P^4 + \cdots$.

   d. When would $I + P$ be a good approximation for $(I - P)^{-1}$?

   e. Let A be nonsingular and assume that $\|A^{-1}E\| < 1$. Show that $(A + E)$ is nonsingular and that

$$(A + E)^{-1} = (I + F)A^{-1} \quad \text{where } \|F\| \leq \frac{\|A^{-1}E\|}{(1 - \|A^{-1}E\|)}.$$

7. Show that solving the matrix equation

$$AX = \begin{bmatrix} 1 & 1 & 2 & 1 \\ -1 & 1 & 0 & 1 \\ 2 & 1 & 1 & 0 \\ 1 & 3 & 1 & 0 \end{bmatrix} X = \begin{bmatrix} 1 & 1.01 & 0.998 \\ 2 & 2.02 & 1.997 \\ 3 & 3 & 3 \\ 4 & 2.4 & 3.5 \end{bmatrix}$$

$$= B = [B_1, B_1 + \Delta B_1, B_1 + \Delta B_2]$$

is equivalent to solving the three systems $AX = B_1$, $AX = B_1 + \Delta B_1$, and $AX = B_1 + \Delta B_2$. Solve the matrix equation by row reducing $[A \,\vdots\, B]$ using either MATMAN or MATLAB. Compare the results with Theorem 2.9 and estimate $\text{cond}_1(A)$.

8. Verify that the exact solution of the following system is $x = -1$, $y = 1$:

$$\begin{aligned} 300x + 400y &= 100 \\ 30001x + 40002y &= 10001. \end{aligned}$$

Suppose that the right-hand side of the second equation is changed to 9999.

Show that the new system has the solution $x = 3$ and $y = -2$. Compute the relative changes $\|\Delta X\|_1/\|X\|_1$ and $\|\Delta B\|_1/\|B\|_1$ and then use Theorem 2.9 to estimate $\text{cond}_1(A)$.

9. For the matrix $H = \begin{bmatrix} 1 & \frac{1}{2} & \frac{1}{3} \\ \frac{1}{2} & \frac{1}{3} & \frac{1}{4} \\ \frac{1}{3} & \frac{1}{4} & \frac{1}{5} \end{bmatrix}$ and the perturbation $\Delta H = \begin{bmatrix} 0 & 0 & 0 \\ 0 & 0 & 0 \\ 0 & 0 & \frac{1}{50} \end{bmatrix}$, compute H^{-1}, $(H + \Delta H)^{-1}$, and $\text{cond}_\infty(H)$.

10. For the following matrices, use MATLAB to compute the inverses of A, B, and C as well as the inverses $(A + \Delta M)^{-1}$, $(B + \Delta M)^{-1}$, $(C + \Delta M)^{-1}$:

$$A = \begin{bmatrix} 1 & 1 & 2 & 1 \\ -1 & 1 & 0 & 1 \\ 2 & 1 & 1 & 0 \\ 1 & 3 & 1 & 0 \end{bmatrix},$$

$$B = \begin{bmatrix} 1 & -1 & 2 & 1 \\ 2 & -1 & 0 & 2 \\ 4 & 1 & -11 & -1 \\ 1 & 2 & 3 & 83 \end{bmatrix}, \quad \text{and} \quad C = \begin{bmatrix} 1 & \frac{1}{2} & \frac{1}{3} & \frac{1}{4} \\ \frac{1}{2} & \frac{1}{3} & \frac{1}{4} & \frac{1}{5} \\ \frac{1}{3} & \frac{1}{4} & \frac{1}{5} & \frac{1}{6} \\ \frac{1}{4} & \frac{1}{5} & \frac{1}{6} & \frac{1}{7} \end{bmatrix}$$

$$\Delta M = \begin{bmatrix} 0.001 & 0 & 0 & 0 \\ 0 & 0 & 0 & 0 \\ 0 & 0 & 0 & 0 \\ 0 & 0 & 0.002 & 0 \end{bmatrix}.$$

Use Theorem 2.10 to estimate the 1-condition numbers of A, B, and C.

11. Find the 1-condition number for each of the matrices of Exercise 10.

12. Use the FORTRAN driver in Appendix E to call appropriate LINPACK subroutines to solve any linear system $AX = B$ where A is no larger than 50×50. Test your subroutine on the linear systems of Exercise 3 in Section 1.1, 9 in Section 2.2, and 7 in this section.

13. Use the program from Exercise 12 above to compare the condition number estimate from LINPACK with the results of Exercise 11 above.

14. Compute the condition number, with respect to each of the three norms, for the matrix

$$J = \begin{bmatrix} 3 & 1 & -2 \\ 0 & 3 & 1 \\ 0 & 0 & 1 \end{bmatrix}.$$

15. Establish the following inequalities for the condition numbers of an $n \times n$ matrix A:

a. $(1/n)\,\text{cond}_2(A) \le \text{cond}_1(A) \le n\text{cond}_2(A)$

b. $(1/n)\,\text{cond}_\infty(A) \le \text{cond}_2(A) \le n\text{cond}_\infty(A)$

c. $(1/n^2)\text{cond}_1(A) \le \text{cond}_\infty(A) \le n^2\text{cond}_1(A)$

16. Let X_0 be an approximate solution of the linear system $AX = B$. Define the residual $R = B - AX_0$ and then show that $R = AE$ where $E = X - X_0$ is the error in the approximate solution. This is the basis for the following iterative refinement algorithm:

For $i = 0, 1, 2, \ldots, nc$
  Compute the residual $R = B - AX_i$ (in double precision).
  Solve $AE = R$ (in single precision) and
  Compute $X_{i+1} = X_i + E$ (in double precision).

2.5 ■ POSITIVE DEFINITE MATRICES

There are many interesting special kinds of matrices. Among those that occur most frequently in applications are the symmetric positive definite matrices. A case can be made for viewing the positive definite matrices as the "proper" generalization, to matrices, of the positive numbers.

■ DEFINITION 2.6 ■

The $n \times n$ symmetric matrix A is **positive definite** if

$$X \ne 0 \Rightarrow X^TAX = \sum_{i=1}^{n} \sum_{j=1}^{n} a_{ij}x_ix_j > 0.$$

A is **positive semidefinite** if

$$X \ne 0 \Rightarrow X^TAX \ge 0.$$

Let A be positive definite. Then for $X = \mathbf{e_i} = [0, 0, \ldots, 1, \ldots, 0]^T$, we have $X^TAX = a_{ii} > 0$. Thus each diagonal element of a positive definite matrix must be positive. A more general result is that every principal submatrix of a positive definite matrix must also be positive definite. In particular, if A is partitioned so that $A = \left[\begin{array}{c|c} A_{11} & A_{12} \\ \hline A_{21} & A_{22} \end{array}\right]$, then A_{11} and A_{22} are both positive definite. To see this, let $X_1 \neq 0$ and set $X = \begin{bmatrix} X_1 \\ 0 \end{bmatrix}$. Since $X \neq 0$, we must have $X^TAX = X_1^TA_{11}X_1 > 0$. It follows that A_{11} is positive definite. A similar argument shows that A_{22} is also positive definite.

The next theorem gives one of the most useful properties of positive definite matrices.

THEOREM 2.13

If the matrix A is positive definite, then Gaussian elimination can proceed without pivoting. After the first k columns have been reduced, the $n - k \times n - k$ unreduced lower-right-hand corner is always positive definite. Moreover, A is nonsingular with $\det(A) > 0$, and the largest element of A, in absolute value, appears on the diagonal.

PROOF Let us first observe that if P is nonsingular, then PAP^T is also positive definite; because $X \neq 0 \Rightarrow P^TX \neq 0 \Rightarrow X^T(PAP^T)X = (P^TX)^TA(P^TX) > 0$. Since $a_{11} \neq 0$, we can reduce the first column of A without any row interchanges. Let P be the matrix that describes the row operations used to reduce the first column of A. Then P is a unit lower triangular matrix of the form $P = \left[\begin{array}{c|c} 1 & 0 \\ \hline \mathbf{c} & I_{n-1} \end{array}\right]$ for which $PA = \left[\begin{array}{c|c} a_{11} & \mathbf{r} \\ \hline 0 & A_{22}^{(2)} \end{array}\right]$. Now $PAP^T = \left[\begin{array}{c|c} a_{11} & 0 \\ \hline 0 & A_{22}^{(2)} \end{array}\right]$ is positive definite, so $A_{22}^{(2)}$, the lower-right corner of PA and of PAP^T, is also positive definite. It follows, by induction on the size of the matrix A, that the Gaussian reduction of A can proceed without pivoting and will yield an LU factorization of A in which the diagonal entries of U are positive. Since $\det(P) = 1$, $\det(A) = a_{11}\det(A_{22}^{(2)})$, and it follows by induction that $\det(A) > 0$.

To see that the largest element of A is on the diagonal, we suppose the contrary. Thus suppose that $a_{ij} = a_{ji} > 0$ is the entry of A with the largest absolute value. Then for $X = \mathbf{e_i} - \mathbf{e_j}$ we have

$$X^TAX = a_{ii} - a_{ij} - a_{ji} + a_{jj} < 0,$$

which is impossible since A is positive definite. If $a_{ij} = a_{ji} < 0$ is the element of A with the largest absolute value, then for $X = \mathbf{e_i} + \mathbf{e_j}$,

$X^TAX = a_{ii} + a_{ij} + a_{ji} + a_{jj} < 0$, which is again a contradiction. Thus the largest element of A must appear on the diagonal. ■

If we have computed the LU factorization of a positive definite matrix A, then we can factor out the positive diagonal elements of U to obtain the factorization $A = LDU = LDL^T = (LD^{1/2})(LD^{1/2})^T = L_1L_1^T$. This decomposition is called the **Cholesky factorization** of A.

THEOREM 2.14

Cholesky: The symmetric matrix A is positive definite if and only if there is a unique lower-triangular matrix L, with positive diagonal entries, such that $A = LL^T$.

PROOF If $A = LL^T$ and $X \neq 0$, then $L^TX \neq 0$, since L is nonsingular. Thus $X^TAX = X^TLL^TX = (L^TX)^T(L^TX) = \|L^TX\|_2^2 > 0$, so A is positive definite. We have already proved the converse, but we will give an alternate proof by induction on n. For $n = 1$, $A = a_{11} = \sqrt{a_{11}}\sqrt{a_{11}} = LL^T$. Now suppose that the theorem is true for $n - 1 \times n - 1$ positive definite matrices. Let $A = \begin{bmatrix} A_{11} & A_{12} \\ A_{21} & a_{nn} \end{bmatrix}$ where A_{11} is $n - 1 \times n - 1$ and positive definite. By inductive hypothesis $A_{11} = L_{11}L_{11}^T$, where L_{11} is lower triangular with positive diagonal entries. If we assume that $L = \begin{bmatrix} L_{11} & 0 \\ L_{21} & x \end{bmatrix}$, then $A = LL^T$ is equivalent to

$$LL^T = \begin{bmatrix} L_{11} & 0 \\ L_{21} & x \end{bmatrix}\begin{bmatrix} L_{11}^T & L_{21}^T \\ 0 & x \end{bmatrix} = \begin{bmatrix} L_{11}L_{11}^T & L_{11}L_{21}^T \\ L_{21}L_{11}^T & L_{21}L_{21}^T + x^2 \end{bmatrix} = \begin{bmatrix} A_{11} & A_{12} \\ A_{21} & a_{nn} \end{bmatrix}.$$

From this last equation we see that L_{21} is determined uniquely as the solution of the triangular system $L_{11}L_{21}^T = A_{12}$, and that x is then determined by $x = +\sqrt{(a_{nn} - L_{21}L_{21}^T)}$. To see that x is well defined we use the fact that $\det(A) > 0$. Taking determinants in the last display line yields $(x \det(L_{11}))(x \det(L_{11}^T)) = x^2(\det(L_{11}))^2 = \det(A) > 0$. It follows that $x^2 > 0$ and hence that x is real and positive. Thus the Cholesky factorization of A is uniquely determined by the preceding equations and the unique Cholesky factorization of A_{11} provided by the inductive hypothesis. ■

The redundant inductive proof of Theorem 2.14 was included because it provides the basis for an efficient algorithm for computing the Cholesky factorization.

CHOLESKY FACTORIZATION

1. INPUT: $A(n, n)$ symmetric and positive definite
2. For $k = 1, 2, \ldots, n$
3. For $i = 1, 2, \ldots, k - 1$
4. $$a_{ki} = l_{ki} = l_{ii}^{-1}\left(a_{ki} - \sum_{j=1}^{i-1} l_{ij}l_{kj}\right) \tag{2.10}$$
5. $$a_{kk} = l_{kk} = \sqrt{a_{kk} - \sum_{j=1}^{k-1} l_{kj}^2}$$
6. OUTPUT: $L(n, n)$ lower triangular with positive diagonal entries

We leave it to the reader to show that this algorithm is only half as costly as Algorithm (2.7); that is, it requires only about $(n^3/6)$ multiplications. The positive definite matrices are sufficiently important, as we shall see in Chapter 6, that the LINPACK and LAPACK packages provide separate subroutines for factoring matrices that are known to be positive definite, thus yielding a reduction of 50% in computing time. The EISPACK package also provides separate routines for finding the eigenvalues of positive definite matrices.

■ DEFINITION *2.7* ■

The $n \times n$ matrix A is **(column) diagonally dominant** if in each column the absolute value of the diagonal element is larger than the sum of the absolute values of the off-diagonal elements. That is,

$$|a_{ii}| > \|\mathrm{Col}_i(A) - a_{ii}\mathbf{e_1}\|_1 = \sum_{k=1, \neq i}^{n} |a_{ki}| \quad \text{for } i = 1, 2, \ldots, n.$$

For nonsymmetric matrices there is a companion notion of (row) diagonal dominance. For symmetric matrices the two notions are the same.

Consider the matrices

$$A = \begin{bmatrix} 5 & 1 & 2.1 & -2 \\ -1 & 3 & 1 & 0.9 \\ -1 & 1.9 & 4 & 0 \\ 1 & 0 & 0.5 & 3 \end{bmatrix} \quad \text{and} \quad B = \begin{bmatrix} 4 & -1 & -1 & 0 & 0 \\ -1 & 4 & -1 & -1 & 0 \\ -1 & -1 & 4 & -1 & -1 \\ 0 & -1 & -1 & 4 & -1 \\ 0 & 0 & -1 & -1 & 4 \end{bmatrix}.$$

The matrix A is column diagonally dominant but not row diagonally dominant. The matrix B just misses being diagonally dominant.

A frequently useful way of establishing that a symmetric matrix A is positive definite and hence nonsingular is given in the next theorem. The converse of Theorem 2.15 is false as shown by the matrices in Exercises 3, 6, and 12.

THEOREM 2.15

If A is symmetric, diagonally dominant, and has positive diagonal entries, then A is positive definite.

PROOF We will show that Gaussian elimination can proceed without pivoting and hence that A has a Cholesky factorization. Since A is symmetric and diagonally dominant, a_{11} is larger than any off-diagonal entry in the first row or the first column of A. Thus one step of Gaussian elimination reduces A to

$$A \to A' = \left[\begin{array}{c|c} a_{11} & a_{12} \cdots a_{1n} \\ \hline 0 & A_1 \end{array}\right].$$

We will now show that A_1 is also symmetric, diagonally dominant, and has positive diagonal entries. The matrix A_1 is symmetric since, by the symmetry of A, for $i, j \neq 1$, $\text{ent}_{ij}(A') = a_{ij} - \dfrac{a_{i1}}{a_{11}} a_{1j} = a_{ji} - \dfrac{a_{j1}}{a_{11}} a_{1i} = \text{ent}_{ji}(A')$. Consider now the sum of the off-diagonal elements in column $k - 1$ of A_1. (These elements are in column k of A'.) We have

$$\sum_{i=2, \neq k}^{n} |\text{ent}_{ik}(A')| = \left| a_{2k} - \frac{a_{21}}{a_{11}} a_{1k} \right| + \left| a_{3k} - \frac{a_{31}}{a_{11}} a_{1k} \right| + \cdots + \left| a_{nk} - \frac{a_{n1}}{a_{11}} a_{1k} \right|$$

which, using the triangle inequality in $\mathcal{R}$ and the diagonal dominance of A,

$$\leq \sum_{i=2, \neq k}^{n} |a_{ik}| + \frac{|a_{1k}|}{a_{11}} \sum_{i=2, \neq k}^{n} |a_{i1}| < (a_{kk} - |a_{1k}|) + \frac{|a_{1k}|}{a_{11}} (a_{11} - |a_{k1}|)$$

$$= a_{kk} - \frac{|a_{1k}|\,|a_{k1}|}{a_{11}} = \text{ent}_{kk}(A') = \text{ent}_{k-1,k-1}(A_1).$$

Therefore, A_1 is diagonally dominant with positive diagonal elements. It now follows easily, by induction, that A can be reduced, without row interchanges, to an upper triangular matrix with positive diagonal entries. Thus A has a Cholesky factorization and by Theorem 2.14 is positive definite as claimed. ■

■ SOFTWARE NOTES ■

The LINPACK subroutines SCHDC and SPOCO will compute the Cholesky factorization.

The MATLAB command **Chol(A)** will compute the Cholesky decomposition.

The MATALG command **M LU A** can be used to compute the LU factorization from which you can easily obtain the Cholesky decomposition.

2.5 ■ Exercise Set

1. Show that Algorithm 2.10 requires about $\left(\frac{n^3}{6}\right)$ multiplications.

2. For any $m \times n$ matrix A, show that A^TA is positive semidefinite, with nonnegative diagonal entries. Further show that $\mathcal{NS}(A) = \mathcal{NS}(A^TA)$ and hence that A^TA has the same rank as A.

3. For $A = \begin{bmatrix} 1 & 2 \\ -3 & -4 \end{bmatrix}$, show that A^TA is positive definite but not diagonally dominant.

4. Show that every eigenvalue of a positive definite matrix is positive.

5. Show that $A = \begin{bmatrix} a & b \\ b & c \end{bmatrix}$ is positive definite if and only if $a > 0$ and $ac - b^2 > 0$. For $X = \begin{bmatrix} x \\ y \end{bmatrix}$, show that the graph of $X^TAX = 1$ is an ellipse if and only if A is positive definite.

6. Show that $A = \begin{bmatrix} 0.1 & b \\ b & 100 \end{bmatrix}$ is positive definite if and only if $b^2 < 10$.

7. For any $n \times n$ positive definite matrix A, show that $(a_{ij})^2 < a_{ii}a_{jj}$ for $i \neq j$.

8. Show that $A = \begin{bmatrix} 4 & -1 & 0 & 0 & 0 \\ -1 & 4 & -1 & 0 & 0 \\ 0 & -1 & 4 & -1 & 0 \\ 0 & 0 & -1 & 4 & -1 \\ 0 & 0 & 0 & -1 & 4 \end{bmatrix}$ is positive definite and find its Cholesky factorization.

9. Show that a diagonally dominant matrix must be nonsingular regardless of whether it is symmetric.

10. Use the decomposition from Exercise 8 to solve the system $AX = [1\ 2\ 3\ 4\ 3]^T$.

11. Assume that A is positive definite. Show how to obtain the Cholesky decomposition of any leading principal minor of A from the Cholesky factorization of A.

12. Find a Cholesky factorization for $B = \begin{bmatrix} 4 & -1 & -1 & 0 & 0 \\ -1 & 4 & -1 & -1 & 0 \\ -1 & -1 & 4 & -1 & -1 \\ 0 & -1 & -1 & 4 & -1 \\ 0 & 0 & -1 & -1 & 4 \end{bmatrix}$.

3

Eigenvalues and Eigenvectors

3.1 ■ REVIEW

In this chapter and the next, we will consider numerical methods for computing some, or all, of the eigenvalues and eigenvectors of a square matrix A. This is one of the most frequently encountered computational problems of linear algebra. We begin with a brief review of the basic facts associated with this problem as treated in most beginning linear algebra texts.

■ **DEFINITION 3.1** ■

Let A be an $n \times n$ matrix. A nonzero vector X is a (right) **eigenvector** of A if there exists a scalar λ, real or complex, such that $AX = \lambda X$. The scalar λ is the **eigenvalue** associated with the eigenvector X.

Motivation for the study of eigenvalues and eigenvectors comes from the geometry of linear operators and the study of systems of linear differential equations of the form $Y'(t) = AY(t)$. It is easy to see that $Y = Xe^{\lambda t}$ is a solution of $Y'(t) = AY(t)$ if and only if $AX = \lambda X$, that is, if and only if X is an eigenvector of A and λ is the associated eigenvalue. If $\lambda = a + bi$ is a complex eigenvalue, then oscillatory terms of the form $e^{at} \cos bt$ appear in the solution of $Y'(t) = AY(t)$.

If A is interpreted as a linear transformation on $\mathcal{R}^n$, then the real eigenvectors determine directions (subspaces) on which the action of A is either an expansion ($\lambda > 1$), a contraction ($0 < \lambda < 1$), or a reflection followed by an expansion ($\lambda < -1$) or a contraction ($-1 < \lambda < 0$). Thus a linear transformation has its simplest representation relative to a basis consisting partly or entirely of eigenvectors of A. Complex eigenvalues are associated with rotations of the space. Recall that two different matrix representatives of the same linear operator are similar matrices (Cullen, 1972).

■ **DEFINITION 3.2** ■

The matrices A and B are said to be **similar** if and only if there exists a nonsingular matrix P such that $B = P^{-1}AP$. The matrices A and B are **orthogonally similar** if P is orthogonal ($P^TP = I$), and **unitarily similar** if P is unitary ($P^HP = \bar{P}^TP = I$).

The basic connection between eigenvectors and similarity is presented in the following theorem.

THEOREM 3.1

An $n \times n$ matrix A is similar to a diagonal matrix D, $P^{-1}AP = D = \text{Dg}\{\lambda_1, \lambda_2, \ldots, \lambda_n\}$, if and only if the columns of P are a linearly independent set of eigenvectors of A.

PROOF If we partition P into its columns, $P = [P_1, P_2, \ldots, P_n]$, then $P^{-1}AP = D = \text{Dg}\{\lambda_1, \lambda_2, \ldots, \lambda_n\}$ is equivalent to $AP = PD$ which, if we partition the matrices AP and PD into columns, can be rewritten as

$$[AP_1, AP_2, \ldots, AP_n] = [\lambda_1 P_1, \lambda_2 P_2, \ldots, \lambda_n P_n].$$

It follows that $AP_i = \lambda_i P_i$, so each column of P is an eigenvector of A. The matrix P will have an inverse if and only if its columns are linearly independent. ■

Note that the basic equation $AX = \lambda X$ is equivalent to the homogeneous system $(\lambda I - A)X = 0$. Thus eigenvectors are nonzero solutions of certain homogeneous linear systems and therefore are never unique. Since, by Theorem 1.2, a square homogeneous system has a nonzero solution if and only if the coefficient matrix is singular, it follows that the system $(\lambda I - A)X = 0$ has a nonzero solution if and only if $\det(\lambda I - A) = 0$. This observation is the basis for the first solution of the problem of finding the eigenvalues and eigenvectors of the matrix A:

1. Find the **characteristic polynomial** $c(\lambda) = \det(\lambda I - A)$.
2. Find the eigenvalues by solving the **characteristic equation** $c(\lambda) = \det(\lambda I - A) = 0$.
3. For each eigenvalue λ_i, find a basis for the corresponding **eigenspace** $\mathscr{ES}(\lambda_i) \equiv \mathscr{NS}(\lambda_i I - A) = \{X \text{ in } \mathscr{R}^n : (\lambda_i I - A)X = 0\}$ by solving the homogeneous system

$$(\lambda_i I - A)X = 0.$$

Note that a real matrix A may have some complex eigenvalues and that some eigenvectors may have complex entries. We will now illustrate the basic method for a 3×3 matrix.

■ EXAMPLE 1

Find the eigenvalues and eigenvectors of the matrix $A = \begin{bmatrix} 2 & 1 & 1 \\ 2 & 3 & 2 \\ 1 & 1 & 2 \end{bmatrix}$ and a matrix P such that $P^{-1}AP$ is a diagonal matrix.

SOLUTION The characteristic polynomial of A is

$$c(\lambda) = \det(\lambda I - A) = \det\begin{bmatrix} \lambda - 2 & -1 & -1 \\ -2 & \lambda - 3 & -2 \\ -1 & -1 & \lambda - 2 \end{bmatrix},$$

which we evaluate using the definition of the determinant. Thus

$$\begin{aligned} c(\lambda) &= (\lambda - 2)\det\begin{bmatrix} \lambda - 3 & -2 \\ -1 & \lambda - 2 \end{bmatrix} - (-1)\det\begin{bmatrix} -2 & -2 \\ -1 & \lambda - 2 \end{bmatrix} \\ &\quad + (-1)\det\begin{bmatrix} -2 & \lambda - 3 \\ -1 & -1 \end{bmatrix} \\ &= (\lambda - 2)[(\lambda - 3)(\lambda - 2) - 2] + [-2(\lambda - 2) - 2] - [2 + (\lambda - 3)] \\ &= \lambda^3 - 7\lambda^2 + 11\lambda - 5. \end{aligned}$$

The eigenvalues of A are thus the roots of the characteristic equation

$$c(\lambda) = \lambda^3 - 7\lambda^2 + 11\lambda - 5 = (\lambda - 5)(\lambda - 1)^2 = 0.$$

To help in factoring such polynomials recall that if a monic polynomial equation, with integer coefficients, is to have any rational roots, they must be integers that are factors of the constant term. Thus the rational roots of the above equation, if there are any, must come from the set $\{1, -1, 5, -5\}$.

To find the eigenvectors for $\lambda = 5$ we solve the homogeneous system $(5I - A)X = 0$. The coefficient matrix is

$$(5I - A) = \begin{bmatrix} 3 & -1 & -1 \\ -2 & 2 & -2 \\ -1 & -1 & 3 \end{bmatrix}, \quad \text{which row reduces to } \begin{bmatrix} 1 & 0 & -1 \\ 0 & 1 & -2 \\ 0 & 0 & 0 \end{bmatrix}.$$

Note that the final matrix has a zero row; we expect this because the eigenvalues are chosen so that the matrix $(\lambda I - A)$ is singular. The system $(5I - A)X = 0$ is thus equivalent to the two equations: $x_1 - x_3 = 0$ and

$x_2 - 2x_3 = 0$. The general solution of the system $(5I - A)X = 0$ is thus

$$X = \begin{bmatrix} c_1 \\ 2c_1 \\ c_1 \end{bmatrix} = c_1 \begin{bmatrix} 1 \\ 2 \\ 1 \end{bmatrix} = c_1 X_1 \quad \text{where } c_1 \text{ is arbitrary.}$$

Thus the eigenspace $\mathscr{E}\mathscr{S}(5)$ is one dimensional and X_1 is a basis for that space.

For $\lambda = 1$ we have

$$(I - A) = \begin{bmatrix} -1 & -1 & -1 \\ -2 & -2 & -2 \\ -1 & -1 & -1 \end{bmatrix}, \quad \text{which row reduces to} \quad \begin{bmatrix} 1 & 1 & 1 \\ 0 & 0 & 0 \\ 0 & 0 & 0 \end{bmatrix}.$$

The system $(I - A)X = 0$ thus reduces to the single equation $x_1 + x_2 + x_3 = 0$, and the general solution is then

$$X = \begin{bmatrix} -c_2 - c_3 \\ c_2 \\ c_3 \end{bmatrix} = c_2 \begin{bmatrix} -1 \\ 1 \\ 0 \end{bmatrix} + c_3 \begin{bmatrix} -1 \\ 0 \\ 1 \end{bmatrix} = c_2 X_2 + c_3 X_3$$

where c_2 and c_3 are arbitrary. In this case the eigenspace $\mathscr{E}\mathscr{S}(1)$ is two-dimensional, with $\{X_2, X_3\}$ as a basis.

If we construct a matrix P whose columns are the basis vectors for the eigenspaces, we obtain

$$P = \begin{bmatrix} 1 & -1 & -1 \\ 2 & 1 & 0 \\ 1 & 0 & 1 \end{bmatrix}$$

for which

$$P^{-1}AP = \begin{bmatrix} 5 & 0 & 0 \\ 0 & 1 & 0 \\ 0 & 0 & 1 \end{bmatrix} \quad \text{is a diagonal matrix.}$$

■

For the matrix of Example 1 we had a double eigenvalue ($\lambda = 1$) for which we found a two-dimensional eigenspace. In general we will say that an eigenvalue λ_i has **algebraic multiplicity** k if $c(\lambda) = (\lambda - \lambda_i)^k q(\lambda)$ and $q(\lambda_i) \neq 0$. The dimension of the eigenspace $\mathscr{S}(\lambda_i)$ is called the **geometric**

multiplicity of λ_i. In general, as the next example illustrates, the geometric multiplicity of an eigenvalue can be any integer between 1 and the algebraic multiplicity.

EXAMPLE 2

The matrix $J = \begin{bmatrix} 2 & a & 0 & 0 \\ 0 & 2 & b & 0 \\ 0 & 0 & 2 & c \\ 0 & 0 & 0 & 2 \end{bmatrix}$ has characteristic polynomial $\det(\lambda I - J) = (\lambda - 2)^4$, so $\lambda = 2$ is an eigenvalue of J of algebraic multiplicity 4.

The matrix $(2I - J) = \begin{bmatrix} 0 & a & 0 & 0 \\ 0 & 0 & b & 0 \\ 0 & 0 & 0 & c \\ 0 & 0 & 0 & 0 \end{bmatrix}$ has rank 3 if $abc \neq 0$, rank 2 if one of the superdiagonal entries is zero, rank 1 if two of the superdiagonal entries are zero, and rank 0 if all of these entries are zero. Thus the geometric multiplicity of $\lambda = 2$ is

1 if $abc \neq 0$,
2 if $c = 0$ and $ab \neq 0$,
3 if $b = c = 0$ and $a \neq 0$, and
4 if $a = b = c = 0$.

If the eigenvalues of A are distinct (that is, they all have algebraic multiplicity 1), then each eigenvalue also has geometric multiplicity 1. In this case the eigenvectors are linearly independent and the matrix A is certain to be similar to a diagonal matrix. A more general result is contained in the following theorem from elementary linear algebra.

THEOREM 3.2

The eigenvectors of A, associated with distinct eigenvalues, must be linearly independent.

PROOF Suppose that $AX_i = \lambda_i X_i$, $X_i \neq 0$, and the λ_i are distinct for $i = 1, 2, \ldots, k$. If the eigenvectors are dependent, then there must exist an integer r such that the set $\{X_1, X_2, \ldots, X_r\}$ is independent while the set $\{X_1, X_2, \ldots, X_r, X_{r+1}\}$ is dependent. It follows that X_{r+1} must be a linear combination of the first r eigenvectors

$$X_{r+1} = a_1X_1 + a_2X_2 + a_3X_3 + \cdots + a_rX_r. \tag{3.1}$$

If we multiply Equation (3.1) on the left by A and use the fact that the X_i are eigenvectors, we obtain

$$\lambda_{r+1}X_{r+1} = a_1\lambda_1X_1 + a_2\lambda_2X_2 + a_3\lambda_3X_3 + \cdots + a_r\lambda_rX_r. \tag{3.2}$$

We now multiply Equation (3.1) by λ_{r+1} to obtain

$$\lambda_{r+1}X_{r+1} = a_1\lambda_{r+1}X_1 + a_2\lambda_{r+1}X_2 + a_3\lambda_{r+1}X_3 + \cdots + a_r\lambda_{r+1}X_r. \tag{3.3}$$

Subtracting Equation (3.3) from Equation (3.2) now yields

$$\begin{aligned} 0 = a_1(\lambda_1 - \lambda_{r+1})X_1 &+ a_2(\lambda_2 - \lambda_{r+1})X_2 \\ &+ a_3(\lambda_3 - \lambda_{r+1})X_3 + \cdots + a_r(\lambda_r - \lambda_{r+1})X_r. \end{aligned}$$

Since the set $\{X_1, X_2, \ldots, X_r\}$ is linearly independent by assumption, we must have

$$a_i(\lambda_i - \lambda_{r+1}) = 0 \quad \text{for } i = 1, 2, 3, \ldots, r.$$

The distinctness of the eigenvalues tells us that $(\lambda_i - \lambda_{r+1}) \neq 0$ and hence that $a_i = 0$. From Equation (3.1) it follows that $X_{r+1} = 0$, which is a contradiction. Therefore the X_i must be independent. ■

The next theorem lists some properties shared by similar matrices.

THEOREM 3.3

Similar matrices have the same

1. characteristic polynomial,
2. eigenvalues (but not the same eigenvectors),
3. trace,
4. determinant,
5. rank, and
6. minimum polynomial (the monic polynomial $m(x)$ of least degree such that $m(A) = 0$).

PROOF Using elementary properties of the determinant we see that if $B = P^{-1}AP$, then

$$\begin{aligned} \det(\lambda I - B) &= \det(\lambda P^{-1}IP - P^{-1}AP) = \det(P^{-1}(\lambda I - A)P) \\ &= \det(P^{-1})\det(\lambda I - A)\det(P) = \det(\lambda I - A). \end{aligned}$$

Thus A and B have the same characteristic polynomial. The remainder of the proof is left to the exercises. ■

A nonzero vector Y such that $Y^TA = \lambda Y^T(A^TY = \lambda Y)$ is called a **left eigenvector** of A. The following theorem gives an orthogonality relationship, between the left and right eigenvectors, that will be useful in some of our later work.

THEOREM 3.4

Let X be a right eigenvector for A, $AX = \lambda X$, and let Y be a left eigenvector for A, $A^TY = \mu Y$. If $\lambda \neq \mu$, then X and Y are orthogonal, that is, $Y^TX = 0$.

PROOF Note that $AX = \lambda X \Rightarrow Y^TAX = \lambda Y^TX$ and that $A^TY = \mu Y \Rightarrow Y^TA = \mu Y^T \Rightarrow Y^TAX = \mu Y^TX$. It follows that $\lambda Y^TX = \mu Y^TX$ and, since $\lambda \neq \mu$, that $Y^TX = 0$. ■

For real symmetric matrices, the eigenvalue problem has several special features. As a corollary of Theorem 3.4, since every left eigenvector of a symmetric matrix is also a right eigenvector, it follows that eigenvectors, for distinct real eigenvalues of a symmetric matrix, must be orthogonal. It is also easy to see that all the eigenvalues of a symmetric matrix must be real. To see this, suppose that $A^T = A$ and that $AX = \lambda X$ where $\lambda = a + bi$ and $X = U + iV$ are complex. It follows that $\overline{AX} = A\bar{X} = \bar{\lambda}\bar{X}$, so $\bar{X} = U - iV$ is an eigenvector of A for $\bar{\lambda} = a - bi$. If we set $X^H = \bar{X}^T$(X^H is called the **Hermitian conjugate** of X), we have

$$AX = \lambda X \Rightarrow X^HAX = \lambda X^HX$$

and

$$A\bar{X} = \bar{\lambda}\bar{X} \Rightarrow X^HA = \bar{\lambda}X^H \Rightarrow X^HAX = \bar{\lambda}X^HX.$$

Since X^HX is real and positive, it follows that $\lambda = \bar{\lambda}$, which implies that λ is real. It now follows, from Theorem 3.4, that any symmetric matrix with distinct real eigenvalues must be orthogonally similar to a diagonal matrix. You can construct the orthogonal transforming matrix P as in Example 1, except that you must choose the constants so the eigenvectors have 2-norm one. It will turn out to be unnecessary that the eigenvalues be distinct, so that the following theorem is true. The proof will be completed in Section 3.2.

THEOREM 3.5

Every real symmetric matrix is orthogonally similar to a diagonal matrix. ■

■ EXAMPLE 3

For $A = \begin{bmatrix} -2 & -3 & -3 \\ -3 & 2 & -1 \\ -3 & -1 & 2 \end{bmatrix}$, find an orthogonal matrix P such that P^TAP is diagonal.

SOLUTION The characteristic polynomial of A is $\det(\lambda I - A) = \lambda^3 - 2\lambda^2 - 23\lambda + 60 = (\lambda - 4)(\lambda + 5)(\lambda - 3)$. Thus the eigenvalues of A are 3, 4, and -5. Simple row reductions, like those in Example 1, show that the eigenvectors are

$$X_1 = c_1\begin{bmatrix} -1 \\ 1 \\ 1 \end{bmatrix} \text{ for } \lambda_1 = 4, \quad X_2 = c_2\begin{bmatrix} 0 \\ -1 \\ 1 \end{bmatrix} \text{ for } \lambda_2 = 3,$$

and

$$X_3 = c_3\begin{bmatrix} 2 \\ 1 \\ 1 \end{bmatrix} \text{ for } \lambda_3 = -5.$$

As in Example 1, the matrix $P = \begin{bmatrix} -1 & 0 & 2 \\ 1 & -1 & 1 \\ 1 & 1 & 1 \end{bmatrix}$, whose columns are independent eigenvectors of A, is such that $P^{-1}AP = \text{Dg}\{4, 3, -5\}$. Note that the columns of P are mutually orthogonal, as guaranteed by Theorem 3.4, but are not unit vectors. If we choose $c_1 = 1/\sqrt{3} = 1\ \|\text{col}_1(P)\|_2$, $c_2 = 1/\sqrt{2}$, and $c_3 = 1/\sqrt{6}$, the resulting vectors are both mutually orthogonal and of unit length (**orthonormal**). The matrix $Q = \begin{bmatrix} \frac{-1}{\sqrt{3}} & 0 & \frac{2}{\sqrt{6}} \\ \frac{1}{\sqrt{3}} & \frac{-1}{\sqrt{2}} & \frac{1}{\sqrt{6}} \\ \frac{1}{\sqrt{3}} & \frac{1}{\sqrt{2}} & \frac{1}{\sqrt{6}} \end{bmatrix}$ is thus an orthogonal matrix ($Q^TQ = I \Rightarrow Q^{-1} = Q^T$) for which $Q^{-1}AQ = Q^TAQ = \text{Dg}\{4, 3, -5\}$. ■

The method we have described and illustrated in Examples 1 and 3 presents computational difficulties at each step. The evaluation of the determinant will be difficult and expensive because the entries of the matrix involve the parameter λ. Even if we could find the characteristic polynomial (see Exercises 14 and 16), the polynomial equation $c(\lambda) = 0$ cannot be solved exactly in most cases (life is not an integer). Thus one would have to resort to some sort of approximation method, for example Newton's method, to approximate the roots. For the approximate roots λ_i, the coefficient matrices of the systems $(\lambda_i I - A)X = 0$ would likely be nonsingular but close to being singular; that is, they would be ill-conditioned. In this case it is unlikely that we would be able to compute accurate eigenvectors. The remainder of this chapter and all of Chapter 4 will be devoted to the design and analysis of algorithms that will solve the algebraic eigenvalue problem in an efficient and stable way.

■ SOFTWARE NOTES ■

The MATLAB command **poly(A)** or the MATALG command **P, A** will compute the characteristic polynomial. If the coefficients are stored in the vector **v**, then the command **polyval(v,x)** in MATLAB or **=poly(v,x)** in MATALG will evaluate the characteristic polynomial at the number x.

The MATLAB command **[P,D]=eig(A)** will compute an $n \times n$ matrix P such that $P^{-1}AP = D$ is a diagonal matrix. The columns of P are the eigenvectors of A and the diagonal entries of D are the eigenvalues. If some eigenvalues are complex, then P will also have complex entries.

The MATALG command **=eval(A)** will compute the real and imaginary parts of the eigenvalues of A. If the eigenvalues are all real, then the command **P=evec(A)** will compute the eigenvectors and store them as the columns of a matrix P such that $P^{-1}AP$ is a diagonal matrix. If some eigenvalues of A are complex, the corresponding columns of P contain the real and imaginary parts of the associated eigenvectors. In this case $P^{-1}AP$ will be block diagonal with some 2×2 diagonal blocks.

3.1 ■ Exercise Set

1. For each of the following matrices find the characteristic polynomial, the eigenvalues, and the eigenvectors. If possible find a nonsingular matrix P so that $P^{-1}AP$ is a diagonal matrix.

a. $A = \begin{bmatrix} 5 & 4 & 2 \\ 4 & 5 & 2 \\ 2 & 2 & 2 \end{bmatrix}$

b. $A = \begin{bmatrix} 5 & -1 & 1 \\ -1 & 5 & -1 \\ 1 & -1 & 5 \end{bmatrix}$

c. $A = \begin{bmatrix} 0 & 1 & 0 & 0 \\ 0 & 0 & 1 & 0 \\ 0 & 0 & 0 & 1 \\ -36 & 0 & 13 & 0 \end{bmatrix}$

2. For each of the following symmetric matrices find the eigenvalues and eigenvectors. Verify that the eigenvectors are orthogonal and then find an orthogonal matrix P such that $P^TAP = P^{-1}AP$ is a diagonal matrix.

a. $A = \begin{bmatrix} 4 & 1 \\ 1 & 4 \end{bmatrix}$

b. $A = \begin{bmatrix} 3 & 2 & 2 \\ 2 & 2 & 0 \\ 2 & 0 & 4 \end{bmatrix}$

c. $A = \begin{bmatrix} 2 & -1 & 1 \\ -1 & 2 & -1 \\ 1 & -1 & 2 \end{bmatrix}$

d. $A = \begin{bmatrix} 2 & -1 & 0 \\ -1 & 2 & -1 \\ 0 & -1 & 2 \end{bmatrix}$

3. a. Verify that $Y(t) = Xe^{\lambda t}$ is a solution of $Y'(t) = AY(t)$ if and only if $AX = \lambda X$.

b. Use the result of Example 3 to find a general solution of

$$Y'(t) = \begin{bmatrix} -2 & -3 & -3 \\ -3 & 2 & -1 \\ -3 & -1 & 2 \end{bmatrix} Y(t).$$

c. Find a solution satisfying $Y(0) = [1\ \ 1\ \ 1]^T$.

4. a. Verify, for any $n \times n$ matrix A, that $\det(\lambda I - A) = \lambda^n - \text{tr}(A)\lambda^{n-1} + \cdots + (-1)^n \det(A)$, where the **trace** of A is defined by $\text{tr}(A) = a_{11} + a_{22} + \cdots + a_{nn}$.

b. Show that for any matrix A, the sum of the eigenvalues of A is $\text{tr}(A)$ and the product of the eigenvalues of A is $\det(A)$.

5. For any eigenvalue of A, verify that the geometric multiplicity is less than or equal to the algebraic multiplicity.

6. Show that a real matrix P is orthogonal if and only if its columns are orthonormal vectors (mutually orthogonal unit vectors). What is the corresponding statement about complex matrices?

7. Prove parts two through six of Theorem 3.3.

8. An $n \times n$ complex matrix A is **Hermitian** if $A^H \equiv \bar{A}^T = A$.

a. Write the most general 3×3 Hermitian matrix.

b. Prove that the eigenvalues of a Hermitian matrix must be real and that eigenvectors associated with distinct eigenvalues are orthogonal ($X^H Y = 0$).

(In general, the properties and applications of Hermitian matrices are similar to those of the real symmetric matrices. Thus the Hermitian matrices are the "proper" way to generalize, to complex matrices, the notion of a real symmetric matrix.)

9. Use the software to find the characteristic polynomial of the matrix

$$A = \begin{bmatrix} 1 & 2 & 3 & 4 \\ 2 & 3 & 5 & 6 \\ 3 & 5 & 2 & 7 \\ 4 & 6 & 7 & 1 \end{bmatrix}.$$

10. Use the software to evaluate the characteristic polynomial, from Exercise 9, for 60 evenly spaced values of λ between -6 and 6. In MATLAB, try the commands **v=poly(A); X=-6:2:6; Y=polyval(v,X); plot(X,Y).** In MATALG, try the commands **P,A; v; t=-6; =poly(v,t); L,t,-6,6,60.** What information does this give you about the graph of the characteristic polynomial and the eigenvalues of A?

11. For $X_1 = [1, 1, 1, 1]^T$ and the matrix A of Exercise 9, compute X_{20} where $X_{k+1} = \dfrac{1}{\|AX_k\|_\infty} AX_k$. The MATLAB looping feature or the MATALG **R** (for repeat) function will be useful in this calculation. Is X_{20} close to an eigenvector of A?

12. How are the eigenvectors of A and $P^{-1}AP$ related?

13. Suppose that A is similar to a diagonal matrix $D = P^{-1}AP$. Show that the rows of P^{-1} are left eigenvectors of A and hence that Theorem 3.4 is a consequence of the identity $P^{-1}P = I$.

14. One method for finding the characteristic polynomial of A on a computer is the SOURIAU-FRAME algorithm, which is used by the P command in MATALG.

$B_0 = I$

For $k = 1, 2, \ldots, n$

$A_k = AB_{k-1}$

$c_k = \dfrac{-1}{k} \operatorname{tr}(A_k)$

$B_k = A_k + c_kI.$

The characteristic equation is $\lambda^n + c_1\lambda^{n-1} + c_2\lambda^{n-2} + \cdots + c_{n-1}\lambda + c_n$.

a. Use the software to work through this algorithm for the matrices

$$A = \begin{bmatrix} 2 & -2 & 3 \\ 1 & 1 & 1 \\ 1 & 3 & -1 \end{bmatrix} \quad \text{and} \quad B = \begin{bmatrix} 2 & 1 & -1 & 2 \\ 1 & 3 & 2 & -3 \\ -1 & 2 & 1 & -1 \\ 2 & -3 & -1 & 4 \end{bmatrix}.$$

b. Show that this method requires about n^4 multiplications for an $n \times n$ matrix A.

15. **a.** Let $p(x) = x^n + a_{n-1}x^{n-1} + a_{n-2}x^{n-2} + \cdots + a_2x^2 + a_1x + a_0$ be any monic polynomial of degree n. The **companion matrix** of $p(x)$ is defined to be

$$C = \text{Comp}(p(x)) = \begin{bmatrix} 0 & 0 & 0 & \cdots & 0 & -a_0 \\ 1 & 0 & 0 & \cdots & 0 & -a_1 \\ 0 & 1 & 0 & \cdots & 0 & -a_2 \\ & & & \ddots & & \\ 0 & 0 & 0 & \cdots & 0 & -a_{n-2} \\ 0 & 0 & 0 & \cdots & 1 & -a_{n-1} \end{bmatrix}.$$

Show that the characteristic polynomial of C is $p(\lambda)$. *Hint*: Use the following sequence of elementary row operations on $\lambda I - C$: $R_1 \leftarrow \lambda R_2 + R_1$, $R_1 \leftarrow \lambda^2 R_3 + R_1$, $R_1 \leftarrow \lambda^3 R_4 + R_1, \ldots, R_1 \leftarrow \lambda^{n-1}R_n + R_1$, and then use the expansion by the first row.

b. Show that each eigenvalue of C has geometric multiplicity one.

16. **a.** Show that the characteristic polynomial $c(\lambda)$ of an $n \times n$ matrix A can be computed as follows:

1. Choose n distinct numbers $t_1, t_2, \ldots, t_n$.
2. Evaluate $\det(t_iI - A)$ for $i = 1, 2, \ldots, n$.
3. Determine scalars $x_1, x_2, \ldots, x_n$ such that the polynomial

$$c(\lambda) = x_1 + x_2\lambda + x_3\lambda^2 + x_4\lambda^3 + \cdots + x_n\lambda^{n-1} + \lambda^n$$

passes through the n points $(t_i, \det(t_iI - A))$, $i = 1, 2, \ldots, n$.

b. Use this method to find the characteristic polynomial for the matrix of Exercise 9.

c. Show that this method takes about $n^4/3$ multiplications.

3.2 ■ CANONICAL FORMS

It is easy to see that the relation of similarity ($\approx$) on the set of all $n \times n$ matrices is **reflexive** ($A \approx A$ for all A), **symmetric** ($A \approx B \Rightarrow B \approx A$), and **transitive** ($A \approx B$, $B \approx C \Rightarrow A \approx C$). Such relations are called **equivalence relations** and are known to partition the set of all $n \times n$ matrices into disjoint equivalence classes (Cullen, 1972). An interesting problem is to determine the simplest element of the equivalence class containing a given matrix A, that is, to find the most nearly diagonal matrix similar to a given matrix. Theorem 3.5 gives the solution to this problem if A is symmetric. For nonsymmetric matrices, Theorem 3.1 tells us that A is similar to a diagonal matrix if and only if A has a full set of eigenvectors. If A does not have a full set of eigenvectors (because at least one eigenvalue has geometric multiplicity less than its algebraic multiplicity), then A is not similar to a diagonal matrix. If A is not similar to a diagonal matrix, then A is said to be **defective** or **nondiagonalizable**. If A is similar to a diagonal matrix, then A is said to be **diagonalizable** or **nondefective**. A solution to the problem we raised above is provided by the following theorem of Jordan. (The proof, which is given in Cullen, 1992, is omitted.)

THEOREM 3.6

Jordan canonical form: An $n \times n$ matrix A, with characteristic polynomial

$$c(\lambda) = \det(\lambda I - A) = \prod_{i=1}^{t} (\lambda - \lambda_i)^{e_i}$$

is similar to a block diagonal matrix of the form $J = \mathrm{Dg}\{J_1, J_2, \ldots, J_t\}$. In this block diagonal matrix, each J_i is an $e_i \times e_i$ matrix, which is itself a block diagonal matrix whose diagonal blocks are **simple Jordan matrices** of the form

$$J_r(\lambda) = \lambda I_r + N = \begin{bmatrix} \lambda & 1 & 0 & \cdots & 0 & 0 \\ 0 & \lambda & 1 & \cdots & 0 & 0 \\ 0 & 0 & \lambda & \cdots & 0 & 0 \\ \vdots & \vdots & \vdots & \ddots & \vdots & \vdots \\ 0 & 0 & 0 & \cdots & \lambda & 1 \\ 0 & 0 & 0 & \cdots & 0 & \lambda \end{bmatrix}.$$

■

matrices that are in Jordan canonical form are

$$\left[\begin{array}{ccc:cc} 2 & 1 & 0 & 0 & 0 \\ 0 & 2 & 1 & 0 & 0 \\ 0 & 0 & 2 & 0 & 0 \\ \hdashline 0 & 0 & 0 & 3 & 1 \\ 0 & 0 & 0 & 0 & 3 \end{array}\right], \quad \left[\begin{array}{c:cc:cc} 5 & 0 & 0 & 0 & 0 \\ \hdashline 0 & 5 & 1 & 0 & 0 \\ 0 & 0 & 5 & 0 & 0 \\ \hdashline 0 & 0 & 0 & 6 & 0 \\ 0 & 0 & 0 & 0 & 7 \end{array}\right],$$

and

$$\left[\begin{array}{c:cc:c:cc} 3 & 0 & 0 & 0 & 0 & 0 \\ \hdashline 0 & 4 & 1 & 0 & 0 & 0 \\ 0 & 0 & 4 & 0 & 0 & 0 \\ \hdashline 0 & 0 & 0 & 4 & 0 & 0 \\ \hdashline 0 & 0 & 0 & 0 & 4 & 1 \\ 0 & 0 & 0 & 0 & 0 & 4 \end{array}\right].$$

Theorem 3.6, although a theoretically satisfying result, is very difficult to implement numerically. To see why this is the case, we note that the characteristic polynomial of the matrix

$$J' = J + \Delta J = \begin{bmatrix} 2 & 1 & 0 & 0 \\ 0 & 2 & 1 & 0 \\ 0 & 0 & 2 & 1 \\ \varepsilon^4 & 0 & 0 & 2 \end{bmatrix}$$

is

$$(\lambda - 2)^4 - \varepsilon^4 = ((\lambda - 2)^2 + \varepsilon^2)((\lambda - 2)^2 - \varepsilon^2)$$

so that the eigenvalues of J' are $2 - \varepsilon i$, $2 + \varepsilon i$, $2 - \varepsilon$, and $2 + \varepsilon$. Since for $\varepsilon \neq 0$, these eigenvalues are distinct, Theorem 3.1 tells us that J' is similar to a diagonal matrix. Thus we see that a small change in just one element of the Jordan matrix

$$J = \begin{bmatrix} 2 & 1 & 0 & 0 \\ 0 & 2 & 1 & 0 \\ 0 & 0 & 2 & 1 \\ 0 & 0 & 0 & 2 \end{bmatrix}$$

has changed the Jordan form to

$$\begin{bmatrix} 2-\varepsilon i & 0 & 0 & 0 \\ 0 & 2+\varepsilon i & 0 & 0 \\ 0 & 0 & 2-\varepsilon & 0 \\ 0 & 0 & 0 & 2+\varepsilon \end{bmatrix},$$

a matrix with a very different structure. The above example suggests that it would be very difficult to compute the Jordan canonical form using the approximate arithmetic of a modern digital computer.

A theoretically less satisfactory result, which is however more suitable for numerical work, is described in the next theorem. This theorem provides a canonical form for unitary similarity and a partial solution to the problem of finding a canonical form for orthogonal similarity.

THEOREM 3.7

Schur: Any $n \times n$ complex matrix A is unitarily similar to an upper triangular matrix. If A is real and all the eigenvalues of A are real, then A is orthogonally similar to an upper triangular matrix.

PROOF (Induction on n) The result is obvious for $n = 1$. Thus we suppose that the theorem is true for $n - 1 \times n - 1$ matrices, and consider an $n \times n$ matrix A. Let λ_1 be an eigenvalue of A and let X_1 be an eigenvector of A for which $\|X_1\|_2 = 1$. Note that if λ_1 is real, then so is X_1. Using the Gram-Schmidt orthogonalization process (Cullen, 1972), or the construction described in Exercise 4, we can construct a unitary matrix P_1 whose first column is X_1. P_1 will be a real orthogonal matrix if λ_1 is real. Now since the columns of P_1 are orthonormal, we have

$$P_1^H A P_1 = \begin{bmatrix} X_1^H \\ \mathrm{Col}_2(P_1)^H \\ \vdots \\ \mathrm{Col}_n(P_1)^H \end{bmatrix} [\lambda_1 X_1 \quad A\,\mathrm{Col}_2(P_1) \cdots A\,\mathrm{Col}_n(P_1)] = \left[\begin{array}{c|c} \lambda_1 & A_{12} \\ \hline 0 & A_{22} \end{array}\right]$$

where A_{22} is an $n - 1 \times n - 1$ matrix whose eigenvalues are those of A, excluding λ_1. By the inductive hypothesis, A_{22} is unitarily similar to an upper triangular matrix. Thus there exists an $n - 1 \times n - 1$ unitary matrix P_2 such that $P_2^H A_{22} P_2 = T_2$ is upper triangular. We now have that $P_3 = \left[\begin{array}{c|c} 1 & 0 \\ \hline 0 & P_2 \end{array}\right]$

is an $n \times n$ unitary matrix for which

$$(P_1P_3)^H A(P_1P_3) = P_3^H(P_1^H AP_1)P_3 = \left[\begin{array}{c:c} 1 & 0 \\ \hdashline 0 & P_2^H \end{array}\right]\left[\begin{array}{c:c} \lambda_1 & A_{12} \\ \hdashline 0 & A_{22} \end{array}\right]\left[\begin{array}{c:c} 1 & 0 \\ \hdashline 0 & P_2 \end{array}\right]$$

$$= \left[\begin{array}{c:c} \lambda_1 & A'_{12} \\ \hdashline 0 & P_2^H A_{22} P_2 \end{array}\right] = \left[\begin{array}{c:c} \lambda_1 & A'_{12} \\ \hdashline 0 & T_2 \end{array}\right]$$

is an upper triangular matrix. Now $P = P_1P_3$ is unitary, so we have established the desired result. Note that P is real, and hence orthogonal, if all the eigenvalues of A are real. ■

Exercises 13 and 14 give some insights on how to further reduce the triangular matrix of Schur's theorem in the direction of the Jordan form.

If we apply the result of Schur's theorem to a real symmetric matrix A, then, because the eigenvalues of A are all real, we have an orthogonal matrix P such that $P^{-1}AP = P^TAP = U$ is an upper triangular matrix. Since $U^T = (P^TAP)^T = P^TAP = U$ we see that U is both symmetric and upper triangular; thus U is a diagonal matrix. This argument completes the proof of Theorem 3.5, which was proven in Section 3.1 for the special case of distinct eigenvalues.

If $AX = \lambda X$, then it follows easily that $A^2X = \lambda^2 X$, and if $p(x)$ is any polynomial, then $p(A)X = p(\lambda)X$. A slightly more general result follows from Theorem 3.7.

THEOREM 3.8

If the eigenvalues of the $n \times n$ matrix A are $\lambda_1, \lambda_2, \ldots, \lambda_n$ and $p(x)$ is any polynomial, then the eigenvalues of $p(A)$ are $p(\lambda_1), p(\lambda_2), \ldots, p(\lambda_n)$.

PROOF By Schur's theorem there is a unitary matrix Q such that $Q^HAQ = U$ is upper triangular. It follows that $A = QUQ^H$, so we have

$$A^2 = (QUQ^H)(QUQ^H) = QU(Q^HQ)UQ^H = QU^2Q^H.$$

It follows, by induction on k, that $A^k = QU^kQ^H$ for any positive integer k. Now if

$$p(x) = a_m x^m + a_{m-1}x^{m-1} + \cdots + a_1 x + a_0 = \sum_{k=0}^{m} a_k x^k$$

is any polynomial, then we have

$$p(A) = \sum_{k=0}^{m} a_k A^k = \sum_{k=0}^{m} a_k(QU^kQ^H) = Q\Big(\sum_{k=0}^{m} a_k U^k\Big)Q^H = Qp(U)Q^H,$$

which shows that $p(A)$ and $p(U)$ are unitarily similar. Now, a direct calculation (Exercise 6) shows that if U is upper triangular then so is $p(U)$ and, moreover, that you can obtain the diagonal elements of $p(U)$ by evaluating p at the diagonal elements of U. Thus we have

$$p(U) = p\left(\begin{bmatrix} \lambda_1 & u_{12} & u_{13} & \cdots & u_{1n} \\ 0 & \lambda_2 & u_{23} & \cdots & u_{2n} \\ 0 & 0 & \lambda_3 & \cdots & u_{3n} \\ \vdots & \vdots & \vdots & \ddots & \vdots \\ 0 & 0 & 0 & \cdots & \lambda_n \end{bmatrix}\right)$$

$$= \begin{bmatrix} p(\lambda_1) & * & * & \cdots & * \\ 0 & p(\lambda_2) & * & \cdots & * \\ 0 & 0 & p(\lambda_3) & \cdots & * \\ \vdots & \vdots & \vdots & \ddots & \vdots \\ 0 & 0 & 0 & \cdots & p(\lambda_n) \end{bmatrix}.$$

Since a triangular matrix displays its eigenvalues on the diagonal, it follows that the eigenvalues of $p(U)$, and hence of $p(A)$, are $p(\lambda_i)$ for $i = 1, 2, \ldots, n$. ■

We will now use Theorem 3.8 to develop a useful characterization of two of the eigenvalues of a symmetric matrix A in terms of the maximum and minimum values of the quadratic form, $q(X) = X^TAX$, determined by the matrix A. The quotient X^TAX/X^TX is frequently called the **Rayleigh quotient** for the matrix A at the vector X.

THEOREM 3.9

Let A be a symmetric matrix with eigenvalues $\lambda_1 \geq \lambda_2 \geq \cdots \geq \lambda_n$. Then $\lambda_1 = \max\{(X^TAX/X^TX): X \neq 0\}$ and $\lambda_n = \min\{(X^TAX/X^TX): X \neq 0\}$.

PROOF The matrix $\lambda_1 I - A$ is a polynomial in A and, by Theorem 3.8, its eigenvalues are $0 = \lambda_1 - \lambda_1$ and $\lambda_1 - \lambda_n \geq \lambda_1 - \lambda_{n-1} \geq \cdots \geq \lambda_1 - \lambda_3 \geq \lambda_1 - \lambda_2 \geq 0$. Thus by Exercise 7, $\lambda_1 I - A$ is a positive semidefinite matrix. So for any $X \neq 0$,

$$\frac{X^T(\lambda_1 I - A)X}{X^TX} = \lambda_1 - \frac{X^TAX}{X^TX} \geq 0.$$

Therefore $\lambda_1 \geq X^TAX/X^TX$ for all $X \neq 0$. That is, λ_1 is an upper bound for the set $\{X^TAX/X^TX : X \neq 0\}$. To see that this upper bound is actually achieved, we let $X = X_1$ be an eigenvector for λ_1. Since $AX_1 = \lambda_1 X_1$, the value of the Rayleigh quotient at X_1 is

$$\frac{X_1^T A X_1}{X_1^T X_1} = \frac{X_1^T(\lambda_1 X_1)}{X_1^T X_1} = \lambda_1.$$

Thus we have

$$\lambda_1 = \max\left\{\frac{X^T A X}{X^T X} : X \neq 0\right\}.$$

Now consider the matrix $A - \lambda_n I$ whose eigenvalues, by Theorem 3.8, are $\lambda_1 - \lambda_n \geq \lambda_2 - \lambda_n \geq \cdots \geq \lambda_{n-1} - \lambda_n \geq \lambda_n - \lambda_n = 0$. Therefore $A - \lambda_n I$ is positive semidefinite. Thus for any $X \neq 0$, we must have

$$\frac{X^T(A - \lambda_n I)X}{X^T X} = \frac{X^T A X}{X^T X} - \lambda_n \geq 0,$$

so that λ_n is a lower bound for the set $\{X^T AX/X^T X : X \neq 0\}$. If we now set $X = X_n$, where $AX_n = \lambda_n X_n$, we have $X_n^T A X_n / X_n^T X_n = X_n^T(\lambda_n X_n)/X_n^T X_n = \lambda_n$. Thus we actually reach the lower bound, and we have $\lambda_n = \min\{X^T AX/X^T X : X \neq 0\}$. ■

3.2 ■ Exercise Set

1. Prove that similarity is an equivalence relation on $\mathscr{R}^{n \times n}$, the set of all $n \times n$ matrices.

2. For each of the following matrices find the characteristic and the minimum polynomial:

$$\begin{bmatrix} 2 & 1 & 0 & 0 & 0 \\ 0 & 2 & 1 & 0 & 0 \\ 0 & 0 & 2 & 0 & 0 \\ 0 & 0 & 0 & 3 & 1 \\ 0 & 0 & 0 & 0 & 3 \end{bmatrix}, \quad \begin{bmatrix} 5 & 0 & 0 & 0 & 0 \\ 0 & 5 & 1 & 0 & 0 \\ 0 & 0 & 5 & 0 & 0 \\ 0 & 0 & 0 & 6 & 0 \\ 0 & 0 & 0 & 0 & 7 \end{bmatrix}, \quad \begin{bmatrix} 3 & 0 & 0 & 0 & 0 & 0 \\ 0 & 4 & 1 & 0 & 0 & 0 \\ 0 & 0 & 4 & 0 & 0 & 0 \\ 0 & 0 & 0 & 4 & 0 & 0 \\ 0 & 0 & 0 & 0 & 4 & 1 \\ 0 & 0 & 0 & 0 & 0 & 4 \end{bmatrix}.$$

Hint: Recall that the minimum polynomial must be a divisor of the characteristic polynomial and must be zero at each eigenvalue of A (Cullen, 1972).

3. Let X be a unit vector in $\mathscr{C}^n$, with respect to the 2-norm. Show that $H = I - 2XX^H$ is both Hermitian and unitary. What if X is real?

4. Let $X_1 = \begin{bmatrix} x_1 \\ Y \end{bmatrix}$ be a unit vector in $\mathscr{C}^n$ with x_1 real. Show that the matrix $Q = \left[\begin{array}{c|c} x_1 & Y^H \\ \hline Y & I - \dfrac{1}{1 - x_1} YY^H \end{array}\right]$ is both Hermitian and unitary by computing the matrix of Exercise 3 for a unit vector X in the direction of $X_1 - \mathbf{e}_1$. Show also that if X_1 is real, then Q is symmetric and orthogonal.

5. A complex matrix A is said to be **normal** if $A^HA = AA^H$. A real matrix A is **normal** if $A^TA = AA^T$.

- **a.** Show that all symmetric, Hermitian, orthogonal, and unitary matrices are normal.
- **b.** Show that if A is normal, then so is Q^HAQ for any unitary matrix Q.
- **c.** Show that an upper triangular normal matrix must be a diagonal matrix.
- **d.** Show, from Theorem 3.7, that any normal matrix is unitarily similar to a diagonal matrix.

6. Show that if U is upper triangular, then so is $p(U)$ for any polynomial $p(x)$, and, moreover, that you can obtain the diagonal elements (eigenvalues) of $p(U)$ by evaluating p at the diagonal elements (eigenvalues) of U.

7. Show that a symmetric matrix A is positive definite (semidefinite) if and only if all the eigenvalues of A are positive (nonnegative).

8. The Cayley-Hamilton theorem states that every square matrix satisfies its characteristic equation. That is, if $c(\lambda) = \det(\lambda I - A) = \lambda^n + a_{n-1}\lambda^{n-1} + \cdots + a_1\lambda + a_0$, then $c(A) = A^n + a_{n-1}A^{n-1} + \cdots + a_1A + a_0I = 0$.

- **a.** Show that this theorem is an easy corollary of Exercise 6 if A is similar to a diagonal matrix.
- **b.** Show that this theorem is also true for the simple Jordan matrix $J_r(\lambda)$ and hence that the general result follows from Theorem 3.6.

9.
- **a.** Use the result of Exercise 8 to show that A^{-1} is a polynomial in A.
- **b.** Show that the eigenvalues of A^{-1} are the reciprocals of the eigenvalues of A and that the eigenvectors of A and A^{-1} are the same.

10. Consider the matrix $A = \begin{bmatrix} 1 & 2 & 3 & 4 \\ 2 & 3 & 5 & 6 \\ 3 & 5 & 2 & 7 \\ 4 & 6 & 7 & 1 \end{bmatrix}$. Choose 10 vectors X in $\mathscr{R}^4$ and use the software to compute the Rayleigh quotient for each vector. What do your calculations and Theorem 3.9 tell you about the eigenvalues of A?

11. Let A be any symmetric matrix. Show that $\max\{X^TAX/X^TX : X \neq 0\} = \max\{X^TAX : \|X\|_2 = 1\}$.

12. Show that $\begin{bmatrix} 2 & 1 \\ 0 & 2 \end{bmatrix}$ is similar to $\begin{bmatrix} 2 & \varepsilon \\ 0 & 2 \end{bmatrix}$ for any $\varepsilon \neq 0$. Is the same thing true for any simple Jordan matrix?

13. Find a matrix $P = \begin{bmatrix} I & X \\ 0 & I \end{bmatrix}$ such that $P^{-1}\left[\begin{array}{cc:cc} 1 & 3 & 4 & 2 \\ 0 & 1 & 3 & 7 \\ \hdashline 0 & 0 & 2 & 5 \\ 0 & 0 & 0 & 2 \end{array}\right]P = \begin{bmatrix} 1 & 3 & 0 & 0 \\ 0 & 1 & 0 & 0 \\ 0 & 0 & 2 & 5 \\ 0 & 0 & 0 & 2 \end{bmatrix}$.

Note that $P^{-1} = \begin{bmatrix} I & -X \\ 0 & I \end{bmatrix}$.

14. a. Determine k so that the operations $kR_3 + R_2$ and $-kC_2 + C_3$, applied to

$$A = \begin{bmatrix} \lambda_1 & a_{12} & a_{13} & a_{14} \\ 0 & \lambda_1 & a_{23} & a_{24} \\ 0 & 0 & \lambda_2 & a_{34} \\ 0 & 0 & 0 & \lambda_2 \end{bmatrix}$$

will yield a matrix with 0 in the $(2, 3)$ position.

b. Show that the resulting matrix is similar to A.

c. Apply this method to the matrix of Exercise 13; produce zeros in positions $(2, 3)$, $(1, 3)$, $(2, 4)$, and $(1, 4)$ in that order. Is the order important?

d. Develop an algorithm for reducing the upper triangular matrix of Theorem 3.7 to a block-diagonal matrix in which each of the diagonal blocks is an upper triangular matrix with equal diagonal entries. Can you state a theorem?

15. Construct a proof, like that of Theorem 3.7, which shows that every A in $\mathscr{C}^{n\times n}$ is similar to an upper triangular matrix. Do not assume that P is unitary.

FIGURE 3.1 ■ GEOMETRY OF THE COMPLEX PLANE

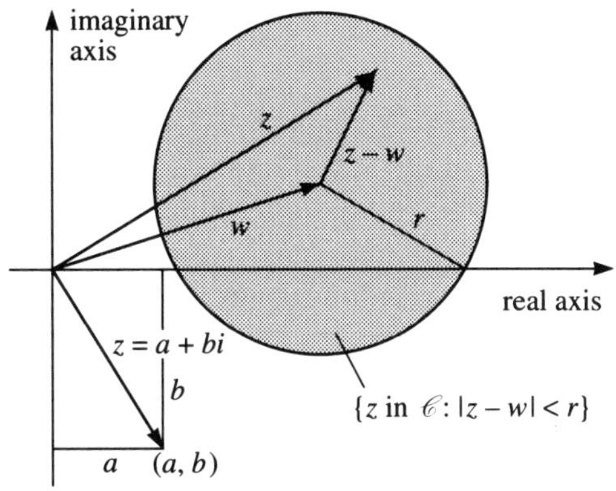

3.3 ■ GERSCHGORIN CIRCLES

Recall that we can associate the complex number $z = a + bi$ with the point in the plane whose coordinates are (a, b), or equivalently with the vector from the origin to the point (a, b). Figure 3.1 illustrates some of the simple geometry of the complex plane.

Note, as a consequence of the definition of the determinant function, that the coefficients of the characteristic polynomial of A are polynomials in the elements of A and hence continuous functions of the elements of A. Moreover, since the roots of a polynomial are continuous functions of the coefficients, it follows that the eigenvalues of a matrix are continuous functions of the elements of the matrix. Sometimes, despite this continuity, small changes in A will generate changes in the eigenvalues that are much larger than the changes in A. (Recall the discussion preceding Theorem 3.7.) In this section, we take a brief look at the stability of the eigenvalue problem.

The eigenvalues of a diagonal matrix are just the diagonal elements of the matrix. The following theorem provides a useful measure of how far the eigenvalues of a matrix are from the diagonal elements of the matrix.

THEOREM 3.10

Gerschgorin: Let A be any $n \times n$ complex matrix and let

$$\mathcal{D}_i = \left\{ z \text{ in } \mathcal{C} : |z - a_{ii}| \le r_i = \sum_{k=1, \neq i}^{n} |a_{ik}| = \|\text{Row}_i(A) - a_{ii}\mathbf{e}_i\|_1 \right\}$$

be the disk in the complex plane with center at a_{ii} and radius r_i. Then all the eigenvalues of A lie within the union of the n disks $\mathcal{D}_i$.

PROOF Let $AX = \lambda X$ with the eigenvector X normalized so that $\|X\|_\infty = 1$. Without loss of generality, let us assume that $x_i = 1$ and $|x_j| \le 1$ for $j \ne i$. Since entry i of AX must equal entry i of λX, we have

$$\sum_{j=1}^{n} a_{ij}x_j = \lambda x_i \Rightarrow \lambda - a_{ii} = \sum_{j=1,\ne i}^{n} a_{ij}x_j \Rightarrow$$

$$|\lambda - a_{ii}| \le \sum_{j=1,\ne i}^{n} |a_{ij}|\,|x_j| \le \sum_{j=1,\ne i}^{n} |a_{ij}| = r_i.$$

Thus λ lies in the disk $\mathscr{D}_i$ and hence in the union $\bigcup_{j=1}^{n} \mathscr{D}_j$. ■

■ EXAMPLE 4

Determine the Gerschgorin region for the matrix

$$A = \begin{bmatrix} 3 & 0.5 & 0 & 1 & -0.5 \\ 0.5 & -3 & 0.5 & -0.5 & 0.5 \\ 0.1 & 0.4 & 5 & -1 & 0 \\ 0.3 & 0.2 & -0.1 & 8 & 0.4 \\ 0.8 & 0.6 & 0.1 & 0 & -1 \end{bmatrix}.$$

SOLUTION The first disk is centered at $z = 3$ and has radius $r_1 = |0.5| + |1| + |-0.5| = 2$. Thus, $\mathscr{D}_1 = \{z \text{ in } \mathscr{C} : |z - 3| \le 2\}$. Similarly, the other Gerschgorin disks are

$$\mathscr{D}_2 = \{z \text{ in } \mathscr{C} : |z + 3| \le 2\},$$
$$\mathscr{D}_3 = \{z \text{ in } \mathscr{C} : |z - 5| \le 1.5\},$$
$$\mathscr{D}_4 = \{z \text{ in } \mathscr{C} : |z - 8| \le 1\},$$

and

$$\mathscr{D}_5 = \{z \text{ in } \mathscr{C} : |z + 1| \le 1.5\}.$$

These five disks are sketched in Figure 3.2. Note that the five disks form

FIGURE 3.2 ■ THE GERSCHGORIN REGION FOR THE MATRIX A OF EXAMPLE 4

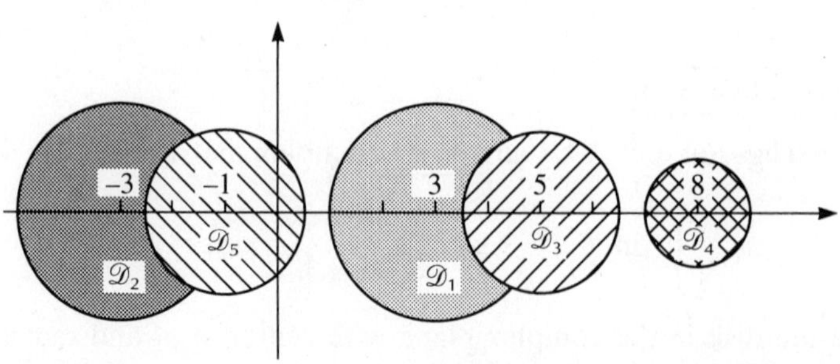

three disjoint regions in the complex plane. The actual values for the eigenvalues of A, computed using the software, are 8.08350, 2.85936, −0.759075, −3.18796, and 5.00417. Note that they are much closer to the diagonal entries of A than predicted by Theorem 3.10.

In Example 4, the eigenvalues were quite close to the centers of the Gerschgorin circles. To show that the eigenvalues can be on or near the boundary of the Gerschgorin region, consider the matrix in Example 5.

EXAMPLE 5

Find the Gerschgorin region and the eigenvalues of $B = \begin{bmatrix} 0 & 0 & 0 & -1 \\ 1 & 0 & 0 & 0 \\ 0 & 1 & 0 & 0 \\ 0 & 0 & 1 & 0 \end{bmatrix}$.

SOLUTION In this case all the Gerschgorin circles have center 0 and radius 1. Thus the Gerschgorin region is just the unit disk $\mathcal{G} = \{z \text{ in } \mathcal{C} : |z| \leq 1\}$. By Exercise 15 of Section 3.1, the characteristic polynomial of A is $\lambda^4 + 1$. Thus the eigenvalues of A are the complex numbers

$$\lambda_1 = e^{(\pi/4)i} = \frac{1+i}{\sqrt{2}}, \qquad \lambda_2 = e^{(3\pi/4)i} = \frac{-1+i}{\sqrt{2}},$$

$$\lambda_3 = e^{(5\pi/4)i} = \frac{-1-i}{\sqrt{2}}, \qquad \lambda_4 = e^{(-\pi/4)i} = \frac{1-i}{\sqrt{2}}$$

all of which lie on the boundary of $\mathcal{G}$ as shown in Figure 3.3.

Because the eigenvalues of A^T are the same as the eigenvalues of A, we can apply Theorem 3.10 to A^T to obtain (possibly) additional information about

FIGURE 3.3 ■ THE GERSCHGORIN REGION FOR THE MATRIX B OF EXAMPLE 5

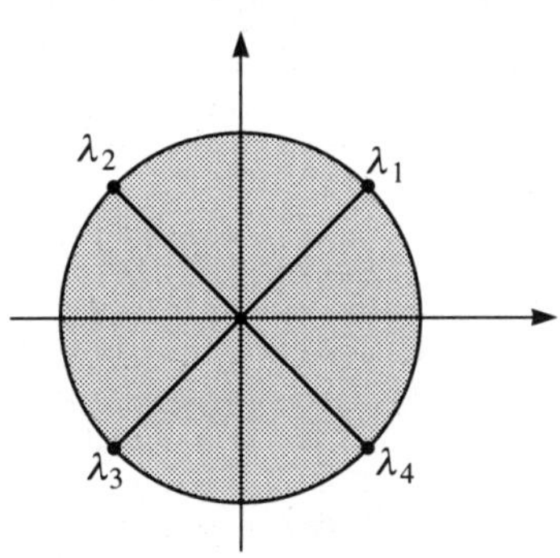

the eigenvalues of A. The resulting region could be called the (column) Gerschgorin region for A while the region described in Theorem 3.10 would be called the (row) Gerschgorin region of A. The (column) Gerschgorin region for the matrix A is

$$\bigcup_{i=1}^{n} \left\{ z \text{ in } \mathscr{C} : |z - a_{ii}| \le c_i = \sum_{j=1, \ne i}^{n} |a_{ji}| \right\}.$$

We leave it to the reader to compare the row and column Gerschgorin regions for the matrices of Examples 4 and 5.

■ DEFINITION *3.3* ■

The set of all eigenvalues of the matrix A is called the **spectrum** of A and is denoted $\Lambda(A)$. The **spectral radius** of A, $\rho(A)$, is the radius of the smallest circle, centered at the origin, which contains the spectrum of A

$$\rho(A) \equiv \max\{|\lambda_i| : \lambda_i \text{ in } \Lambda(A)\}.$$

You can obtain an upper bound for the spectral radius of A by taking the radius of any circle, centered at the origin, which contains the Gerschgorin region of A. The smallest such circle has radius

$$\max\{|a_{ii}| + r_i : i = 1, 2, \ldots, n\} = \max\{\|\text{Row}_i(A)\|_1\} = \|A\|_\infty.$$

Thus $\rho(A) \le \|A\|_\infty$. A similar argument, using the column Gerschgorin region, establishes that $\rho(A) \le \|A\|_1$. It is easy to see, independent of Theorem 3.10, that the spectral radius of A is less than any induced matrix norm of A. (See Exercise 11.)

We can obtain a useful corollary of Theorem 3.10 by considering the Gerschgorin region for a diagonally dominant matrix. For such a matrix, none of the Gerschgorin circles can contain the origin. Thus $\lambda = 0$ is not an eigenvalue of A, so the matrix A is nonsingular. If the diagonally dominant matrix A is symmetric, with positive diagonal entries, then the Gerschgorin circles lie completely in the right half-plane. Since the eigenvalues of a symmetric matrix are always real, it follows that they are all positive and hence, from Exercise 7 of Section 3.2, that A is positive definite. We state this useful result as Theorem 3.11. Note that this argument provides an easy proof of Theorem 2.15.

THEOREM *3.11*

Any diagonally dominant matrix is nonsingular. A symmetric, diagonally dominant matrix with positive diagonal entries is positive definite. ■

Next we present a stronger version of Theorem 3.10. The proof, which depends on topological properties (Wilkinson, 1965), will be omitted.

THEOREM 3.12

If the Gerschgorin region splits into k disjoint subsets, then the ith subset contains as many eigenvalues as there are circles in that subset. ■

The Gerschgorin region pictured in Figure 3.2 splits into three disjoint subsets. Theorem 3.12 states that there are two eigenvalues in $\mathcal{D}_2 \cup \mathcal{D}_5$, two eigenvalues in $\mathcal{D}_1 \cup \mathcal{D}_3$, and one eigenvalue in $\mathcal{D}_4$.

Let A be an $n \times n$ matrix and let ΔA be an $n \times n$ matrix of perturbations. If the elements of ΔA are small, then we expect the eigenvalues of $A + \Delta A$ to be near the eigenvalues of A. We also expect the eigenvectors of $A + \Delta A$ to be near the eigenvectors of A. A complete analysis of this situation is beyond the scope of this text. The interested reader is referred to the *The Algebraic Eigenvalue Problem* by James Wilkinson (Wilkinson, 1965), which treats the problem very thoroughly. We will restrict our attention in this text to a few of the simpler results.

We can gain some insight into how small changes in the matrix affect the eigenvalues by differentiating the basic equation $AX = \lambda X$. Let us assume that differentiable changes in A induce differentiable changes in the simple eigenvalue λ and in the corresponding unit right eigenvector X, $\|X\|_2 = 1$. Let Y be a unit left eigenvector of A for λ ($Y^TA = \lambda Y^T$, $\|Y\|_2 = 1$). To see how a change in a_{ij} affects λ, we will use the product rule and implicit differentiation to compute the partial derivative of λ with respect to a_{ij}. Thus we have

$$\frac{\partial(AX)}{\partial a_{ij}} = \frac{\partial(\lambda X)}{\partial a_{ij}} \Rightarrow \frac{\partial A}{\partial a_{ij}} X + A\frac{\partial X}{\partial a_{ij}} = \frac{\partial \lambda}{\partial a_{ij}} X + \lambda \frac{\partial X}{\partial a_{ij}} \tag{3.4}$$

Now $\partial A/\partial a_{ij} = E_{ij}$, the matrix whose only nonzero entry is a 1 in position (i, j). If we multiply Equation (3.4) on the left by Y^T and use the fact that $Y^TA = \lambda Y^T$, we obtain

$$Y^T(E_{ij}X) + Y^TA\frac{\partial X}{\partial a_{ij}} = \frac{\partial \lambda}{\partial a_{ij}} Y^TX + \lambda Y^T\frac{\partial X}{\partial a_{ij}} \Rightarrow Y^T(x_j\mathbf{e_i}) = \frac{\partial \lambda}{\partial a_{ij}} Y^TX.$$

It follows that

$$\frac{\partial \lambda}{\partial a_{ij}} = \frac{y_ix_j}{Y^TX} \quad \text{with } |y_ix_j| < 1. \tag{3.5}$$

Thus we see that $\partial\lambda/\partial a_{ij}$ will be large if and only if Y^TX is small. If $n = 3$, $Y^TX = \cos\theta$, where θ is the angle between X and Y. Thus we see that the problem of computing the eigenvalue λ is ill-conditioned if the left eigenvector X and the right eigenvector Y are close to orthogonal. The number $1/|Y^TX|$ is thus a condition number for the eigenvalue λ. It is quite possible that some of the eigenvalues of A are well-conditioned while other eigenvalues of A are ill-conditioned. If A is symmetric (or Hermitian), then

$X = Y$ and $Y^T X = X^T X = 1$. This shows that the eigenvalue problem for a symmetric matrix is always well-conditioned.

Our next theorem yields a global condition number for the eigenvalue problem for a diagonalizable matrix A. The theorem gives an upper bound on the distance of an eigenvalue of $A + \Delta A$ from the spectrum of A. This result is known as the Bauer-Fike theorem.

THEOREM 3.13

Bauer-Fike: If the matrices A and $A + \Delta A$ are both diagonalizable, with $P^{-1}AP = D = \mathrm{Dg}\{\lambda_1, \lambda_2, \ldots, \lambda_n\}$, and if μ is an eigenvalue of $A + \Delta A$, then

$$\min_{\lambda \varepsilon \Lambda(A)} \{|\lambda - \mu|\} \leq \|P\| \, \|P^{-1}\| \, \|\Delta A\| = \mathrm{cond}(P) \, \|\Delta A\|.$$

PROOF Let μ be an eigenvalue of $A + \Delta A$. If μ is an eigenvalue of A, the theorem is obvious. Thus we assume that μ is not an eigenvalue of A so that the diagonal matrix $(D - \mu I)$ is nonsingular. Since μ is an eigenvalue of $A + \Delta A$, it follows that $(A + \Delta A - \mu I)$ is singular and so is

$$\begin{aligned} P^{-1}(A + \Delta A - \mu I)P &= D + P^{-1}\Delta A P - \mu I \\ &= (D - \mu I)(I + (D - \mu I)^{-1}P^{-1}\Delta A P). \end{aligned}$$

Since $(D - \mu I)$ is nonsingular by assumption, $I + (D - \mu I)^{-1}P^{-1}\Delta A P$ must be singular. It follows from Exercise 10, or Exercise 6 in Section 2.4, that

$$\begin{aligned} 1 &\leq \|(D - \mu I)^{-1}P^{-1}\Delta A P)\| \leq \|(D - \mu I)^{-1}\| \, \|P^{-1}\| \, \|P\| \, \|\Delta A\| \\ &= \rho((D - \mu I)^{-1}) \, \|P^{-1}\| \, \|P\| \, \|\Delta A\| = \frac{1}{\min\{|\lambda_i - \mu|\}} \mathrm{cond}(P) \, \|\Delta A\|, \end{aligned}$$

since for any of our three norms, the norm of a diagonal matrix is the same as its spectral radius. Finally we have the desired result

$$\min_{\lambda \varepsilon \Lambda(A)} \{|\lambda - \mu|\} \leq \|P\| \, \|P^{-1}\| \, \|\Delta A\| = \mathrm{cond}(P) \, \|\Delta A\| .$$

■

Theorem 3.13 shows that the condition number of the eigenvector matrix P is an overall condition number for the eigenvalue problem for the matrix A. If the matrix A is symmetric, then the diagonalizing matrix P can be assumed to be orthogonal. For an orthogonal matrix P it is easy to see that $\|P\| = 1$ for any of our three norms. Thus we see again that the eigenvalue problem for symmetric matrices is always well-conditioned.

3.3 ■ Exercise Set

1. a. Sketch the Gerschgorin regions for each of the following matrices:

$$T = \begin{bmatrix} 2 & -1 & 0 & 0 \\ -1 & 2 & -1 & 0 \\ 0 & -1 & 2 & -1 \\ 0 & 0 & -1 & 2 \end{bmatrix}, \quad F = \begin{bmatrix} 4 & 1 & 1 \\ 0 & 2 & 1 \\ -2 & 0 & 9 \end{bmatrix},$$

$$S = \begin{bmatrix} 3 & 1 & -1 \\ -1 & -2 & 0.5 \\ 1 & 0.5 & 0 \end{bmatrix}, \quad U = \begin{bmatrix} 1 & -1 & 0 & -1 \\ 1 & 1 & 1 & 0 \\ 0 & -1 & 1 & -1 \\ 1 & 0 & 1 & 1 \end{bmatrix}.$$

b. Use the software to find the eigenvalues of each of the matrices in part a. Show them on the sketches.

2. What does the Gerschgorin theorem tell you about the eigenvalues of

$$M = \begin{bmatrix} 1 & 10^{-3} & 10^{-3} & 10^{-3} \\ 10^{-3} & 1 & 10^{-3} & 10^{-3} \\ 10^{-3} & 10^{-3} & 2 & 1 \\ 10^{-3} & 10^{-3} & 10^{-3} & 2 \end{bmatrix}?$$

3. Find the eigenvectors of the matrix $J = \begin{bmatrix} 2 & 1 \\ 0 & 2+\varepsilon \end{bmatrix}$ and thus find a nonsingular matrix P such that $P^{-1}JP$ is a diagonal matrix. Find the limit of P as $\varepsilon \to 0$.

4. What is the strongest statement you can make about the eigenvalues of the matrix

$$C = \begin{bmatrix} 1 & -0.5 & -0.5 & 0 \\ -0.5 & 1.5 & i & 0 \\ 0 & -0.5i & 5 & 0.5i \\ -1 & 0 & 0 & 5i \end{bmatrix}$$

using Theorems 3.10 and 3.12.

5. Prove that if a real matrix has disjoint Gerschgorin disks, then the eigenvalues must all be real.

6. Let A be an $n \times n$ matrix that is similar to a diagonal matrix; $P^{-1}AP = D = \mathrm{Dg}\{\lambda_1, \ldots, \lambda_n\}$. Describe the Gerschgorin region for the matrix $P^{-1}(A + \Delta A)P$ and interpret your result as a perturbation theorem for the eigenvalues of A.

7. Let $A = \begin{bmatrix} 101 & 110 \\ -90 & -98 \end{bmatrix}$. Find the eigenvalues and the eigenvectors of A and find a nonsingular matrix P such that $P^{-1}AP$ is diagonal. Find cond(P) and use Theorem 3.13 to estimate how changing a_{11} to 100 might change the eigenvalues. Compute the eigenvalues of the new matrix and compare the actual perturbations in the eigenvalues with the above estimates.

8. Consider the 8×8 matrix $F = \begin{bmatrix} 8 & 7 & 6 & 5 & 4 & 3 & 2 & 1 \\ 7 & 7 & 6 & 5 & 4 & 3 & 2 & 1 \\ 0 & 6 & 6 & 5 & 4 & 3 & 2 & 1 \\ 0 & 0 & 5 & 5 & 4 & 3 & 2 & 1 \\ 0 & 0 & 0 & 4 & 4 & 3 & 2 & 1 \\ 0 & 0 & 0 & 0 & 3 & 3 & 2 & 1 \\ 0 & 0 & 0 & 0 & 0 & 2 & 2 & 1 \\ 0 & 0 & 0 & 0 & 0 & 0 & 1 & 1 \end{bmatrix}$.

a. Show that the determinant of $F = 1$, and hence that the product of the eigenvalues of F, is 1.

b. Use the software to find the eigenvalues of F.

c. Change the $(1, 8)$ entry of F to 1.0001 and compute the eigenvalues of the perturbed matrix $F + \Delta F$. Compute the relative change in each eigenvalue.

d. Find $\det(F + \Delta F)$.

e. Use the software to find the right eigenvectors X_i of F and construct the matrix P of right eigenvectors. Construct the matrix $Q = P^{-1}$ of left eigenvectors of F (right eigenvectors of F^T). Scale the columns of Q so that they are unit vectors. Then compute the product Q^TP and thus find the condition numbers $s_i = 1/Y_i^TX_i$ for each of the eigenvalues.

f. Compute $\text{cond}_1(P) = \|P\|_1 \|P^{-1}\|_1$.

9. Let $B = \begin{bmatrix} 2 & -1 & 0 & 0 & 0 \\ -1 & 2 & -1 & 0 & 0 \\ 0 & -1 & 2 & -1 & 0 \\ 0 & 0 & -1 & 2 & -1 \\ 0 & 0 & 0 & -1 & 2 \end{bmatrix}$ and let $\Delta B = \begin{bmatrix} 0 & 0 & 0 & 0 & 0 \\ 0 & 0 & \varepsilon & 0 & 0 \\ 0 & \varepsilon & 0 & 0 & 0 \\ 0 & 0 & 0 & 0 & \varepsilon \\ 0 & 0 & 0 & \varepsilon & 0 \end{bmatrix}$. For $\varepsilon = 0.0001$ compute the eigenvalues of B and of $B + \Delta B$. What are the relative changes in the eigenvalues generated by the perturbation ΔB?

10. **a.** Let $\|A\|_\alpha$ be the matrix norm induced by the vector norm $\|\mathbf{x}\|_\alpha$, $\alpha = 1, 2, \infty$. Show that $\|A\|_\alpha \geq \rho(A)$ with equality if A is a diagonal matrix.

b. Show that if $I + B$ is singular then $\|B\|_\alpha \geq 1$.

11. Let $\|A\|$ be any induced matrix norm. Show, without reference to Theorem 3.10, that $\rho(A) < \|A\|$.

12. Let A be $n \times n$ and diagonalizable; $P^{-1}AP = D = \text{Dg}\{\lambda_1, \ldots, \lambda_n\}$. Let μ and Y be an approximate eigenvalue, eigenvector pair with $\|Y\|_2 = 1$. Show that $Y = (A - \mu I)^{-1}\mathbf{r} = P(D - \mu I)^{-1}P^{-1}\mathbf{r}$ and hence that the residual vector $\mathbf{r} = AY - \mu Y$ satisfies $\min_i \{|\lambda_i - \mu|\} \leq \text{cond}(P)\,\|\mathbf{r}\|_2$.

13. Let A be an $n \times n$ diagonalizable matrix, and let λ be an eigenvalue of A with right eigenvector X $(AX = \lambda X)$ and left eigenvector Y $(A^TY = \lambda Y \Rightarrow Y^TA = \lambda Y^T)$. Let the perturbation matrix $\Delta A = \varepsilon B$ with $\|B\|_2 = \|A\|_2$ and let $\lambda(\varepsilon)$ and $X(\varepsilon)$ satisfy

$$(A + \Delta A)X(\varepsilon) = \lambda(\varepsilon)X(\varepsilon) \quad \text{where} \quad \lambda(0) = \lambda \quad \text{and} \quad X(0) = X.$$

Show that the relative error in the eigenvalue satisfies

$$\left|\frac{\lambda(\varepsilon) - \lambda}{\lambda}\right| \approx |\varepsilon| \left|\frac{\lambda'(0)}{\lambda}\right| \leq \frac{\|\Delta A\|_2}{|\lambda|} \frac{\|X\|_2\,\|Y\|_2}{|Y^TX|}.$$

Hint: Differentiate $(A + \Delta A)X(\varepsilon) = \lambda(\varepsilon)X(\varepsilon)$ with respect to λ, set $\lambda = 0$, and then multiply by Y^T. Then use the Cauchy-Schwarz inequality.

14. Compare the Gerschgorin regions for the matrices B and $X^{-1}BX$ where

$$B = \begin{bmatrix} 1 & \frac{1}{2} & \frac{1}{2} \\ \frac{1}{2} & 4 & \frac{1}{2} \\ \frac{3}{4} & \frac{1}{2} & 6 \end{bmatrix} \quad \text{and} \quad X = \begin{bmatrix} 1 & 0 & 0 \\ 0 & \frac{5}{3} & 0 \\ 0 & 0 & \frac{15}{7} \end{bmatrix}.$$

3.4 ■ THE POWER METHOD

All of the machine methods for computing eigenvalues and eigenvectors are iterative in nature. That is, they compute a sequence of approximations to the eigenvector and/or the related eigenvalue. In discussing such methods we will need to consider questions of convergence of the related sequences and criteria for stopping the machine calculations.

The method we will consider in this section can be viewed as a generalization of Picard's method, or functional iteration, for finding a solution of the equation $x = f(x)$. This method finds an approximate solution to the equation as the limit of the sequence $x_k = f(x_{k-1})$, where x_0 is an initial approximation to a solution. The sequence is known to converge to a solution if f is a contraction, that is, if $|f'(x)| < 1$ in an appropriate interval containing x_0 and the solution.

EXAMPLE 6

For the matrix $A = \begin{bmatrix} 1 & 2 & 3 & 4 \\ 2 & 3 & 5 & 6 \\ 3 & 5 & 2 & 7 \\ 4 & 6 & 7 & 1 \end{bmatrix}$ and the vector $X_0 = \begin{bmatrix} 1 \\ 1 \\ 1 \\ 1 \end{bmatrix}$ we compute a few terms of the vector sequence defined by $X_k = AX_{k-1}$:

$$X_1 = AX_0 = \begin{bmatrix} 10 \\ 16 \\ 17 \\ 18 \end{bmatrix}, \quad X_2 = AX_1 = \begin{bmatrix} 165 \\ 261 \\ 270 \\ 273 \end{bmatrix},$$

$$X_3 = AX_2 = \begin{bmatrix} 2589 \\ 4101 \\ 4251 \\ 4389 \end{bmatrix}, \quad X_4 = AX_3 = \begin{bmatrix} 41100 \\ 65070 \\ 67497 \\ 69108 \end{bmatrix}.$$

(A sequence like this is often called a **Krylov sequence**.)

We can check to see if X_3 is close to an eigenvector of A by comparing the ratios of the components of X_3 and $X_4 = AX_3$. Thus we have

$$\frac{41100}{2589} = 15.8749, \quad \frac{65070}{4101} = 15.8669, \quad \frac{67497}{4251} = 15.8779,$$

and

$$\frac{69108}{4389} = 15.7457.$$

From these calculations we conclude that X_3 is quite close to an eigenvector of A for an eigenvalue that is approximately $\lambda = 15.8$. ■

To better understand what is going on in Example 6, we assume that the $n \times n$ matrix A is similar to a diagonal matrix with eigenvectors Y_i, $i = 1, 2, \ldots, n$, satisfying $AY_i = \lambda_i Y_i$. Let us also assume that the eigenvalues are ordered so that λ_1 is the **dominant eigenvalue** in the sense that

$$|\lambda_1| > |\lambda_2| \geq |\lambda_3| \geq \cdots \geq |\lambda_n|.$$

Since A is diagonalizable, the eigenvectors Y_i form a basis for $\mathcal{R}^n$. Thus every vector in $\mathcal{R}^n$, including X_0, is uniquely expressible as a linear combination of the Y_i. Thus there exist scalars a_i such that

$$X_0 = a_1 Y_1 + a_2 Y_2 + a_3 Y_3 + \cdots + a_n Y_n.$$

Now since $AY_i = \lambda_i Y_i$, we have

$$\begin{aligned} X_1 = AX_0 &= a_1 A Y_1 + a_2 A Y_2 + \cdots + a_n A Y_n \\ &= a_1 \lambda_1 Y_1 + a_2 \lambda_2 Y_2 + \cdots + a_n \lambda_n Y_n \end{aligned}$$

and

$$X_2 = AX_1 = a_1 \lambda_1^2 Y_1 + a_2 \lambda_2^2 Y_2 + \cdots + a_n \lambda_n^2 Y_n.$$

It follows easily by induction that

$$X_k = AX_{k-1} = a_1 \lambda_1^k Y_1 + a_2 \lambda_2^k Y_2 + a_3 \lambda_3^k Y_3 + \cdots + a_n \lambda_n^k Y_n.$$

If we rewrite the last expression for X_k by factoring out λ_1^k, we obtain

$$X_k = \lambda_1^k \left\{ a_1 Y_1 + a_2 \left(\frac{\lambda_2}{\lambda_1}\right)^k Y_2 + a_3 \left(\frac{\lambda_3}{\lambda_1}\right)^k Y_3 + \cdots + a_n \left(\frac{\lambda_n}{\lambda_1}\right)^k Y_n \right\} \tag{3.6}$$

and we begin to see what is happening. Since $|\lambda_1| > |\lambda_i|$, each of the fractions λ_i/λ_1 is, in absolute value, less than 1. Thus, each of the terms $(\lambda_i/\lambda_1)^k \to 0$, as k gets large and X_k approaches a scalar multiple of Y_1, an eigenvector for the dominant eigenvalue. It is apparent from Equation (3.6) that the sequence $\{X_k\}$ will actually converge to a finite nonzero vector only if $\lambda_1 = 1$; otherwise it will diverge or approach zero. It is however correct to say that $\lim_{k\to\infty} (X_k - a_1\lambda_1^k Y_1) = 0$.

We can interpret the above calculation geometrically by observing that each time we multiply by A, the approximate eigenvector X_k "rotates" so that it more nearly points in the direction of the dominant eigenvector Y_1. This point of view is helpful as it suggests that we can scale the vectors in some convenient manner to keep the sequence from diverging. Ideally we would like to take advantage of the fact that

$$\lim_{k\to\infty} (X_k - \lambda_1^k a_1 Y_1) = 0 \quad \text{is equivalent to} \quad \lim_{k\to\infty} \frac{1}{\lambda_1^k} X_k = a_1 Y_1. \tag{3.7}$$

In general we cannot use Equation (3.7) as the basis for an iteration, because we do not know λ_1. Since all we are interested in is the direction of the vectors, we can use any convenient scale factor to keep the sequence from diverging. Thus we could replace the sequence $X_k = AX_{k-1}$ with a sequence of the form $X_k = (1/s_k)AX_{k-1}$. The most commonly used scale factors are $s_k = \|AX_{k-1}\|_p$, for $p = 1, 2$, or ∞. The eigenvalue λ_1 can then be approximated from the eigenvector approximation X_k via any one of the following observations:

$$\frac{\text{ent}_i(AX_k)}{\text{ent}_i(X_k)} \to \lambda_1, \quad \|AX_k\|_p \to \lambda_1, \quad \frac{X_k^T A X_k}{X_k^T X_k} \to \lambda_1. \tag{3.8}$$

We will show in Chapter 5 that the Rayleigh quotient (the last of the above methods) provides the best estimate of the eigenvalue λ_1 available from the approximate eigenvector X_k.

EXAMPLE 7

Let us compute a few terms of the scaled sequence for the matrix A of Example 6. We scale by the ∞-norm and do the calculations using the software:

MATLAB	MATALG
for r=1:11 **Y=A*X;, X=Y/norm(Y,inf)** **end**	**X=(1/NORMO(A*X))*A*X** **R 10**

The output for the first 11 iterates is

k	x_1	x_2	x_3	x_4
0	1	1	1	1
1	0.555556	0.888889	0.944444	1
2	0.604396	0.956044	0.989011	1
3	0.589884	0.934381	0.968558	1
4	0.594721	0.941570	0.976689	1
5	0.592994	0.939015	0.973544	1
6	0.593624	0.939945	0.974734	1
7	0.593391	0.939601	0.974286	1
8	0.593478	0.939729	0.974454	1
9	0.593445	0.939681	0.974391	1
10	0.593458	0.939699	0.974415	1
11	0.593453	0.939693	0.974406	1

The sequence certainly seems to be converging, and X_{11} should be a good approximation to an eigenvector for λ_1. The Rayleigh quotient estimate for the associated eigenvalue can be computed using the software and yields the approximation $\lambda_1 = X_{11}^T A X_{11} / X_{11}^T X_{11} = 15.8328$. ■

The results of the above discussion are summarized in the following theorem. This method of estimating the dominant eigenvector and its eigenvalue is called the **power method**.

THEOREM 3.14

Let A be an $n \times n$ diagonalizable matrix with eigenvalues λ_i such that

$$|\lambda_1| > |\lambda_2| \geq |\lambda_3| \geq \cdots \geq |\lambda_n|.$$

Let $\{Y_i : i = 1, 2, \ldots, n\}$ be a basis for $\mathscr{R}^n$ consisting of eigenvectors of A with $AY_i = \lambda_i Y_i$, and let $X_0 = a_1 Y_1 + a_2 Y_2 + \cdots + a_n Y_n$ be arbitrary except that $a_1 \neq 0$. Then, the sequence defined by

$$X_k = \frac{1}{\|AX_{k-1}\|} AX_{k-1} \tag{3.9}$$

either converges to a scalar multiple of Y_1 or, if λ_1 is negative, flip-flops between Y_1 and $-Y_1$. The sequence of Rayleigh quotients converges to λ_1, that is,

$$\lim_{k\to\infty} \frac{X_k^T A X_k}{X_k^T X_k} = \lambda_1.$$

■

A more efficient implementation of the basic power method algorithm, which stops when two successive eigenvector estimates are closer than some user-specified tolerance TOL or when the number of iterations exceeds a preset value MAX, follows. Note that if the dominant eigenvalue is negative, then $Z + X$ gets small. The algorithm provides for this case by testing $\|Z + X\| < \text{TOL}$.

POWER METHOD ALGORITHM

(3.10)

1. INPUT: $A(n, n)$, $X(n, 1)$ (arbitrary), TOL, MAX
2. For $i = 1, 2, \ldots,$ MAX
3. $\quad Z = AX$
4. $\quad s = \|Z\|$
5. $\quad Z = (1/s)Z$
6. $\quad$ If $\|Z - X\| < \text{TOL}$ or $\|Z + X\| < \text{TOL}$, Exit loop
7. $\quad X = Z$
8. $\lambda = \dfrac{Z^T A Z}{Z^T Z}$
9. OUTPUT: Dominant eigenvalue λ and its eigenvector X
10. Stop

The rate of the convergence of the sequences generated by Algorithm (3.10) is controlled by the ratio $|\lambda_2/\lambda_1|$, which is called the **dominance ratio** of the matrix A. If the dominance ratio is close to 1, the convergence will be quite slow, while if the dominance ratio is small, the convergence will be rapid. Note that

$$(0.5)^{10} \approx (0.9)^{66} \approx (0.99)^{690} \approx 0.97 \times 10^{-3},$$

which gives some indication of the relative number of iterations required, to provide comparable accuracy, for matrices with dominance ratios of 0.5, 0.9, and 0.99.

Theorem 3.14 has several hypotheses that are difficult to verify for a given matrix A: It will not be easy to decide if A is similar to a diagonal matrix (unless A is symmetric), it will not be apparent if A has a dominant eigenvalue, and it will certainly not be clear that the starting vector X_0 has a nonzero component in the direction of the dominant eigenvector Y_1 ($a_1 \neq 0$).

If we were so unfortunate as to choose the starting vector X_0 such that $a_1 = 0$, then, for the unscaled iteration, we would have

$$\begin{aligned} X_k &= AX_{k-1} = a_2\lambda_2^k Y_2 + a_3\lambda_3^k Y_3 + \cdots + a_n\lambda_n^k Y_n \\ &= \lambda_2^k\left\{a_2 Y_2 + a_3\left(\frac{\lambda_3}{\lambda_2}\right)^k Y_3 + \cdots + a_n\left(\frac{\lambda_n}{\lambda_2}\right)^k\right\}, \end{aligned}$$

which suggests that we could expect to obtain an eigenvector for the second-largest eigenvalue, provided that $a_2 \neq 0$ and $|\lambda_2| > |\lambda_3|$. In practice it may still happen that the sequence will converge to Y_1. This is because the round-off error associated with the machine arithmetic may introduce a small component in the direction of Y_1; in this case the sequence may begin to converge to Y_2 and then slowly change direction and converge to Y_1. Exercise 4 illustrates this case. Essentially the same behavior would be expected if a_1 were much smaller than a_2. If you suspect that this situation is occurring, it may be prudent to restart the iteration with a different initial vector X_0.

The matrix A could fail to have a dominant eigenvalue because $\lambda_1 = \lambda_2$, because $\lambda_1 = -\lambda_2$, or because $\lambda_2 = \overline{\lambda_1}$. In the first case the scaled sequence (3.9) would converge to a scalar multiple of $a_1 Y_1 + a_2 Y_2$, which is another eigenvector for the dominant eigenvalue $\lambda_1 = \lambda_2$. Thus this case presents no difficulty.

If $\lambda_1 = -\lambda_2$, then $(\lambda_2/\lambda_1)^k = (-1)^k = \pm 1$, and the sequence (3.9) will eventually flip-flop between a scalar multiple of $a_1 Y_1 + a_2 Y_2$ and a scalar multiple of $a_1 Y_1 - a_2 Y_2$, each of which is an eigenvector for A^2 associated with the eigenvalue λ_1^2. We can easily detect this situation (see Exercise 3, matrix B), and in this case we get a bonus: two eigenvectors and two eigenvalues. To see how to find the eigenvectors of A from the iterates, we note that if $X_{2k} = b_1 Y_1 + b_2 Y_2$ then,

$$AX_{2k} = \lambda_1 b_1 Y_1 - \lambda_1 b_2 Y_2 \quad \text{and} \quad \lambda_1 X_{2k} = \lambda_1 b_1 Y_1 + \lambda_1 b_2 Y_2.$$

It follows that

$$\lambda_1 X_{2k} + AX_{2k} = 2b_1\lambda_1 Y_1 \quad \text{and} \quad \lambda_1 X_{2k} - AX_{2k} = 2b_2\lambda_1 Y_2$$

are eigenvectors of A for λ_1 and $\lambda_2 = -\lambda_1$.

If $\lambda_2 = \overline{\lambda_1}$, then $(\lambda_2/\lambda_1)^k$ is a complex number of unit modulus that moves about the unit circle as the iteration proceeds. In this case the iterates will not converge and we will observe rather wild oscillations. This case is at least detectable. In Chapter 5 we will discuss one way of estimating the complex eigenvalues from three successive terms of the Krylov sequence. There is no way that the power method, applied to a real matrix with a real starting vector, can converge to a complex eigenvector. Exercise 3, matrix C, illustrates how the power method behaves when the dominant eigenvalue of a real matrix is complex.

If A is not similar to a diagonal matrix, we can still show that the sequence (3.9) will converge. The analysis requires a closer look at the Jordan canonical form and is omitted. In this case the convergence is apt to be quite slow. Exercise 3, matrix A, illustrates this case and Exercise 11 discusses the convergence in a special case.

If the eigenvalues of A are $\{\lambda_1, \lambda_2, \lambda_3, \ldots, \lambda_n\}$, then the eigenvalues of $A - kI$ will be $\{\lambda_1 - k, \lambda_2 - k, \lambda_3 - k, \ldots, \lambda_n - k\}$. If k is near λ_1, then the dominant eigenvalue of $A - kI$ will not be $\lambda_1 - k$, and the power method, applied to $A - kI$, will produce an eigenvector of $A - kI$, which is also an eigenvector of A. This new eigenvector will be different from the dominant eigenvector of A. In this case k is called an origin shift. The use of origin shifts is explored in several of the exercises. If the dominant eigenvalue is complex, then the use of a complex shift and a complex starting vector will produce the dominant complex eigenvector. Exercise 12 illustrates the use of complex shifts.

■ EXAMPLE 8

Use the shift $k = 15$ to find another eigenvalue and eigenvector for the matrix of Example 6.

SOLUTION If we use the starting vector $X_0 = [1\ 2\ 3\ 4]^T$ and infinity norm scaling, we obtain, using the software, the following table of values:

k	x_1	x_2	x_3	x_4	$\|X_k\|_\infty$
1	0.7895	0.8947	0.1053	−1.0000	19.0000
2	−0.5566	−0.6290	−0.0656	1.0000	23.2632
3	0.5053	0.5917	0.1485	−1.0000	20.4593
4	−0.4582	−0.5506	−0.2162	1.0000	20.6110
5	0.4196	0.5138	0.2752	−1.0000	20.6498
6	−0.3878	−0.4810	−0.3263	1.0000	20.6877
7	0.3614	0.4519	0.3703	−1.0000	20.7207
8	−0.3395	−0.4265	−0.4082	1.0000	20.7495
9	0.3213	0.4042	0.4407	−1.0000	20.7744
10	−0.3062	−0.3849	−0.4686	1.0000	20.7958
11	0.2936	0.3682	0.4924	−1.0000	20.8143
12	−0.2831	−0.3538	−0.5127	1.0000	20.8302
13	0.2743	0.3414	0.5300	−1.0000	20.8438
14	−0.2670	−0.3308	−0.5447	1.0000	20.8554
15	0.2608	0.3217	0.5572	−1.0000	20.8653
16	−0.2557	−0.3139	−0.5678	1.0000	20.8762

From the preceding table, we can conclude that $X_{16} = [-0.2557 \ -0.3139 \ -0.5678 \ 1]^T$ is an approximate eigenvector of $A - 15I$ and of A. The corresponding eigenvalue of $A - 15I$ is -20.88. For the original matrix A it follows that X_{16} is an approximate eigenvector for the eigenvalue $\lambda_4 = -20.88 + 15 = -5.88$.

■ SOFTWARE NOTES ■

The MATALG command **R n** will repeat the last formula n times. If that formula is of the form $x = f(x)$, then the result will be n steps of functional iteration. There is no facility in MATALG for repeating a series of steps.

MATLAB provides for the repetition of any number of steps. Looping can be accomplished by beginning a series of steps with the command **for k=1:n** and ending with the command **end**. Loops can also begin with a **while** command.

MATLAB will do complex arithmetic. Set **i=sqrt(-1)** and then enter complex numbers in the form **a+b*i**.

In MATLAB it is often helpful to use the fact that ending a command with a semicolon will suppress the printing of that value. This is useful in controlling the appearance of the output.

3.4 ■ Exercise Set

1. For a 4 × 4 matrix A, the power method, as defined by Equation (3.9), with ∞-norm scaling, yields the vectors

$$X_0 = \begin{bmatrix} 1 \\ 1 \\ 1 \\ 1 \end{bmatrix}, \quad \ldots, \quad X_{10} = \begin{bmatrix} 0.99929 \\ -0.99950 \\ 0.99980 \\ 1 \end{bmatrix}, \quad X_{11} = \begin{bmatrix} 0.99977 \\ -1 \\ 0.99973 \\ 0.99996 \end{bmatrix},$$

and

$$AX_{11} = \begin{bmatrix} 11.9990 \\ -11.9990 \\ 11.9978 \\ 11.9978 \end{bmatrix}.$$

a. What can you conclude about the eigenvalues and eigenvectors of A?

b. The power method, applied to the matrix $B = A - 10I$, yields the vectors

$$X_0 = \begin{bmatrix} 1 \\ 2 \\ 3 \\ 4 \end{bmatrix}, \quad \ldots, \quad X_9 = \begin{bmatrix} 0.91226 \\ 0.74353 \\ 0.83127 \\ -1 \end{bmatrix}, \quad X_{10} = \begin{bmatrix} -0.94556 \\ -0.15163 \\ -0.20607 \\ 1 \end{bmatrix},$$

and

$$X_{11} = \begin{bmatrix} 0.955860 \\ -0.351881 \\ -0.307741 \\ -1 \end{bmatrix}.$$

What can you say about the eigenvalues and eigenvectors of the matrix $B = A - 10I$? What can you say about the eigenvalues and eigenvectors of A?

c. The power method, applied to the matrix $C = (A - 3I)^{-1}$, yields the vectors

$$X_0 = \begin{bmatrix} 1 \\ 1 \\ 1 \\ 1 \end{bmatrix}, \quad \ldots, \quad X_4 = \begin{bmatrix} 1 \\ 0.997442 \\ -0.997211 \\ 0.995261 \end{bmatrix}, \quad X_5 = \begin{bmatrix} -0.999693 \\ -1 \\ 0.999165 \\ -0.999405 \end{bmatrix},$$

and

$$X_6 = \frac{1}{0.999629} \begin{bmatrix} 0.999509 \\ 0.999617 \\ -0.999510 \\ 0.999629 \end{bmatrix} = \begin{bmatrix} 0.999876 \\ 0.999988 \\ -0.999881 \\ 1 \end{bmatrix}.$$

What can you conclude about the eigenvalues and eigenvectors of C? What can you say about the eigenvalues and eigenvectors of A?

2. Use the power method to find the dominant eigenvalue and eigenvector of

$$A = \begin{bmatrix} 1 & 1 & -1 \\ 1 & 2 & -1 \\ 1 & -1 & 3 \end{bmatrix}, \quad B = \begin{bmatrix} 3 & 1 & 2 & 5 \\ 2 & 1 & 3 & 7 \\ 3 & 1 & 2 & 4 \\ 4 & 1 & 3 & 2 \end{bmatrix},$$

and

$$C = \begin{bmatrix} 2 & -1 & 0 & 0 & 0 \\ -1 & 2 & -1 & 0 & 0 \\ 0 & -1 & 2 & -1 & 0 \\ 0 & 0 & -1 & 2 & -1 \\ 0 & 0 & 0 & -1 & 2 \end{bmatrix}.$$

3. For each of the following matrices use $X_0 = [1\ 1\ 1\ 1]^T$ and compute 20 terms of the sequence (3.9). Interpret the results and find as many eigenvalues and eigenvectors as possible from the iteration.

$$A = \begin{bmatrix} 7 & -2 & 2 & 2 \\ 5 & 0 & 2 & 2 \\ -4 & 2 & 2 & -3 \\ -4 & 2 & -1 & 0 \end{bmatrix}, \quad B = \begin{bmatrix} -3 & 6 & -12 & 0 \\ -5 & 8 & -12 & 0 \\ -5 & 4 & -3 & -2 \\ 1 & -2 & 6 & 1 \end{bmatrix},$$

$$C = \begin{bmatrix} -10 & 5 & 12 & -8 \\ -8 & 4 & 12 & -6 \\ -3 & 0 & 5 & -3 \\ 9 & -5 & -6 & 7 \end{bmatrix}.$$

4. For $A = \begin{bmatrix} 30 & -4 & -2 & 34 \\ 34 & -8 & -2 & 34 \\ -20 & 4 & 10 & -32 \\ -20 & 4 & 1 & -23 \end{bmatrix}$ and $X_0 = \begin{bmatrix} 1 \\ 2 \\ 1 \\ 0 \end{bmatrix}$ compute 10 terms of the sequence (3.9) using 2-norm scaling, and then estimate the dominant eigenvalue. Compute another 400 terms of the sequence and estimate the dominant eigenvalue again. Now take $X_0 = [1\ 1\ 1\ 1]^T$, compute 25 terms of the sequence (3.9), and interpret the results. Explain what is going on here.

5. Let μ_1 be an approximation to the dominant eigenvalue of A. What is the dominant eigenvalue of $A - \mu_1 I$? Use this observation and the power method to find a second eigenvalue and eigenvector for the matrices A and C of Exercise 3. What happens if we apply the power method to $A + I$?

6. Describe how Exercise 4 in Section 3.2 and the power method could be used to construct the orthogonal triangularizing matrix of Schur's theorem. Do you think that this would be an efficient algorithm?

7. Let A be $n \times n$ and symmetric with eigenvalues $\lambda_1 > \lambda_2 \geq \lambda_3 \geq \cdots \geq \lambda_n$. Let $AX_1 = \lambda_1 X_1$ with $\|X_1\|_2 = 1$. Show that the eigenvalues of $A_1 = A - \lambda_1 X_1 X_1^T$ are $0, \lambda_2, \lambda_3, \ldots, \lambda_n$. What are the eigenvectors? Use this deflation method to find a second eigenvalue and eigenvector for the matrix A of Exercise 1.

8. **a.** Suppose that the eigenvalues of A are $\{9, 8, 3, 2\}$. Find the dominance ratio for $A - kI$; $k = 1, 2, 3, \ldots, 10$ and for $k = 5.5$. Sketch the dominance ratio versus k.

b. Show that in general the dominance ratio is smallest for $A - kI$ if $k = (\lambda_2 + \lambda_n)/2$.

9. Find a way to change the scaling in sequence (3.9) so that it will converge even if the dominant eigenvalue of A is negative.

10. Let A be an $n \times n$ symmetric matrix with eigenvalues $\{\lambda_1, \lambda_2, \ldots, \lambda_n\}$. Show that if $\|X\|_2 = 1$ and $\|AX - \lambda X\|_2 \leq \varepsilon$, then $\min\{|\lambda_i - \lambda| : i = 1, 2, \ldots, n\} \leq \varepsilon$. Explain how this could be used as a stopping criterion for a power method iteration for symmetric matrices. Use the Bauer-Fike theorem to explain why the above stopping criterion may be unsatisfactory if the matrix is not symmetric. What would be an acceptable stopping criterion in that case?

11. Suppose that the 3×3 matrix A is not diagonalizable and that $P^{-1}AP = \begin{bmatrix} \lambda_1 & 1 & 0 \\ 0 & \lambda_1 & 0 \\ 0 & 0 & \lambda_3 \end{bmatrix}$. Let the matrix $P = [Y_1\ \ Y_2\ \ Y_3]$, and assume $|\lambda_1| > |\lambda_3|$.

a. Show that $AY_1 = \lambda_1 Y_1$, $AY_2 = Y_1 + \lambda_1 Y_2$, and $AY_3 = \lambda_2 Y_3$.

b. If $X_0 = a_1 Y_1 + a_2 Y_2 + a_3 Y_3$ is the starting vector for (3.6), then show that

$$X_k = AX_{k-1} = (a_1\lambda_1^k + ka_2\lambda_1^{k-1})Y_1 + a_2\lambda_1^k Y_2 + a_3\lambda_3^k Y_3.$$

c. Show that $X_k = k\lambda_1^{k-1}\left\{a_2 Y_1 + \dfrac{a_1\lambda_1}{k} Y_1 + \dfrac{a_2\lambda_1}{k} Y_2 + \dfrac{a_2\lambda_3}{k}\left(\dfrac{\lambda_3}{\lambda_1}\right)^{k-1} Y_3\right\}$.

d. Discuss the behavior of the sequences X_k and $X_k/\|X_k\|$ as k gets large.

12. Consider the matrix C of Exercise 3 and the complex shift $k = 2 + i$. Choose a complex starting vector X_0 and apply the power method to the matrix $C - kI$. Do the calculations using MATLAB and use the scale factor $s = Z(1)$. What conclusions can you make about the eigenvectors and eigenvalues of C?

13. Consider an $n \times n$ matrix A and the Krylov sequence $X_0, X_1 = AX_0, X_2 = AX_1, \ldots, X_n = AX_{n-1}$ generated by an arbitrary nonzero vector X_0. Show that if there exist unique scalars c_i such that

$$X_n = c_0 X_0 + c_1 X_1 + c_2 X_2 + \cdots + c_{n-1} X_{n-1}$$

then the characteristic polynomial of A is $\lambda^n - c_{n-1}\lambda^{n-1} - c_{n-2}\lambda^{n-2} - \cdots - c_1\lambda - c_0$.

a. Use this method to find the characteristic polynomial of the matrix A of Example 6.

b. Count the number of multiplications required to compute the characteristic polynomial using this method. Compare with the operation counts for the methods in Exercises 14 and 16 in Section 3.1.

3.5 ■ THE INVERSE POWER METHOD

Let A be an $n \times n$ nonsingular matrix whose eigenvalues, λ_i, satisfy

$$|\lambda_1| > |\lambda_2| \geq |\lambda_3| \geq \cdots \geq |\lambda_n| > 0.$$

Exercises 1, 5, and 8 in Section 3.4 suggest ways that we can use the fact that the eigenvalues of $A - kI$ are $\{\lambda_i - k : i = 1, 2, \ldots, n\}$ to improve the convergence of the power method and to find a second eigenvector and the corresponding eigenvalue. Another useful observation is that the eigenvalues of A^{-1} are $1/\lambda_1, 1/\lambda_2, \ldots, 1/\lambda_n$ and that the dominant eigenvalue of A^{-1} is $1/\lambda_n$. Applying the power method to A^{-1} would thus yield $1/\lambda_n$ and the eigenvector Y_n, which is an eigenvector for both A and A^{-1}. The algorithm in this case would look like

> X_0 arbitrary
> Compute A^{-1}
> For $i = 1, 2, \ldots,$ convergence
> $\quad Z = A^{-1}X_{i-1}$
> $\quad s = \|Z\|$
> $\quad X_i = \left(\frac{1}{s}\right)Z$
> Estimate λ_n

This algorithm is unattractive because it requires the calculation of A^{-1}. A better way would be to replace "$Z = A^{-1}X_{i-1}$" by "Solve $AZ = X_{i-1}$ for Z." What we would have then is the necessity of solving, at each step, a linear system with coefficient matrix A. Thus we have a set of linear systems, $AZ = X_{i-1}$, where only one of the right-hand sides is available initially. Recall Example 2 in Section 2.2. As discussed in Section 2.2, the most efficient way to handle the solution of these systems is to first compute the LU factorization of A, thus reducing the solution of each system to the solution of two triangular systems. Recall that the solution of two triangular systems requires only about (n^2) multiplications, which is no more work than computing a matrix vector product. This approach replaces the inversion, which takes n^3 multiplications, with the LU factorization, which only takes about $(n^3/3)$ multiplications. The reduction in calculation time could be quite significant for a large problem. This method is called the inverse power method. The algorithm is

INVERSE POWER METHOD ALGORITHM

1. INPUT: $A(n, n)$, $X(n)$ arbitrary, TOL, MAX
2. Factor A, $(PA = LU)$
3. For $i = 1, 2, \ldots, \text{MAX}$
4. SOLVE $LY = PX$ for Y
5. SOLVE $UZ = Y$ for Z
6. $s = \|Z\|$ (3.11)
7. $Z = (1/s)Z$
8. If $\|Z - X\| < \text{TOL}$ or $\|Z + X\| < \text{TOL}$, Exit loop
9. $X = Z$
10. $\lambda = \dfrac{Z^T A Z}{Z^T Z}$
11. OUTPUT: smallest eigenvalue of A and its eigenvector
12. Stop

■ EXAMPLE 9

Find the smallest eigenvalue of $A = \begin{bmatrix} 1 & 2 & 3 & 4 \\ 2 & 3 & 5 & 6 \\ 3 & 5 & 2 & 7 \\ 4 & 6 & 7 & 1 \end{bmatrix}$ using Algorithm (3.11).

SOLUTION We choose as the starting vector $X_0 = [1\ 1\ 1\ 1]^T$, and use Algorithm (3.11) to compute the following iterates:

k	x_1	x_2	x_3	x_4
0	1.0	1.0	1.0	1.0
1	−1.00000	0.661538	0.123077	0.0461538
2	1.00000	−0.630152	−0.0348879	0.0354302
3	−1.00000	0.635237	0.0332009	−0.0357860
4	1.00000	−0.635544	−0.0329597	0.0358799
5	−1.00000	0.635572	0.0329400	0.0358858
6	1.00000	−0.635574	−0.0329383	0.0358863

The sequence has converged very quickly; the alternating signs indicate that the dominant eigenvalue of A^{-1} is negative. The Rayleigh quotient estimate of the smallest eigenvalue of A is $X_6^TAX_6/X_6^TX_6 = -0.226417$. The vector X_6 is an approximate eigenvector of A as well as an approximate eigenvector of A^{-1} for the eigenvalue $\mu = -4.41663 = 1/(-0.226417)$.

Origin shifts work very well with the inverse power method. If the method is applied to the matrix $A - kI$, then it will produce the eigenvector for the dominant eigenvalue of $(A - kI)^{-1}$; say

$$\mu = \frac{1}{\lambda_i - k}.$$

This vector will also be an eigenvector of A and the corresponding eigenvalue of A can be computed from the Rayleigh quotient. This will give us λ_i, the eigenvalue of A closest to k, as well as the eigenvector for λ_i. If k is reasonably close to an eigenvalue of A, this iteration will converge very quickly, often in two or three steps.

Now we are really getting somewhere. By varying k we have the ability to find many, perhaps all, of the eigenvalues of A. In many applications the user may be able to predict, perhaps from a priori knowledge of how the model should behave, reasonable estimates of the eigenvalues of A. At any rate the Gerschgorin theorem will give you some information, or you could simply use $k = a_{ii}$. Another approach is to use a few iterations of the power method to get a rough estimate of the eigenvalue and then switch to the inverse power method to pin it down. If some of the eigenvalues are complex, you can find them by using complex shifts and complex starting vectors. Recall that MATLAB accepts complex matrices without any special programming considerations.

INVERSE POWER METHOD ALGORITHM WITH SHIFTS

(3.12)

```
1. INPUT: A(n, n), TOL, and MAX
2. For j = 1, 2, . . . , r (r = number of eigenvalues desired)
3.      CHOOSE shift k
4.      CHOOSE X arbitrary or an approximate eigenvector
5.      FACTOR A − kI, (P(A − kI) = LU)
6.      For i = 1, 2, . . . , MAX
7.          SOLVE LY = PX for Y
8.          SOLVE UZ = Y for Z
9.          s = ||Z||
10.         Z = (1/s)Z
```

```
        If ||Z − X|| < TOL or ||Z + X|| < TOL, Stop
        X = Z
    λ_j = (Z^T A Z)/(Z^T Z)
OUTPUT: Eigenvalue of A nearest k and its eigenvector
Stop
```

3.5 ■ Exercise Set

1. Use the inverse power method to find the smallest eigenvalue of $B = \begin{bmatrix} 5 & -2 & -0.5 & 1.5 \\ -2 & 5 & 1.5 & -0.5 \\ -0.5 & 1.5 & 5 & -2 \\ 1.5 & -0.5 & -2 & 5 \end{bmatrix}$.

2. Take $A = \begin{bmatrix} 3 & 1 & 2 & 5 \\ 2 & 1 & 3 & 7 \\ 3 & 1 & 2 & 4 \\ 4 & 1 & 3 & 2 \end{bmatrix}$ and $X_0 = \begin{bmatrix} 1 \\ 2 \\ 3 \\ 4 \end{bmatrix}$.

a. Use two iterations of the power method to get a rough estimate μ_1 for λ_1.

b. Use five iterations of the inverse power method on $A - \mu_1 I$ to refine your estimate of λ_1 and its eigenvector X_1.

c. Choose a new starting vector and then use four iterations of the power method on $A - \lambda_1 I$ to get an estimate μ_4 for λ_4.

d. Now use five iterations of the inverse power method on $A - \mu_4 I$ to refine your estimate of λ_4 and find its eigenvector X_4.

e. Choose a new starting vector and use the inverse power method on $A - kI$, where $k = (\lambda_1 + \lambda_4)/2$ to find another eigenvalue and eigenvector of A.

f. Use the fact that $\mathrm{tr}(A)$ is the sum of the eigenvalues to get an estimate of the remaining eigenvalue.

g. Choose a new starting vector and use the inverse power method to refine your estimate of the remaining eigenvalue and its eigenvector.

3. Discuss the storage requirements for Algorithms (3.11) and (3.12).

4. How should Algorithms (3.11) and (3.12) be modified so that it is not necessary to save A?

5. a. For a 4×4 matrix A, the power method with ∞-norm scaling yields the following vectors:

$$X_0 = \begin{bmatrix} 1 \\ 1 \\ 1 \\ 1 \end{bmatrix}, \quad \ldots, \quad X_9 = \begin{bmatrix} 1.0000 \\ 0.54365 \\ -0.67624 \\ 0.11284 \end{bmatrix}, \quad X_{10} = \begin{bmatrix} 1.0000 \\ 0.49966 \\ -0.64458 \\ 0.078723 \end{bmatrix},$$

$$X_{11} = \begin{bmatrix} 1.0000 \\ 0.49567 \\ -0.68512 \\ 0.049847 \end{bmatrix}, \quad X_{12} = \begin{bmatrix} 1.0000 \\ 0.51436 \\ -0.67770 \\ 0.076750 \end{bmatrix}.$$

What can you conclude about the eigenvalues and eigenvectors of A?

b. The inverse power method on $A - 2I$ yields the following vectors:

$$X_0 = \begin{bmatrix} 1 \\ 1 \\ 1 \\ 1 \end{bmatrix}, \quad \ldots, \quad X_3 = \begin{bmatrix} 0.27800 \\ 0.38177 \\ -0.32480 \\ 1.0000 \end{bmatrix}, \quad X_4 = \begin{bmatrix} -0.27995 \\ -0.38166 \\ 0.32483 \\ -1.0000 \end{bmatrix},$$

$$X_5 = \begin{bmatrix} 0.27995 \\ 0.38166 \\ -0.32484 \\ 1.0000 \end{bmatrix}.$$

What can you conclude about the eigenvalues and eigenvectors of A?

c. The inverse power method on $A + 2I$ yields the following vectors:

$$X_0 = \begin{bmatrix} 1 \\ 1 \\ 1 \\ 1 \end{bmatrix}, \quad \ldots, \quad X_9 = \begin{bmatrix} 0.33059 \\ 0.73556 \\ 0.68974 \\ 1.0000 \end{bmatrix}, \quad X_{10} = \begin{bmatrix} -0.98976 \\ 1.0000 \\ -0.80736 \\ 0.33557 \end{bmatrix},$$

$$X_{11} = \begin{bmatrix} -0.409890 \\ -0.544901 \\ -0.42577 \\ -1.0000 \end{bmatrix}.$$

What can you conclude about the eigenvalues and eigenvectors of A?

6. Find all the eigenvalues and eigenvectors of the matrix $A = \begin{bmatrix} 2 & 6 & 1 & 3 \\ 6 & 4 & 3 & 2 \\ 1 & 3 & 2 & 6 \\ 3 & 2 & 6 & 4 \end{bmatrix}$ by using a combination of the power method and the inverse power method as in Exercise 2.

7. Solve the following polynomial equations by finding the eigenvalues of the companion matrix (recall Exercise 15, Section 3.1):

a. $x^3 - 5x^2 + 11x + 14$

b. $x^4 - x^3 - 3x + 8 = 0$

c. $4x^5 - 3x^4 - 7x^3 - 11x^2 + 7x + 9.$

8. The following matrices have some, or all, complex eigenvalues. Use Algorithm (3.12) with complex shifts to find the eigenvalues.

$$A = \begin{bmatrix} 2 & -1 & 3 & 0 \\ 1 & 5 & 2 & -7 \\ -3 & -2 & 0 & 6 \\ 0 & 7 & 3 & 4 \end{bmatrix} \qquad B = \begin{bmatrix} 0 & 0 & 0 & 0 \\ 0 & 0 & 1 & 0 \\ 0 & 0 & 0 & 1 \\ 60 & -200 & 32 & 4 \end{bmatrix}$$

$$C = \begin{bmatrix} 1 & 1 & 2 & 2 \\ -1 & 1 & -2 & 2 \\ -2 & -2 & 1 & 1 \\ 2 & -2 & -1 & 1 \end{bmatrix}$$

4 Transformation Methods for the Eigenvalue Problem

4.1 ■ ELEMENTARY TRANSFORMATIONS

In this chapter we will consider transformation methods for finding the eigenvalues and eigenvectors of a matrix A. These methods attempt to solve the eigenvalue problem for the matrix A by constructing the triangularizing matrix of Schur's theorem. That is, these methods attempt to construct a nonsingular matrix P such that $P^{-1}AP$ is upper triangular,

$$P^{-1}AP = \begin{bmatrix} \lambda_1 & t_{12} & t_{13} & \cdots & t_{1n} \\ 0 & \lambda_2 & t_{23} & \cdots & t_{2n} \\ 0 & 0 & \lambda_3 & \cdots & t_{3n} \\ \vdots & \vdots & \vdots & \ddots & \vdots \\ 0 & 0 & 0 & \cdots & \lambda_n \end{bmatrix} = T.$$

All of the methods we will consider construct P as an infinite product $P = \prod_{i=1}^{\infty} P_i$ where the P_i are relatively simple transforming matrices. Thus to construct $P^{-1}AP$ we construct a sequence $\{A_i\}$ of matrices similar to A, where $A_0 = A$ and $A_i = P_i^{-1}A_{i-1}P_i$ for $i = 1, 2, \ldots$. At each step we choose P_i so that A_i is more nearly triangular than A_{i-1}. In practice we cannot compute an infinite product, so we stop when the transformed matrix is suitably close to an upper triangular matrix. Some of the methods use orthogonal transforming matrices while others do not. We will consider three different kinds of transforming matrices: elementary lower triangular matrices, plane rotations, and elementary reflectors (Householder matrices).

Elementary Lower Triangular Matrices

It follows directly from the definition of matrix multiplication that the rows of the product AB are linear combinations of the rows of B with coefficients from the appropriate row of A. Similarly, the columns of AB are linear combinations of the columns of A with coefficients from the appropriate column of B. It follows, as noted in Section 1.1, that if $\mathscr{E}$ is an elementary row operation, then $\mathscr{E}(A) = \mathscr{E}(I)A$, and if $\mathscr{C}$ is an elementary column operation, then $\mathscr{C}(A) = A\mathscr{C}(I)$. In particular, if $\mathscr{E}$ is the operation $R_j \leftarrow kR_i + R_j$, with $j > i$, then the lower triangular matrix

$$\mathscr{E}(I) = L_{ji}(k) = \begin{array}{c@{}l} \begin{array}{ccccccc} & & \downarrow i & & \downarrow j & & \end{array} & \\ \left[\begin{array}{ccccccc} 1 & \cdots & 0 & \cdots & 0 & \cdots & 0 \\ \vdots & \ddots & \vdots & \vdots & \vdots & \vdots & \vdots \\ 0 & \cdots & 1 & \cdots & 0 & \cdots & 0 \\ \vdots & \cdots & \vdots & \ddots & \vdots & \cdots & \vdots \\ 0 & \cdots & k & \cdots & 1 & \cdots & 0 \\ \vdots & \cdots & \vdots & \cdots & \vdots & \ddots & \vdots \\ 0 & \cdots & 0 & \cdots & 0 & \cdots & 1 \end{array}\right] & \begin{array}{l} \\ \\ \leftarrow i \\ \\ \leftarrow j \\ \\ \\ \end{array} \end{array}$$

differs from the identity matrix only in position (j, i) where the multiplier k occurs. Note that this matrix is also associated with the column operation $C_i \leftarrow kC_j + C_i$. The inverse of this matrix is $L_{ji}(-k)$, which represents both the row operation $R_j \leftarrow -kR_i + R_j$ and the column operation $C_i \leftarrow -kC_j + C_i$.

The **elementary lower triangular matrices** we wish to consider here are products of matrices of the preceding type. We encountered these matrices in Chapter 2 when we developed the LU factorization of the matrix A. Consider the matrix

$$L_i = \begin{array}{c} \begin{array}{ccccccc} & & & \downarrow i & & & \end{array} \\ \left[\begin{array}{ccccccc} 1 & 0 & \cdots & 0 & 0 & \cdots & 0 \\ 0 & 1 & \cdots & 0 & 0 & \cdots & 0 \\ \vdots & \vdots & \ddots & \vdots & \vdots & \vdots & \vdots \\ 0 & 0 & \cdots & 1 & 0 & \cdots & 0 \\ 0 & 0 & \cdots & -r_{i+1,i} & 1 & \cdots & 0 \\ \vdots & \vdots & \vdots & \vdots & \vdots & \ddots & \vdots \\ 0 & 0 & \cdots & -r_{ni} & 0 & \cdots & 1 \end{array}\right] \end{array}$$

$$= L_{ni}(-r_{ni}) \cdots L_{i+2,i}(-r_{i+2,i})L_{i+1,i}(-r_{i+1,i}),$$

which corresponds to the sequence of elementary row operations

$$R_{i+1} \leftarrow -r_{i+1,i}R_i + R_{i+1},$$
$$R_{i+2} \leftarrow -r_{i+2,i}R_i + R_{i+2}, \ldots, R_n \leftarrow -r_{ni}R_i + R_n$$

or to the sequence of elementary column operations

$$C_i \leftarrow -r_{ni}C_n + C_i, C_i \leftarrow -r_{n-1,i}C_{n-1} + C_i, \ldots, C_i \leftarrow -r_{i+1,i}C_{i+1} + C_i.$$

It is easy to verify that

$$L_i^{-1} = 2I - L_i = \begin{bmatrix} 1 & 0 & \cdots & 0 & 0 & \cdots & 0 \\ 0 & 1 & \cdots & 0 & 0 & \cdots & 0 \\ \vdots & \vdots & \ddots & \vdots & \vdots & \vdots & \vdots \\ 0 & 0 & \cdots & 1 & 0 & \cdots & 0 \\ 0 & 0 & \cdots & r_{i+1,i} & 1 & \cdots & 0 \\ \vdots & \vdots & \vdots & \vdots & \vdots & \ddots & \vdots \\ 0 & 0 & \cdots & r_{ni} & 0 & \cdots & 1 \end{bmatrix}.$$

Note that multiplying a matrix A on the left by L_i (premultiplying) changes the last $n - i$ rows of A, while multiplying a matrix A on the right (postmultiplying) by L_i changes only the ith column of A.

Plane Rotations

In $\mathscr{R}^2$ the matrix that rotates the plane through the angle θ is the matrix $R = \begin{bmatrix} \cos\theta & \sin\theta \\ -\sin\theta & \cos\theta \end{bmatrix}$. This matrix satisfies $\begin{bmatrix} x' \\ y' \end{bmatrix} = \begin{bmatrix} \cos\theta & \sin\theta \\ -\sin\theta & \cos\theta \end{bmatrix}\begin{bmatrix} x \\ y \end{bmatrix}$ as one can see by examining the diagram in Figure 4.1. Using elementary trigonometry we have

$$x = OS = OR - SR = OR - TQ = x'\cos\theta - y'\sin\theta$$
$$y = PS = TS + TP = QR + TP = x'\sin\theta + y'\cos\theta.$$

It follows that $\begin{bmatrix} x \\ y \end{bmatrix} = \begin{bmatrix} \cos\theta & -\sin\theta \\ \sin\theta & \cos\theta \end{bmatrix}\begin{bmatrix} x' \\ y' \end{bmatrix}$ and hence that

FIGURE *4.1* ■ PLANE ROTATION

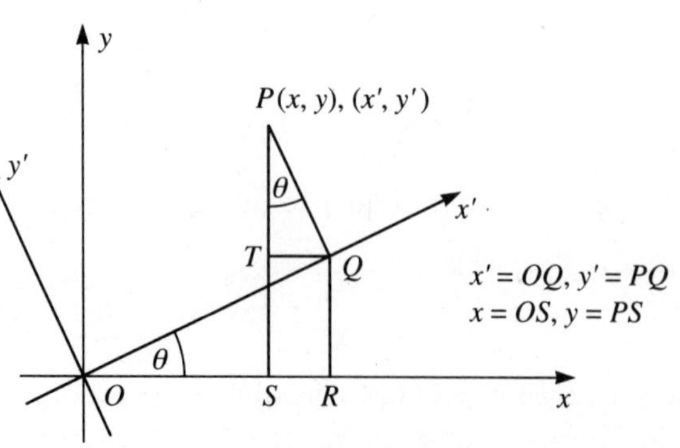

$\begin{bmatrix} x' \\ y' \end{bmatrix} = \begin{bmatrix} \cos\theta & \sin\theta \\ -\sin\theta & \cos\theta \end{bmatrix}\begin{bmatrix} x \\ y \end{bmatrix} = R\begin{bmatrix} x \\ y \end{bmatrix}$. The matrix R has orthonormal columns and hence is an orthogonal matrix. Rotation matrices can be used, as were the elementary lower triangular matrices, to introduce zeros into a matrix. The key observation is that

$$\begin{bmatrix} \cos\theta & \sin\theta \\ -\sin\theta & \cos\theta \end{bmatrix}\begin{bmatrix} a \\ c \end{bmatrix} = \begin{bmatrix} \sqrt{a^2 + c^2} \\ 0 \end{bmatrix} \quad \text{if} \quad \theta = \tan^{-1}\left(\frac{c}{a}\right) = \cot^{-1}\left(\frac{a}{c}\right). \tag{4.1}$$

In $\mathscr{R}^3$ a plane rotation in the $(1,3)$ plane is the transformation determined by the matrix

$$R = R_{13}(\theta) = \begin{bmatrix} \cos\theta & 0 & \sin\theta \\ 0 & 1 & 0 \\ -\sin\theta & 0 & \cos\theta \end{bmatrix}.$$

It follows from Equation (4.1) that

$$R\begin{bmatrix} x_1 \\ x_2 \\ x_3 \end{bmatrix} = \begin{bmatrix} \sqrt{x_1^2 + x_3^2} \\ x_2 \\ 0 \end{bmatrix} \quad \text{if} \quad \theta = \tan^{-1}\left(\frac{x_3}{x_1}\right).$$

Note that this transformation leaves x_2 fixed (geometrically the x_2 axis is left invariant).

More generally, in $\mathscr{R}^n$ a **plane rotation in the (j, k) plane** is the transformation determined by the matrix

$$R = R_{jk}(\theta) = \left[\begin{array}{ccccccc} 1 & \cdots & \overset{\downarrow j}{0} & \cdots & \overset{\downarrow k}{0} & \cdots & 0 \\ \vdots & \ddots & \vdots & \vdots & \vdots & \vdots & \vdots \\ 0 & \cdots & \cos\theta & \cdots & \sin\theta & \cdots & 0 \\ \vdots & \vdots & \vdots & \ddots & \vdots & \vdots & \vdots \\ 0 & \cdots & -\sin\theta & \cdots & \cos\theta & \cdots & 0 \\ \vdots & \vdots & \vdots & \vdots & \vdots & \ddots & \vdots \\ 0 & \cdots & 0 & \cdots & 0 & \cdots & 1 \end{array}\right]\begin{array}{l} \\ \\ \leftarrow j \\ \\ \leftarrow k \\ \\ \\ \end{array}.$$

This matrix differs from the identity matrix I only in positions (j, j), (j, k), (k, j), and (k, k). Premultiplying a matrix A by R changes only rows j and k, while postmultiplying A by R changes only columns j and k. If X is a column vector and we want entry k of RX to be 0, then we choose $\theta = \tan^{-1}(x_k/x_j)$. For any n, the rotation matrix R is orthogonal, so $R^{-1} = R^T$. The matrix R^TAR differs from A only in rows j and k and in columns j and k. Thus the only diagonal elements that change are in positions (j, j) and (k, k).

■ EXAMPLE 1

Determine the rotation matrices R_1, R_2, and R_3 so that $R_3R_2R_1\begin{bmatrix} 4 & 2 & -1 \\ 3 & 2 & 3 \\ 12 & 6 & 2 \end{bmatrix} = \begin{bmatrix} * & * & * \\ 0 & * & * \\ 0 & 0 & * \end{bmatrix}$ is an upper triangular matrix.

SOLUTION We first do a rotation in the $(1, 2)$ plane, which will produce a zero in position $(2, 1)$. By Equation (4.1), we must choose $\theta_1 = \tan^{-1}(a_{21}/a_{11}) = \tan^{-1}(3/4)$; it follows that $\cos\theta_1 = 4/5$ and $\sin\theta_1 = 3/5$. See Figure 4.2. Thus $R_1 = R_{12}(\theta_1) = \begin{bmatrix} \frac{4}{5} & \frac{3}{5} & 0 \\ -\frac{3}{5} & \frac{4}{5} & 0 \\ 0 & 0 & 1 \end{bmatrix}$, and by direct computation we find $R_1A = \begin{bmatrix} 5 & \frac{14}{5} & 1 \\ 0 & \frac{2}{5} & 3 \\ 12 & 6 & 2 \end{bmatrix}$. Note that this transformation leaves the third row of the matrix unchanged.

To produce a zero in position $(3, 1)$ we will do a rotation in the $(1, 3)$ plane through an angle $\theta_2 = \tan^{-1}(12/5)$. It follows that $\cos\theta_2 = 5/13$ and $\sin\theta_2 = 12/13$. See Figure 4.2. Thus $R_2 = R_{13}(\theta_2) = \begin{bmatrix} \frac{5}{13} & 0 & \frac{12}{13} \\ 0 & 1 & 0 \\ -\frac{12}{13} & 0 & \frac{5}{13} \end{bmatrix}$, and by direct computation $R_2R_1A = \begin{bmatrix} 13 & \frac{86}{13} & \frac{29}{13} \\ 0 & \frac{2}{5} & 3 \\ 0 & -\frac{18}{65} & -\frac{2}{13} \end{bmatrix}$. Finally a rotation in the $(2, 3)$ plane can be used to produce a zero in position $(3, 2)$. The proper angle is $\theta_3 = \tan^{-1}(-18/65/2/5) = \tan^{-1}(-9/13)$ for which $\cos\theta_3 = 13/\sqrt{250}$ and

FIGURE 4.2 ■ REFERENCE TRIANGLES FOR EXAMPLE 1

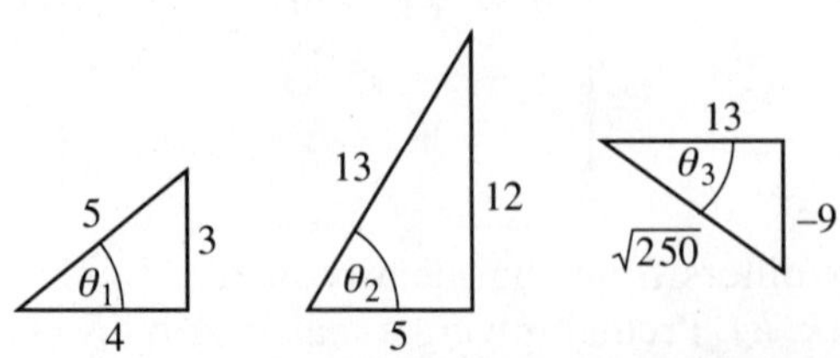

$\sin\theta_3 = -9/\sqrt{250}$. The third rotation matrix is

$$R_3 = R_{23}(\theta_3) = \frac{1}{\sqrt{250}}\begin{bmatrix} \sqrt{250} & 0 & 0 \\ 0 & 13 & -9 \\ 0 & 9 & 13 \end{bmatrix}$$

for which

$$R_3R_2R_1A = \begin{bmatrix} 13 & 6.61538 & 2.23077 \\ 0 & 0.4856504 & 2.55415 \\ 0 & 0 & 1.58114 \end{bmatrix} = U.$$

It is worth observing that our calculations yield a factorization $A = QU$ where $Q = R_1^TR_2^TR_3^T$ is orthogonal and U is upper triangular. ■

Elementary Reflectors (Householder Matrices)

The **elementary reflectors**, or **Householder matrices**, are matrices of the form $H = I - 2\mathbf{w}\mathbf{w}^T$ where $\mathbf{w}$ is a unit vector in $\mathcal{R}^n$. It is easy to see that these matrices are both symmetric and orthogonal (Exercise 3, Section 3.2). To see why these matrices are called elementary reflectors we compute

$$H\mathbf{w} = (I - 2\mathbf{w}\mathbf{w}^T)\mathbf{w} = \mathbf{w} - 2\mathbf{w}(\mathbf{w}^T\mathbf{w}) = \mathbf{w} - 2\mathbf{w} = -\mathbf{w}$$

and, for any vector $\boldsymbol{x}$ orthogonal to $\mathbf{w}$,

$$H\boldsymbol{x} = (I - 2\mathbf{w}\mathbf{w}^T)\boldsymbol{x} = \boldsymbol{x} - 2\mathbf{w}(\mathbf{w}^T\boldsymbol{x}) = \boldsymbol{x}.$$

Thus in $\mathcal{R}^3$ the matrix H can be interpreted as a reflection in the plane perpendicular to $\mathbf{w}$. In $\mathcal{R}^n$ the matrix H can be interpreted as a reflection in the $(n-1)$ dimensional subspace orthogonal to $\mathbf{w}$. To see that this is so we note that if $\boldsymbol{y}$ is any vector in $\mathcal{R}^n$, we can write $\boldsymbol{y} = b\mathbf{w} + \boldsymbol{x}$, where $\boldsymbol{x}$ is some vector in $\mathbf{w}^\perp$, the subspace of $\mathcal{R}^n$ orthogonal to $\mathbf{w}$. It then follows that

$$H\boldsymbol{y} = H(b\mathbf{w} + \boldsymbol{x}) = bH\mathbf{w} + H\boldsymbol{x} = -b\mathbf{w} + \boldsymbol{x},$$

as depicted in Figure 4.3.

FIGURE 4.3 ■ ELEMENTARY REFLECTORS

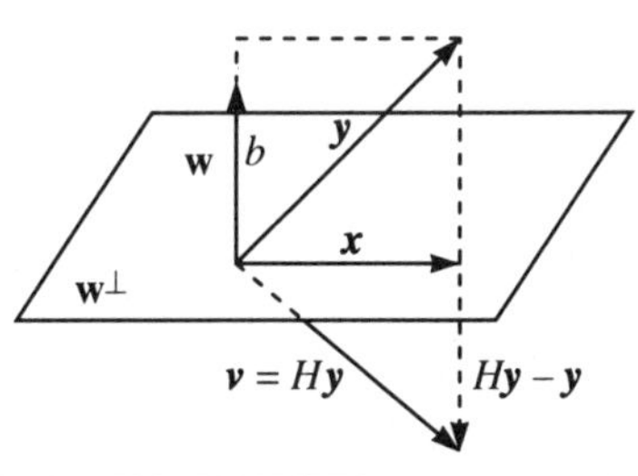

As Figure 4.3 suggests, if $\boldsymbol{y}$ and $\boldsymbol{v}$ are vectors of the same length, $\|\boldsymbol{y}\|_2 = \|\boldsymbol{v}\|_2$, then the elementary reflector determined by $\mathbf{w} = (1/\|\boldsymbol{y} - \boldsymbol{v}\|_2)(\boldsymbol{y} - \boldsymbol{v})$ is such that $H\boldsymbol{y} = \boldsymbol{v}$. In this case the subspace $\mathbf{w}^{\perp}$ bisects the angle between the vectors $\boldsymbol{y}$ and $\boldsymbol{v} = H\boldsymbol{y}$. The general result, suggested by the preceding geometric argument, is the content of our next theorem.

THEOREM 4.1

If $\boldsymbol{y}$ and $\boldsymbol{v}$ are any two unit vectors in $\mathcal{R}^n$, then the elementary reflector $H = I - 2\mathbf{w}\mathbf{w}^T$, with $\mathbf{w} = (1/\|\boldsymbol{y} - \boldsymbol{v}\|_2)(\boldsymbol{y} - \boldsymbol{v})$, is such that $H\boldsymbol{y} = \boldsymbol{v}$.

PROOF If $H\boldsymbol{y} = \boldsymbol{y} - 2\mathbf{w}\mathbf{w}^T\boldsymbol{y} = \boldsymbol{y} - 2(\mathbf{w}^T\boldsymbol{y})\mathbf{w} = \boldsymbol{v}$, then $\mathbf{w} = k(\boldsymbol{y} - \boldsymbol{v})$, a scalar multiple of $\mathbf{w}$. Since $\mathbf{w}$ is to be a unit vector we must choose $k = \pm 1/\|\boldsymbol{y} - \boldsymbol{v}\|_2$. ■

■ EXAMPLE 2

Find elementary reflectors H_1 and H_2 that will satisfy

$$H_1\begin{bmatrix} 3 \\ 1 \\ 1 \\ -5 \end{bmatrix} = \begin{bmatrix} 6 \\ 0 \\ 0 \\ 0 \end{bmatrix} \quad \text{and} \quad H_2\begin{bmatrix} 5 \\ 8 \\ 1 \\ 4 \end{bmatrix} = \begin{bmatrix} 5 \\ 9 \\ 0 \\ 0 \end{bmatrix}.$$

SOLUTION By Theorem 4.1, $H_1 = I - 2\mathbf{w}\mathbf{w}^T$, where $\mathbf{w}$ is a unit vector in the direction of

$$\begin{bmatrix} 3 \\ 1 \\ 1 \\ -5 \end{bmatrix} - \begin{bmatrix} 6 \\ 0 \\ 0 \\ 0 \end{bmatrix} = \begin{bmatrix} -3 \\ 1 \\ 1 \\ -5 \end{bmatrix}.$$

Thus

$$\mathbf{w} = \frac{1}{6}\begin{bmatrix} -3 \\ 1 \\ 1 \\ -5 \end{bmatrix} \quad \text{and} \quad H_1 = I - 2\mathbf{w}\mathbf{w}^T = \frac{1}{18}\begin{bmatrix} 9 & 3 & 3 & -15 \\ 3 & 17 & -1 & 5 \\ 3 & -1 & 17 & 5 \\ -15 & 5 & 5 & -7 \end{bmatrix}.$$

Similarly $H_2 = I - 2\mathbf{z}\mathbf{z}^T$, where $\mathbf{z}$ is a unit vector in the direction of

$$\begin{bmatrix} 5 \\ 8 \\ 1 \\ 4 \end{bmatrix} - \begin{bmatrix} 5 \\ 9 \\ 0 \\ 0 \end{bmatrix} = \begin{bmatrix} 0 \\ -1 \\ 1 \\ 4 \end{bmatrix}.$$

Thus,

$$\mathbf{z} = \frac{1}{\sqrt{18}} \begin{bmatrix} 0 \\ -1 \\ 1 \\ 4 \end{bmatrix} \quad \text{and} \quad H_2 = I - 2\mathbf{z}\mathbf{z}^T = \frac{1}{9} \begin{bmatrix} 9 & 0 & 0 & 0 \\ 0 & 8 & 1 & 4 \\ 0 & 1 & 8 & -4 \\ 0 & 4 & -4 & -7 \end{bmatrix}.$$

■

The elementary transformation matrices we have introduced in this section are all perturbations of the identity matrix. The elementary lower triangular matrices and the rotation matrices are relatively sparse. The calculation of any of the products P_iA can be accomplished with significantly fewer multiplications than the n^3 required to compute the product of two $n \times n$ matrices using a general purpose matrix multiplication subroutine. In general these matrices will not actually be constructed; the products $P_iA \times (AP_i)$ are computed by forming linear combinations of the rows (columns) of A.

We saw in Chapter 2 that computing the product $L_{n-1}L_{n-2} \cdots L_2L_1A = U$, which yielded the LU factorization of A, required only about $(n^3/3)$ multiplications. The product L_iA requires only $(n - i)^2$ multiplications if the $n - i \times i$ lower-left corner of A is zero, and $n(n - i)$ multiplications if the lower-left corner of A is not zero.

For a plane rotation $R = R_{jk}(\theta)$, the product RA requires only $4n$ multiplications. To see this we note that only rows j and k are changed and that each of these rows is replaced by a linear combination of rows j and k of A; that is, if we set $c = \cos\theta$ and $s = \sin\theta$, then

$$\text{Row}_j(RA) = c\text{Row}_j(A) + s\text{Row}_k(A)$$

$$\text{and} \quad \text{Row}_k(RA) = -s\text{Row}_j(A) + c\text{Row}_k(A).$$

For an elementary reflector $H = I - 2\mathbf{w}\mathbf{w}^T$, the product HA can be rewritten as $HA = A - (2\mathbf{w})(\mathbf{w}^TA)$. The computation of the row vector $\mathbf{w}^TA$ takes n^2 multiplications and the product of the column vector $2\mathbf{w}$ and the row vector $\mathbf{w}^TA$ takes another n^2 multiplications. Thus the product HA requires only $2n^2$ multiplications and is much cheaper than general matrix multiplication, which takes n^3 multiplications.

4.1 ■ Exercise Set

1. Show that if $\|\mathbf{w}\|_2 = 1$, then $H = I - 2\mathbf{w}\mathbf{w}^T$ is symmetric and orthogonal.

2. Verify that $\begin{bmatrix} \cos\theta & \sin\theta \\ -\sin\theta & \cos\theta \end{bmatrix}\begin{bmatrix} a & b \\ c & d \end{bmatrix} = \begin{bmatrix} a' & b' \\ 0 & d' \end{bmatrix}$ if $\theta = \tan^{-1}(c/a) = \cot^{-1}(a/c)$. If $-(\pi/2) \le \theta \le \pi/2$, show that $\cos\theta = a/\sqrt{a^2 + c^2}$ and $\sin\theta = c/\sqrt{a^2 + c^2}$.

3. Determine an elementary reflector H so that $H\begin{bmatrix} 12 & 1 & 4 \\ 3 & 2 & 5 \\ 4 & 3 & 6 \end{bmatrix} = \begin{bmatrix} 13 & * & * \\ 0 & * & * \\ 0 & * & * \end{bmatrix}$. What is HAH^{-1} in this case?

4. Determine an elementary reflector H so that $H\begin{bmatrix} 12 & 1 & 4 \\ 3 & 2 & 5 \\ 4 & 3 & 6 \end{bmatrix} = \begin{bmatrix} 12 & * & * \\ 5 & * & * \\ 0 & * & * \end{bmatrix}$. What is HAH^{-1} in this case?

5. Find an elementary lower triangular matrix L so that $L\begin{bmatrix} 2 & 1 & 6 & 4 \\ -4 & 5 & 2 & 0 \\ 6 & 2 & 3 & 1 \\ 1 & 2 & 3 & 4 \end{bmatrix} = \begin{bmatrix} 2 & 1 & 6 & 4 \\ 0 & * & * & * \\ 0 & * & * & * \\ 0 & * & * & * \end{bmatrix}$. What is LAL^{-1} in this case?

6. Find an elementary lower triangular matrix L so that $L\begin{bmatrix} 0 & 1 & 6 & 4 \\ -4 & 5 & 2 & 0 \\ 6 & 2 & 3 & 1 \\ 1 & 2 & 3 & 4 \end{bmatrix}$

$$= \begin{bmatrix} 0 & 1 & 6 & 4 \\ -4 & 5 & 2 & 0 \\ 0 & * & * & * \\ 0 & * & * & * \end{bmatrix}.$$ What is LAL^{-1} in this case?

7. Show that a 2×2 elementary reflector is of the form

$$\begin{bmatrix} \cos\theta & \sin\theta \\ \sin\theta & -\cos\theta \end{bmatrix} = \begin{bmatrix} 1 & 0 \\ 0 & -1 \end{bmatrix} \begin{bmatrix} \cos\theta & \sin\theta \\ -\sin\theta & \cos\theta \end{bmatrix}$$

and thus is the product of a rotation and a reflection in the x-axis.

8. Show that $R^T \begin{bmatrix} a & b \\ b & d \end{bmatrix} R = \begin{bmatrix} a' & 0 \\ 0 & d' \end{bmatrix}$ if R is a plane rotation with $\theta = 1/2 \cot^{-1}((d-a)/2b)$.

9. Consider the 3×3 symmetric matrix $A = \begin{bmatrix} 2 & -5 & 4 \\ -5 & 3 & 2 \\ 4 & 2 & -3 \end{bmatrix}$.

a. Construct, using the result of Exercise 8 in this section, a plane rotation R_1 so that $A^{(1)} = R_1^T A R_1$ has 0 in positions $(2,1)$ and $(1,2)$.

b. Construct a plane rotation R_2 so that $A^{(2)} = R_2^T A^{(1)} R_2$ has 0 in positions $(3,1)$ and $(1,3)$.

c. Construct a plane rotation R_3 so that $A^{(3)} = R_3^T A^{(2)} R_3$ has 0 in positions $(3,2)$ and $(2,3)$.

d. Compare $|a_{21}^{(3)}|^2 + |a_{31}^{(3)}|^2 + |a_{32}^{(3)}|^2$ with $a_{21}^2 + a_{31}^2 + a_{32}^2$.

10. **a.** Show that $\|A\|_E \equiv \sqrt{\sum_{i,j=1}^{n} a_{ij}^2}$ is a matrix norm (the Euclidean matrix norm).

b. Show that $\|A\|_E^2 = \operatorname{tr}(A^T A)$.

c. Show that if P is an orthogonal matrix, then $\|P^T A P\|_E = \|A\|_E$. That is, the Euclidean matrix norm is invariant under orthogonal similarity.

d. Suppose θ is determined as in Exercise 8. Show, as a consequence of part c, that $a'^2 + d'^2 = a^2 + d^2 + 2b^2$.

e. Suppose that A is $n \times n$ and symmetric, $R = R_{ik}(\theta)$ is a plane rotation, $A^{(1)} = R^T A R$, and θ is chosen so that $a_{ik}^{(1)} = 0$. Show that the sum of the squares of the diagonal entries of $A^{(1)}$ is greater than the sum of the squares of the diagonal elements of A.

f. The observations in parts a–e are the basis for the Jacobi method for estimating the eigenvalues of a symmetric matrix. The algorithmic description of the Jacobi method is as follows:

$A^{(0)} = A$

For $s = 0, 1, 2, \ldots$

$A^{(s+1)} = R_s^T A^{(s)} R_s$ where R_s is a plane rotation chosen so that the largest off-diagonal entry of $A^{(s)}$ is reduced to zero.

Show that the sequence $A^{(s)}$ approaches a diagonal matrix whose diagonal entries are the eigenvalues of A.

g. The MATMAN **J** command can be used to implement this method in an interactive way. Try this method on the matrix of Example 5 in Section 3.5.

4.2 ■ THE QR FACTORIZATION

In the last section we saw that, given a matrix A, we could find a plane rotation $R_{ij}(\theta)$ such that $R_{ij}(\theta)A$ has a zero in position (i, j). If $i > j$, the element zeroed is in the lower triangle of the matrix. The product requires only $4n$ multiplications since only rows i and j are changed. We will now show how a matrix can be reduced to upper triangular form using a sequence of plane rotations.

Consider now a 4×4 matrix $A = \begin{bmatrix} a_{11} & a_{12} & a_{13} & a_{14} \\ a_{21} & a_{22} & a_{23} & a_{24} \\ a_{31} & a_{32} & a_{33} & a_{34} \\ a_{41} & a_{42} & a_{43} & a_{44} \end{bmatrix}$ and choose $\theta_1 = \tan^{-1}(a_{21}/a_{11})$. Now $R_{21}(\theta_1) = \begin{bmatrix} \cos\theta_1 & \sin\theta_1 & 0 & 0 \\ -\sin\theta_1 & \cos\theta_1 & 0 & 0 \\ 0 & 0 & 1 & 0 \\ 0 & 0 & 0 & 1 \end{bmatrix}$ is such that

$R_{21}(\theta_1)A = \begin{bmatrix} a_{11}^{(1)} & a_{12}^{(1)} & a_{13}^{(1)} & a_{14}^{(1)} \\ 0 & a_{22}^{(1)} & a_{23}^{(1)} & a_{24}^{(1)} \\ a_{31} & a_{32} & a_{33} & a_{34} \\ a_{41} & a_{42} & a_{43} & a_{44} \end{bmatrix}$. If we now choose $\theta_2 = \tan^{-1}(a_{31}/a_{11}^{(1)})$, the rotation $R_{31}(\theta_2)$ will yield a zero in position $(3, 1)$. The resulting matrix is

$$R_{31}(\theta_2)(R_{21}(\theta_1)A) = \begin{bmatrix} a_{11}^{(2)} & a_{12}^{(2)} & a_{13}^{(2)} & a_{14}^{(2)} \\ 0 & a_{22}^{(1)} & a_{23}^{(1)} & a_{24}^{(1)} \\ 0 & a_{32}^{(1)} & a_{33}^{(1)} & a_{34}^{(1)} \\ a_{41} & a_{42} & a_{43} & a_{44} \end{bmatrix}.$$

Note that the rotation in the $(1, 3)$ plane has not changed rows 2 and 4; in particular, it did not change the zero in the $(2,1)$ position. Here we use the superscript to indicate how many times the entry in that position has changed. The rotation $R_{41}(\theta_3)$, where $\theta_3 = \tan^{-1}(a_{41}/a_{11}^{(2)})$, will now produce a zero in the $(4,1)$ position without changing rows 2 or 3. This completes the reduction of the first column. We now repeat the procedure on the 3×3 lower-right-hand corner of the current matrix. Thus we determine rotations $R_{32}(\theta_4)$ and $R_{42}(\theta_5)$, which will produce zeros in positions $(3,2)$ and $(4,2)$. These rotations will not change the first row or the first column of the matrix. Finally a rotation $R_{43}(\theta_6)$ will produce a zero in position $(4,3)$. Putting all this together we have a reduction of A to an upper triangular matrix R,

$$(R_{43}(\theta_6)R_{42}(\theta_5)R_{32}(\theta_4)R_{41}(\theta_3)R_{31}(\theta_2)R_{21}(\theta_1))A$$
$$= R = \begin{bmatrix} r_{11} & r_{12} & r_{13} & r_{14} \\ 0 & r_{22} & r_{23} & r_{24} \\ 0 & 0 & r_{33} & r_{34} \\ 0 & 0 & 0 & r_{44} \end{bmatrix}.$$

The product of the six plane rotation matrices is an orthogonal matrix Q^T satisfying $Q^TA = R$. This is equivalent to the factorization $A = QR$ where Q is orthogonal and R is upper triangular. It is easy to arrange the computation so that Q is constructed as A is being reduced; the individual rotation matrices need not be formed explicitly. In some applications it suffices to store just enough information about the rotations so that the matrix products like QB or BQ can be formed from the factored form of Q. (See Exercise 11.)

The above argument extends easily to any square matrix and thus establishes the existence of the factorization described in the next theorem.

THEOREM 4.2

If A is any $n \times n$ matrix, then A can be written as $A = QR$ where Q is orthogonal and R is upper triangular. Moreover, if A is nonsingular, then Q and R are unique if R is restricted to have positive diagonal entries.

PROOF To establish the uniqueness we suppose that $A = Q_1R_1 = Q_2R_2$ where Q_1 and Q_2 are orthogonal and R_1 and R_2 are upper triangular with positive diagonal entries. Then,

$$Q_1R_1 = Q_2R_2 \Rightarrow Q_2^TQ_1R_1 = R_2 \Rightarrow Q_2^TQ_1 = R_2R_1^{-1}$$

where $Q_2^TQ_1$ is orthogonal and $R_2R_1^{-1}$ is upper triangular with positive diagonal entries. It is easy to see that the identity matrix is the only orthogonal matrix that is also upper triangular with positive diagonal entries. Thus, $Q_2^T Q_1 = R_2 R_1^{-1} = I$, and it follows that $Q_2 = Q_1$ and $R_2 = R_1$. ■

EXAMPLE 3

Use plane rotations to find the QR factorization of $A = \begin{bmatrix} 3 & -1 & 5 \\ 4 & 7 & 5 \\ -5 & 1 & 1 \end{bmatrix}$.

SOLUTION The first step is to produce a zero in the (2, 1) position of A by a rotation through an angle θ_1 in the (1, 2) plane. As noted in Equation (4.1), the proper angle is $\theta_1 = \tan^{-1}(4/3)$. From elementary right-triangle trigonometry, it follows that $\cos\theta_1 = 3/5$ and $\sin\theta_1 = 4/5$. See Figure 4.4. Now we have

$$R_{21}(\theta_1)A = \begin{bmatrix} \frac{3}{5} & \frac{4}{5} & 0 \\ -\frac{4}{5} & \frac{3}{5} & 0 \\ 0 & 0 & 1 \end{bmatrix} \begin{bmatrix} 3 & -1 & 5 \\ 4 & 7 & 5 \\ -5 & 1 & 1 \end{bmatrix} = \begin{bmatrix} 5 & 5 & 7 \\ 0 & 5 & -1 \\ -5 & 1 & 1 \end{bmatrix} = A_2.$$

We now zero the (3, 1) position of A_2 with a rotation through an angle $\theta_2 = \tan^{-1}(-5/5)$ in the (1, 3) plane. It follows that $\cos\theta_2 = 1/\sqrt{2}$ and $\sin\theta_2 = -1/\sqrt{2}$, so that

$$R_{31}(\theta_2)A_2 = \begin{bmatrix} \frac{1}{\sqrt{2}} & 0 & -\frac{1}{\sqrt{2}} \\ 0 & 1 & 0 \\ \frac{1}{\sqrt{2}} & 0 & \frac{1}{\sqrt{2}} \end{bmatrix} A_2 = \begin{bmatrix} 5\sqrt{2} & 2\sqrt{2} & 3\sqrt{2} \\ 0 & 5 & -1 \\ 0 & 3\sqrt{2} & 2\sqrt{2} \end{bmatrix} = A_3.$$

To obtain a zero in position (3, 2), we need to do a rotation in the (2, 3) plane through an angle $\theta_3 = \tan^{-1}(3\sqrt{2}/5)$, for which $\cos\theta_3 = 5/\sqrt{43}$ and

FIGURE 4.4 ■ REFERENCE TRIANGLES FOR EXAMPLE 4

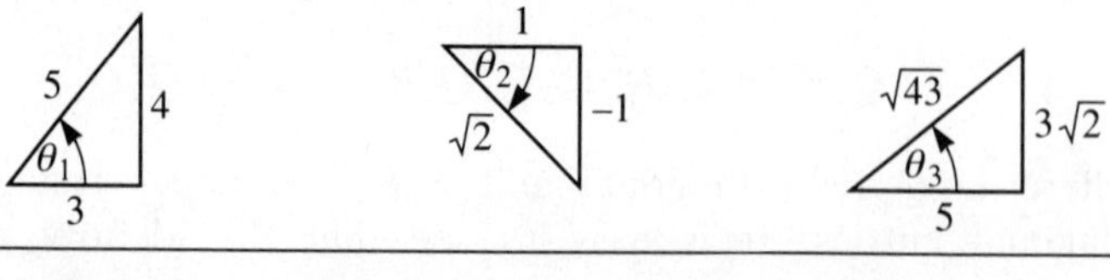

$\sin \theta_3 = 3\sqrt{2}/\sqrt{43}$. Thus, we have

$$R_{32}(\theta_3)A_3 = R_{32}(\theta_3)R_{31}(\theta_2)R_{21}(\theta_1)A$$

$$= \begin{bmatrix} 1 & 0 & 0 \\ 0 & \dfrac{5}{\sqrt{43}} & \dfrac{3\sqrt{2}}{\sqrt{43}} \\ 0 & \dfrac{3\sqrt{2}}{\sqrt{43}} & \dfrac{5}{\sqrt{43}} \end{bmatrix} A_3 = \begin{bmatrix} 4\sqrt{2} & 2\sqrt{2} & 3\sqrt{2} \\ 0 & \sqrt{43} & \dfrac{19}{\sqrt{43}} \\ 0 & 0 & \dfrac{23\sqrt{2}}{\sqrt{43}} \end{bmatrix} = R.$$

If we set $Q^T = R_{32}R_{31}R_{21}$, we have $Q^TA = R$ or $A = QR$ with

$$Q = IR_{21}^TR_{31}^TR_{32}^T = \begin{bmatrix} 0.424264 & -0.335497 & 0.841096 \\ 0.565685 & 0.823492 & 0.0431331 \\ -0.707107 & 0.457496 & 0.539164 \end{bmatrix}.$$

■

The QR factorization of A can be used, as was the LU factorization, to solve a set of linear systems $AX = K_i$. To see how this works we note that $AX = K$ can be written as $(QR)X = K$, which is equivalent to $RX = Q^TK$. We can then solve $AX = K$ by solving the triangular system $RX = Q^TK$. Once the factorization has been obtained, the solution is only a little more costly, in terms of the number of multiplications required, than a solution based on the LU factorization. For a single system the implementation could take the form

$$[A \mid K] \xrightarrow[\text{rotations}]{} [Q^TA \mid Q^TK] = [R \mid K'].$$

That is, doing the rotations on the augmented matrix is equivalent to doing both the factorization of A and the modification of the right-hand side (the solution of $QK' = K$). As in the case of the LU factorization, the explicit factorization would be required if there were multiple right-hand sides, not all of which were available initially.

QR FACTORIZATION USING PLANE ROTATIONS

```
INPUT: A(n, 2n) = [B(n, n), I(n, n)]
For j = 1, 2, . . . , n
    For i = j + 1, j + 2, . . . , n
        d = √(a_jj² + a_ij²)
        c = a_jj/d, s = a_ij/d                      (4.2)
        For k = j, j + 1, . . . , 2n
```

7. $t = a_{jk}$
8. $a_{jk} = c*t + s*a_{ik}$
9. $a_{ik} = -s*t + c*a_{ik}$
10. OUTPUT: $A = [R, Q^T]$ such that $Q^TB = R$.
11. Stop

The QR factorization, obtained using plane rotations, requires about $(4n^3/3)$ multiplications to compute R (Exercise 3), and is thus about four times more expensive than the LU factorization. If Q is computed explicitly, as in Algorithm (4.2), the number of multiplications doubles. Note however that no pivoting is required in the QR factorization; in some applications this is of great importance, and in those cases the QR factorization is more attractive than Gaussian elimination without pivoting.

The QR factorization of A can also be obtained using elementary reflectors. We will again carry out the details for a 4×4 matrix A. We first choose a unit vector $\mathbf{w_1}$, and hence an elementary reflector $H_1 = I - 2\mathbf{w_1}\mathbf{w_1}^T$, such that $H_1\text{Col}_1(A) = k_1\mathbf{e_1}$. By Theorem 4.1, if we take $k_1 = \pm\|\text{Col}_1(A)\|_2$, then we must choose $\mathbf{w_1}$ to be a unit vector in the direction of $\text{Col}_1(A) - k_1\mathbf{e_1}$. Thus

$$\mathbf{w_1} = \frac{1}{c_1}\begin{bmatrix} a_{11} - k_1 \\ a_{21} \\ a_{31} \\ a_{41} \end{bmatrix} \quad \text{and} \quad H_1A = \begin{bmatrix} k_1 & a_{12}^{(1)} & a_{13}^{(1)} & a_{14}^{(1)} \\ 0 & a_{22}^{(1)} & a_{23}^{(1)} & a_{24}^{(1)} \\ 0 & a_{32}^{(1)} & a_{33}^{(1)} & a_{34}^{(1)} \\ 0 & a_{42}^{(1)} & a_{43}^{(1)} & a_{44}^{(1)} \end{bmatrix}$$

where the constant c_1 is to be chosen so that $\mathbf{w_1}$ is a unit vector.

We next choose a unit vector $\mathbf{w_2}$ so that the associated elementary reflector $H_2 = I - 2\mathbf{w_2}\mathbf{w_2}^T$ is such that

$$H_2\,\text{Col}_2(H_1A) = H_2\begin{bmatrix} a_{12}^{(1)} \\ a_{22}^{(1)} \\ a_{32}^{(1)} \\ a_{42}^{(1)} \end{bmatrix} = \begin{bmatrix} a_{12}^{(1)} \\ k_2 \\ 0 \\ 0 \end{bmatrix} = a_{12}^{(1)}\mathbf{e_1} + k_2\mathbf{e_2}.$$

Thus

$$\mathbf{w_2} = \frac{1}{c_2}\begin{bmatrix} 0 \\ a_{22}^{(1)} - k_2 \\ a_{32}^{(1)} \\ a_{42}^{(1)} \end{bmatrix} \quad \text{where} \quad k_2 = \pm\|\text{Col}_2(H_1A) - a_{12}^{(1)}\mathbf{e_1}\|_2$$

and c_2 is chosen so that $\mathbf{w_2}$ is a unit vector. Note that the first entry of $\mathbf{w_2}$ is zero so that H_2 is of the form $H_2 = \left[\begin{array}{c|c} 1 & 0 \\ \hline 0 & * \end{array}\right]$ and premultiplying H_1A by H_2 will not change the first row of H_1A. It follows that

$$H_2H_1A = \begin{bmatrix} k_1 & a_{12}^{(1)} & a_{13}^{(1)} & a_{14}^{(1)} \\ 0 & k_2 & a_{23}^{(2)} & a_{24}^{(2)} \\ 0 & 0 & a_{33}^{(2)} & a_{34}^{(2)} \\ 0 & 0 & a_{43}^{(2)} & a_{44}^{(2)} \end{bmatrix}.$$

Finally we choose $H_3 = I - 2\mathbf{w_3}\mathbf{w_3}^T$ so that

$$H_3 \operatorname{Col}_3(H_2H_1A) = \begin{bmatrix} a_{13}^{(1)} \\ a_{23}^{(2)} \\ k_3 \\ 0 \end{bmatrix}.$$

To accomplish this we choose

$$\mathbf{w_3} = \frac{1}{c_3} \begin{bmatrix} 0 \\ 0 \\ a_{33}^{(2)} - k_3 \\ a_{43}^{(2)} \end{bmatrix}.$$

Then H_3 has the form

$$H_3 = \left[\begin{array}{c|c} I_2 & 0 \\ \hline 0 & * \end{array}\right] \quad \text{and} \quad H_3H_2H_1A = \begin{bmatrix} k_1 & a_{12}^{(1)} & a_{13}^{(1)} & a_{14}^{(1)} \\ 0 & k_2 & a_{23}^{(2)} & a_{24}^{(2)} \\ 0 & 0 & k_3 & a_{34}^{(3)} \\ 0 & 0 & 0 & a_{44}^{(3)} \end{bmatrix}$$

is upper triangular. Thus $A = QR$ where $Q = (H_3H_2H_1)^T = H_1H_2H_3 = IH_1H_2H_3$. Thus we see that Q is the result of postmultiplying I by the elementary reflectors that premultiply A. This observation allows the computation of Q while A is being reduced to R. The procedure we have just described generalizes easily to the $n \times n$ case, so we have an alternate way of computing the QR factorization of A as well as an alternate proof of

Theorem 4.2. The number of multiplications required to compute R in the QR factorization, using elementary reflectors, is about $(2n^3/3)$. This is a 50% reduction over the multiplication count using plane rotations, but it is still greater than for the LU factorization.

A very important observation is that the QR factorization is equivalent to, and provides an efficient implementation of, the Gram-Schmidt orthogonalization process, which was most likely included in your beginning linear algebra course. (See Exercise 10.) If $A = QR$ and R is nonsingular, then A and Q have precisely the same column space. Thus the columns of Q form an orthonormal basis for the column space of A whose first vector is a scalar multiple of the first column of A.

If A is not square, but has independent columns, then it is easy to modify the above observations. Suppose that A is $n \times s$ of rank s. As above, we can find an $n \times n$ orthogonal matrix Q, either as a product of elementary reflectors or of plane rotations, such that

$$\underset{n\times n}{Q^T}\ \underset{n\times s}{A} = \begin{bmatrix} R_1 \\ 0 \end{bmatrix}\}s, \quad \text{or} \quad A = Q\begin{bmatrix} R_1 \\ 0 \end{bmatrix} = [\underset{n\times s}{Q_1}\ \ Q_2]\begin{bmatrix} R_1 \\ 0 \end{bmatrix} = \underset{n\times s}{Q_1}\ \underset{s\times s}{R_1}.$$

In this case the columns of Q_1 form an orthonormal basis for the column space of A whose first vector is a scalar multiple of the first column of A. This is the same orthonormal basis one would obtain with the classical implementation of the Gram-Schmidt process. Note that the columns of Q_2 form an orthonormal basis for the orthogonal complement of the column space of A.

Although the details of the development are beyond the scope of this text, it is important to note that the Q and the R computed by either of the algorithms of this section are very stable. A precise statement of the stability result is given in the next theorem (Wilkinson, 1965).

THEOREM 4.3

Let A be an $n \times n$ matrix and let Q_c and R_c be the output computed by the QR factorization algorithm. Then, there exists a matrix E such that $Q_cR_c = A + E$, and the matrix E is small in the sense that $\|E\|_E \le \Phi(n)\mu\ \|A\|_E$, where $\Phi(n)$ is a slowly growing function of n and μ is the unit round off for the machine. ■

■ SOFTWARE NOTES ■

The MATALG command **M QR A** will generate the QR factorization of A and the user is given the opportunity to name and save the factors. The matrix Q is always square.

The MATLAB command **[Q,R]=qr(A)** will compute the QR factorization and save the factors. The matrix Q is always square.

The LINPACK subroutine **SQRDC** will also compute the QR factorization.

4.2 ■ Exercise Set

1. Show that the product of orthogonal matrices is an orthogonal matrix.

2. Use plane rotations to compute the QR factorizations of the following matrices by hand:

$$A = \begin{bmatrix} 3 & 5 \\ 4 & 10 \end{bmatrix}, \quad B = \begin{bmatrix} 12 & -20 & 41 \\ 9 & -15 & -63 \\ 20 & 50 & 35 \end{bmatrix}.$$

Compute the same factorizations using elementary reflectors.

3. Verify that the QR factorization of an $n \times n$ matrix A, using plane rotations, takes about $(4n^3/3)$ multiplications to compute R and another $(4n^3/3)$ to compute Q explicitly.

4. Verify that the QR factorization of an $n \times n$ matrix A, using elementary reflectors, takes about $(2n^3/3)$ multiplications to compute R and another $(2n^3/3)$ to compute Q explicity.

5. a. Use the software to find the QR factorization of $A = \begin{bmatrix} 1 & 2 & 3 & 4 \\ 2 & 3 & 5 & 6 \\ 3 & 5 & 2 & 7 \\ 4 & 6 & 7 & 1 \end{bmatrix}$.

b. Compute $A_1 = RQ$ and show that it is similar to A.

c. For $i = 1, 2, 3, 4, 5$

Factor $A_i = Q_iR_i$

Compute $A_{i+1} = R_iQ_i$

d. Show that A_5 is similar to A.

e. What seems to be happening to the sequence $\{A_i\}$?

6. Find an orthonormal basis for the subspace of $\mathscr{R}^4$ spanned by the vectors

$$\begin{bmatrix}1\\0\\1\\0\end{bmatrix}, \begin{bmatrix}1\\1\\1\\1\end{bmatrix}, \text{ and } \begin{bmatrix}3\\0\\4\\12\end{bmatrix}.$$

7. Find the QR factorization of $A = \begin{bmatrix}1 & 2 & 1\\1 & -1 & -1\\0 & 3 & 1\end{bmatrix}$.

8. Find an orthonormal basis for the column space of
$B = \begin{bmatrix}3 & 1 & 1\\0 & 0 & 1\\4 & 1 & 1\\12 & 0 & 1\end{bmatrix}$. Compare the result with that of Exercise 6.

9. Use the QR factorization from Exercise 5 to find the solution of $AX = \mathbf{e_2}$.

10. **a.** For X_1 and Y_2 in $\mathscr{R}^n$, determine a so that $X_2 = aX_1 + Y_2$ is orthogonal to X_1.

b. Suppose that $X_1, X_2, \ldots, X_k$ are mutually orthogonal vectors in $\mathscr{R}^n$ and that X_{k+1} is not in the subspace spanned by $\{X_1, X_2, \ldots, X_k\}$. Determine scalars a_i so that $X_{k+1} = a_1X_1 + a_2X_2 + \cdots + a_kX_k + Y_{k+1}$ is orthogonal to each of $X_1, X_2, \ldots, X_k$.

c. Prove that every subspace of $\mathscr{R}^n$ has an orthonormal basis.

11. Consider a compact storage scheme for computing the QR factorization using plane rotations. As in the LU factorization, we want to use the positions in the lower triangle of the matrix to store enough information about the rotations to determine them completely.

a. Which of the following would be best to store: θ, $\tan(\theta)$, $\cos\theta$, $\sin\theta$? Explain.

b. How many multiplications would be required to compute Q^TB, from the compact storage matrix, for an arbitrary vector B in $\mathscr{R}^n$.

12. Develop a compact storage scheme, for computing the QR factorization using elementary reflectors, by storing the $n \times n$ matrix to be factored in the first n rows of an $n + 1 \times n$ matrix A. Then store $\mathbf{w}_1$ in the last n positions of the first column, store $\mathbf{w}_2$ in the last $n - 1$ positions of the second column, and so on. Construct an algorithm to compute RQ from this compact storage matrix.

4.3 ■ HESSENBERG MATRICES

In this section we will show how to reduce a given $n \times n$ matrix A to a reduced form for which the eigenvalue problem is significantly easier.

■ DEFINITION 4.1 ■

An $n \times n$ matrix A is in **upper Hessenberg form** if every element below the first subdiagonal is zero, that is, $a_{ij} = 0$ for $i > j + 1$. An upper Hessenberg matrix is **unreduced** if each subdiagonal entry is different from zero.

Some examples of matrices in upper Hessenberg form are

$$A = \begin{bmatrix} 1 & 1 & -1 \\ 1 & 2 & -1 \\ 0 & -1 & 3 \end{bmatrix}, \quad B = \begin{bmatrix} 2 & 1 & 0 & 0 & 0 & 0 \\ 1 & 2 & 1 & 0 & 0 & 0 \\ 0 & 1 & 2 & 1 & 0 & 0 \\ 0 & 0 & 1 & 2 & 1 & 0 \\ 0 & 0 & 0 & 1 & 2 & 1 \\ 0 & 0 & 0 & 0 & 1 & 2 \end{bmatrix},$$

and

$$C = \left[\begin{array}{cc:ccc} 5 & 4 & 3 & 2 & 1 \\ 9 & 8 & 7 & 6 & 5 \\ \hdashline 0 & 0 & 5 & 3 & 1 \\ 0 & 0 & 2 & 4 & 6 \\ 0 & 0 & 0 & 7 & 8 \end{array}\right].$$

A symmetric upper Hessenberg matrix is **tridiagonal**; the matrix B above is an example of such a matrix. The matrices A and B are unreduced, while C is not; note that C is block upper triangular and that each diagonal block is an unreduced upper Hessenberg matrix. Note also that the companion matrices, described in Exercise 15 in Section 3.1, are special types of upper Hessenberg matrices.

THEOREM 4.4

Every $n \times n$ real matrix A is orthogonally similar to an upper Hessenberg matrix U; that is, there exists an orthogonal matrix P such that $P^TAP = U$ is upper Hessenberg. If A is symmetric, then U is tridiagonal.

PROOF We will construct P as a product of plane rotations. The method is called **Given's method.** We first choose a rotation R_{23} so that the entry in the $(3, 1)$ position of $R_{23}A$ is zero; the appropriate angle is $\theta_1 = \tan^{-1}(a_{31}/a_{21})$. Since R_{23} differs from I only in columns 2 and 3, the product $(R_{23}A)R_{23}^T$ differs from $R_{23}A$ only in columns 2 and 3. Thus the zero in the $(3, 1)$ position of $R_{23}A$ survives the postmultiplication. Recall Exercises 3 through 6 in Section 4.1. If A is 4×4, we have

$$R_{23}A = \begin{bmatrix} a_{11} & a_{12} & a_{13} & a_{14} \\ a_{21}^{(1)} & a_{22}^{(1)} & a_{23}^{(1)} & a_{24}^{(1)} \\ 0 & a_{32}^{(1)} & a_{33}^{(1)} & a_{34}^{(1)} \\ a_{41} & a_{42} & a_{43} & a_{44} \end{bmatrix} \quad \text{and} \quad R_{23}AR_{23}^T = \begin{bmatrix} a_{11} & a_{12}^{(1)} & a_{13}^{(1)} & a_{14} \\ a_{21}^{(1)} & a_{22}^{(2)} & a_{23}^{(2)} & a_{24}^{(1)} \\ 0 & a_{32}^{(2)} & a_{33}^{(2)} & a_{34}^{(1)} \\ a_{41} & a_{42}^{(1)} & a_{43}^{(1)} & a_{44} \end{bmatrix}$$

where the superscript indicates how many times the entry in that position has been changed. We now do a rotation, R_{24} in the $(2, 4)$ plane, to produce a zero in position $(4, 1)$. The premultiplication by R_{24} will change only rows 2 and 4, and the postmultiplication by R_{24}^T will change only columns 2 and 4, thus preserving the zeros in column 1. Thus the matrix $R_{24}(R_{23}AR_{23}^T)R_{24}^T$ will have zeros in positions $(3, 1)$ and $(4, 1)$. In the 4×4 case,

$$R_{24}(R_{23}AR_{23}^T)R_{24}^T = \begin{bmatrix} a_{11} & a_{12}^{(2)} & a_{13}^{(1)} & a_{14}^{(1)} \\ a_{21}^{(2)} & a_{22}^{(4)} & a_{23}^{(3)} & a_{24}^{(3)} \\ 0 & a_{32}^{(3)} & a_{33}^{(2)} & a_{34}^{(2)} \\ 0 & a_{42}^{(3)} & a_{43}^{(2)} & a_{44}^{(2)} \end{bmatrix}$$

and a rotation R_{34} can then be constructed to zero the $(4, 2)$ entry. Since postmultiplying by R_{34}^T will change only columns 3 and 4, it follows that the matrix $U = R_{34}(R_{24}R_{23}AR_{23}^TR_{24}^T)R_{34}^T$ is upper Hessenberg and that the matrix $P = (R_{34}R_{24}R_{23})^T$ is an orthogonal matrix such that $P^TAP = U$. This proves the theorem for the 4×4 case. The generalization to the $n \times n$ case should be evident. If A is symmetric, then so is the upper Hessenberg matrix $U = P^TAP$. In that case U is tridiagonal. ■

For an $n \times n$ matrix A, Given's method will require $(n-2)(n-1)/2$ plane rotations. The total number of multiplications required to compute U is about $(8n^3/3)$; in addition there are about $(n^2/2)$ square roots required.

The reduction to upper Hessenberg form can also be carried out using elementary reflectors. In this case the method is called the **Householder reduction.** We first choose an elementary reflector $H_1 = I - 2\mathbf{w_1}\mathbf{w_1}^T$ so that

$$H_1\text{Col}_1(A) = \begin{bmatrix} a_{11} \\ \pm k_1 \\ 0 \\ \vdots \\ 0 \end{bmatrix} \quad \text{with} \quad k_1 = \sqrt{a_{21}^2 + a_{31}^2 + \cdots + a_{n1}^2}.$$

Thus

$$\mathbf{w_1} = \frac{1}{c_1}\begin{bmatrix} 0 \\ a_{21} \mp k_1 \\ a_{31} \\ \vdots \\ a_{n1} \end{bmatrix} \quad \text{where} \quad c_1 = \sqrt{(a_{21} \mp k_1)^2 + a_{31}^2 + \cdots + a_{n1}^2}$$

is chosen so that $\mathbf{w_1}$ is a unit vector. Now, because the first entry of $\mathbf{w_1} = 0$, $H_1 = \left[\begin{array}{c|c} 1 & 0 \\ \hline 0 & H_1' \end{array}\right]$ and $H_1AH_1^T = H_1AH_1$ has the same first column as H_1A. Thus

$$A_2 = H_1AH_1 = \begin{bmatrix} a_{11} & a_{12}^{(1)} & \cdots \\ \pm k_1 & a_{22}^{(1)} & \cdots \\ 0 & a_{32}^{(1)} & \cdots \\ \vdots & \vdots & \vdots \\ 0 & a_{n2}^{(1)} & \cdots \end{bmatrix}.$$

We now choose $H_2 = I - 2\mathbf{w_2}\mathbf{w_2}^T$ so that

$$H_2\text{Col}_2(A_2) = \begin{bmatrix} a_{12}^{(1)} \\ a_{22}^{(1)} \\ \pm k_2 \\ 0 \\ \vdots \\ 0 \end{bmatrix}.$$

Thus

$$\mathbf{w_2} = \frac{1}{c_2}\begin{bmatrix} 0 \\ 0 \\ a_{32}^{(1)} \mp k_2 \\ a_{42}^{(1)} \\ \vdots \\ a_{n2}^{(1)} \end{bmatrix}$$

where c_2 is chosen so that $\mathbf{w_2}$ is a unit vector. In this case, because the first two entries of $\mathbf{w_2}$ are zero,

$$H_2 = \left[\begin{array}{cc|c} 1 & 0 & \\ 0 & 1 & 0 \\ \hline \multicolumn{2}{c|}{0} & H_2' \end{array}\right],$$

so that $H_2A_2H_2^T$ and H_2A_2 have the same first two columns. Thus $A_3 = H_2A_2H_2$ has its first two columns in upper Hessenberg form. Continuing in this way we obtain, after $n - 1$ steps, the upper Hessenberg matrix

$$U = P^TAP = (H_{n-1}H_{n-2}\cdots H_2H_1)A(H_1H_2\cdots H_{n-1}).$$

The preceding reduction, using elementary reflectors, takes about $(4n^3/3)$ multiplications, which is about half as many as Given's method. Thus, as in the case of the QR factorization, elementary reflectors are a more efficient tool than plane rotations. In Exercise 3 you are asked to show that the work can be rearranged so that the Householder reduction requires only one square root per step. It is important to note that the results of the Householder reduction and the Given's reduction may be different.

A subtle point about the Householder reduction is that the k_i need not be taken to be positive. If a_{21} is positive, it is best to take k_1 to be negative, and if a_{21} is negative, then it is best to take k_1 to be positive. This helps to avoid catastrophic cancellation in the calculation of the $\mathbf{w}_i$.

■ **EXAMPLE 4**

Use Householder's method to reduce the symmetric matrix

$$A = \begin{bmatrix} 1 & 2 & 1 & 2 \\ 2 & 2 & -1 & 1 \\ 1 & -1 & 1 & 1 \\ 2 & 1 & 1 & 1 \end{bmatrix}$$ to upper Hessenberg (in this case tridiagonal) form.

SOLUTION The first step is to choose an elementary reflector H_1 such that

$$H_1 \begin{bmatrix} 1 \\ 2 \\ 1 \\ 2 \end{bmatrix} = \begin{bmatrix} 1 \\ k_1 \\ 0 \\ 0 \end{bmatrix}.$$

We must choose

$$k_1 = \pm \left\| \begin{bmatrix} 2 \\ 1 \\ 2 \end{bmatrix} \right\|_2 = \pm 3.$$

We choose the minus sign because $a_{21} > 0$, so that

$$\mathbf{w_1} = \frac{1}{c_1} \begin{bmatrix} 0 \\ 2+3 \\ 1 \\ 2 \end{bmatrix} = \frac{1}{\sqrt{30}} \begin{bmatrix} 0 \\ 5 \\ 1 \\ 2 \end{bmatrix}.$$

The elementary reflector

$$H_1 = I - 2\mathbf{w_1}\mathbf{w_1}^T = I - \frac{2}{30} \begin{bmatrix} 0 \\ 5 \\ 1 \\ 2 \end{bmatrix} [0 \quad 5 \quad 1 \quad 2]$$

$$= \frac{1}{15} \begin{bmatrix} 15 & 0 & 0 & 0 \\ 0 & -10 & -5 & -10 \\ 0 & -5 & 14 & -2 \\ 0 & -10 & -2 & 11 \end{bmatrix}$$

and

$$A_2 = H_1AH_1^T = H_1AH_1 = \frac{1}{225}\begin{bmatrix} 225 & -675 & 0 & 0 \\ -675 & 525 & 105 & -15 \\ 0 & 105 & 354 & 303 \\ 0 & -15 & 303 & 21 \end{bmatrix}$$

$$= \begin{bmatrix} 1 & -3 & 0 & 0 \\ -3 & \frac{7}{3} & \frac{7}{15} & -\frac{1}{15} \\ 0 & \frac{7}{15} & \frac{354}{225} & \frac{303}{225} \\ 0 & -\frac{1}{15} & \frac{303}{225} & \frac{21}{225} \end{bmatrix}.$$

The next step is to choose H_2 such that

$$H_2\text{Col}_2(A_2) = \begin{bmatrix} -3 \\ \frac{7}{3} \\ k_2 \\ 0 \end{bmatrix} \quad \text{where} \quad k_2 = \pm\frac{\sqrt{50}}{15} = \pm\frac{\sqrt{2}}{3}.$$

Thus

$$\mathbf{w_2} = \frac{1}{c_2}\begin{bmatrix} 0 \\ 0 \\ \frac{7}{15} + \frac{\sqrt{2}}{3} \\ -\frac{1}{15} \end{bmatrix} = \begin{bmatrix} 0 \\ 0 \\ 0.997484208 \\ -0.07088902 \end{bmatrix} \quad \text{and}$$

$$H_2 = I - 2\mathbf{w_2}\mathbf{w_2}^T = \begin{bmatrix} 1 & 0 & 0 & 0 \\ 0 & 1 & 0 & 0 \\ 0 & 0 & -0.989949 & 0.141421 \\ 0 & 0 & 0.141421 & 0.989949 \end{bmatrix}.$$

Finally

$$A_3 = H_2A_2H_2 = (H_2H_1)A(H_1H_2) = (H_1H_2)^TA(H_1H_2)$$

$$= \begin{bmatrix} 1 & -3 & 0 & 0 \\ -3 & 2.3333 & -0.4714 & 0 \\ 0 & -0.4714 & 1.16666 & -1.5 \\ 0 & 0 & -1.5 & 0.5 \end{bmatrix}$$

is the tridiagonal matrix similar to A and

$$H = H_1H_2 = \begin{bmatrix} 1 & 0 & 0 & 0 \\ 0 & -0.6667 & 0.2357 & -0.7071 \\ 0 & -0.3333 & -0.9428 & 0 \\ 0 & -0.6667 & 0.2357 & 0.7071 \end{bmatrix}$$

is the orthogonal transforming matrix. ■

It would certainly be easier to do the QR factorization for an upper Hessenberg matrix, or a symmetric tridiagonal matrix, than for a full matrix. In this case only $n - 1$ plane rotations [in planes $(1, 2), (2, 3), \ldots, (n - 1, n)$] are required and the number of multiplications is reduced to about $(2n^2)$. In this important case there is no advantage in using elementary reflectors because the elementary reflectors are just 2×2 orthogonal matrices.

In the case where A is upper Hessenberg, if $A = QR$, then the matrix $A_2 = RQ = Q^TAQ$ is also upper Hessenberg. To see this we note that

$$A_2 = RQ = \begin{bmatrix} x & x & x & \cdots & x \\ 0 & x & x & \cdots & x \\ 0 & 0 & x & \cdots & x \\ \vdots & \vdots & \vdots & \ddots & \vdots \\ 0 & 0 & 0 & \cdots & x \end{bmatrix} R_{12}^T R_{23}^T \cdots R_{n-1,n}^T$$

$$= \begin{bmatrix} x & x & x & \cdots & x \\ y & x & x & \cdots & x \\ 0 & 0 & x & \cdots & x \\ \vdots & \vdots & \vdots & \ddots & \vdots \\ 0 & 0 & 0 & \cdots & x \end{bmatrix} R_{23}^T \cdots R_{n-1,n}^T$$

$$= \begin{bmatrix} x & x & x & \cdots & x \\ y & x & x & \cdots & x \\ 0 & y & x & \cdots & x \\ \vdots & \vdots & \vdots & \ddots & \vdots \\ 0 & 0 & 0 & \cdots & x \end{bmatrix} R_{34}^T \cdots R_{n-1,n}^T$$

$$= \cdots = \begin{bmatrix} x & x & x & \cdots & x & x \\ y & x & x & \cdots & x & x \\ 0 & y & x & \cdots & x & x \\ \vdots & \vdots & \ddots & \ddots & \vdots & \vdots \\ 0 & 0 & 0 & \cdots & x & x \\ 0 & 0 & 0 & \cdots & y & x \end{bmatrix}$$

where x indicates an element that may be nonzero, and y indicates an element that may become nonzero because of the postmultiplication by the rotation matrices.

■ SOFTWARE NOTES ■

The MATLAB command **[P,H]=hess(A)** will return an upper Hessenberg matrix H and an orthogonal matrix P such that $P^TAP = H$. The MATLAB algorithms work from the lower-right corner rather than from the upper-left corner as described in the text. MATLAB yields a different, but equally valid, result for Example 4.

4.3 ■ Exercise Set

1. Find the QR factorization of the upper Hessenberg matrix $A = \begin{bmatrix} 3 & 5 & 10 & 0 \\ 4 & 15 & -10 & 5 \\ 0 & 12 & 13 & 26 \\ 0 & 0 & 1 & 2 \end{bmatrix}$. Compute RQ and observe that it is still upper Hessenberg.

2. Show that the QR factorization of an upper Hessenberg requires only about $(2n^2)$ multiplications, using either plane rotations or elementary reflectors.

3. Show that the calculation can be organized so that only one square root is required per step in the Householder reduction to upper Hessenberg form. *Hint*: Express H_1 in terms of $V_1 = c_1\mathbf{w}_1$.

4. Use plane rotations to find a tridiagonal matrix similar to
$A = \begin{bmatrix} 0 & 12 & 16 & -15 \\ 12 & 288 & 309 & 185 \\ 16 & 309 & 312 & 80 \\ -15 & 185 & 80 & -600 \end{bmatrix}$. The numerical work for this problem can be carried out by hand.

5. Find an upper Hessenberg matrix similar to $A = \begin{bmatrix} 0 & 12 & 16 & -15 \\ 12 & 288 & 309 & 185 \\ 16 & 309 & 312 & 90 \\ -15 & 185 & 80 & -600 \end{bmatrix}$.

6. Write a subroutine, or a MATLAB m-file, to implement the Householder reduction of an $n \times n$ ($n \leq 30$) matrix to upper Hessenberg form.

7. Use elementary lower triangular matrices to show that an $n \times n$ matrix A is similar to an upper Hessenberg matrix if no zero elements occur on the subdiagonal at inopportune times. How many multiplications does this reduction require?

8. Let H be an unreduced upper-Hessenberg matrix and let λ be an eigenvalue. Show that the dimension of the associated eigenspace is 1. What does this say about the Jordan canonical form of H?

9. Show that an unreduced symmetric tridiagonal matrix has only simple eigenvalues.

10. Show that every unreduced upper-Hessenberg matrix is diagonally similar to an unreduced upper-Hessenberg matrix with all subdiagonal elements equal to 1.

11. Let H be an unreduced upper-Hessenberg matrix and let $\mathbf{x}_1 = \mathbf{e}_1$. Define $H\mathbf{x}_i = \mathbf{x}_{i+1}$ for $i = 1, 2, \ldots, n - 1$ and let $X = [\mathbf{x}_1\ \mathbf{x}_2 \cdots \mathbf{x}_n]$. Show that $X^{-1}HX = C$ is a companion matrix. What are the diagonal entries of X?

4.4 ■ THE QR ALGORITHM

In this section we will study a remarkably effective algorithm for computing all the eigenvalues and eigenvectors of an arbitrary matrix A. The algorithm we will study was developed by J. G. F. Francis in 1961 and was based in part on work by H. Rutishauser in 1958. The software packages EISPACK and LAPACK contain a number of FORTRAN subroutines that provide finely tuned implementations of this method and its refinements. The EISPACK package is frequently cited as a model for software developers. A path chart for the EISPACK subroutines is given in Appendix F. For a detailed discussion of this package and a listing of the source code, see (Smith et al., 1976). Both MATLAB and MATALG use EISPACK subroutines to compute eigenvalues and eigenvectors.

In its simplest form the QR algorithm generates a sequence of matrices orthogonally similar to the given matrix A that converges to an upper triangular matrix.

BASIC QR ALGORITHM

1. INPUT: $A(n, n)$, m
2. $A_1 = A$
3. For $i = 1, 2, \ldots, m$
4. Factor $A_i = Q_i R_i$ (Determine Q_i^T such that $Q_i^T A_i = R_i$ where Q_i is orthogonal and R_i is upper triangular. (4.3)
5. Compute $A_{i+1} = R_i Q_i$.
6. OUTPUT: A_m orthogonally similar to A and close to upper triangular.

EXAMPLE 5

Find the first five elements of the sequence generated by Algorithm (4.3) for the matrix

$$A = \begin{bmatrix} 1 & 2 & 2 & 0 \\ 3 & 1 & 3 & 3 \\ 2 & 2 & 1 & 2 \\ 1 & 2 & 3 & 1 \end{bmatrix}.$$

SOLUTION The MATLAB commands **[Q,R]=qr(A), A1=R*Q** yield the following results:

$$Q = \begin{bmatrix} -0.2582 & 0.5703 & 0.0487 & -0.7783 \\ -0.7746 & -0.5403 & 0.3062 & -0.1197 \\ -0.5164 & 0.2401 & -0.7656 & 0.2993 \\ 0.2382 & 0.5703 & 0.5637 & 0.5388 \end{bmatrix},$$

$$R = \begin{bmatrix} -3.8730 & -0.28402 & -4.1312 & -3.6148 \\ 0 & 2.2211 & 1.4707 & -0.05703 \\ 0 & 0 & 1.9417 & -0.0487 \\ 0 & 0 & 0 & 0.7783 \end{bmatrix},$$

and

$$A_1 = RQ = \begin{bmatrix} 6.2667 & -3.7277 & 0.0665 & 0.1700 \\ -2.3327 & -1.1721 & -0.7673 & -0.1330 \\ -0.9901 & 0.4385 & -1.5139 & 0.5550 \\ -0.2010 & 0.4438 & 0.4387 & 0.4194 \end{bmatrix}.$$

If you repeat the above commands you will get the following matrices, all similar to A:

$$A_2 = \begin{bmatrix} 7.2638 & -0.6008 & 1.5200 & -0.6141 \\ 0.9519 & -2.2394 & -0.5000 & 0.5544 \\ 0.2036 & -0.0796 & -1.5378 & -0.1104 \\ 0.0159 & 0.0773 & 0.1236 & 0.5133 \end{bmatrix},$$

$$A_3 = \begin{bmatrix} 7.1862 & -1.7801 & 1.2814 & 0.5447 \\ -0.2977 & -2.1653 & -0.7051 & -0.4872 \\ -0.0437 & -0.0347 & -1.5492 & 0.2378 \\ -0.0011 & 0.0196 & 0.0317 & 0.5283 \end{bmatrix},$$

$$A_4 = \begin{bmatrix} 7.2480 & -1.3667 & 1.3810 & -0.5758 \\ 0.0968 & -2.2449 & -0.6286 & 0.4940 \\ 0.0092 & -0.0328 & -1.5316 & -0.2153 \\ 0.0001 & 0.0046 & 0.0090 & 0.5285 \end{bmatrix},$$

$$A_5 = \begin{bmatrix} 7.2310 & -1.4731 & 1.3848 & 0.5656 \\ -0.0306 & -2.2366 & -0.6396 & -0.4897 \\ -0.0019 & -0.0208 & -1.5229 & 0.2318 \\ -0.0000 & 0.0011 & 0.0026 & 0.5284 \end{bmatrix}.$$

Clearly the matrices are approaching upper triangular form and the diagonal entries of A_5 are good approximations for the eigenvalues of A. Additional iterations would yield more accurate approximations to the eigenvalues.

■

Some additional computational experience with this algorithm was given in Exercise 5 in Section 4.2. For the symmetric matrix of that exercise the sequence A_i seemed to be converging to a diagonal matrix. It is however too much to hope for that this sequence will always converge to an upper triangular matrix. If A is an orthogonal matrix, then the unique QR factorization of A is $A = AI$. In this case each $A_i = A$ and the sequence goes nowhere at all. Thus the basic QR algorithm is not universally convergent to an upper triangular matrix. Note that the eigenvalues of an orthogonal matrix are all of absolute value 1. In general, as for the power method, repeated eigenvalues and eigenvalues of equal modulus will cause the most difficulty.

It is easy to see that $A_1 = Q_1R_1$ and $A_2 = R_1Q_1$ imply that $A_2 = Q_1^TA_1Q_1$. Thus the terms of the sequence generated by Algorithm (4.3) are all orthogonally similar to the original matrix A. The algorithm also preserves several important kinds of structure that the matrix A may have. In particular:

if A is symmetric, so is each A_i;

if A is upper Hessenberg, so is each A_i;

if A is symmetric and tridiagonal, so is each A_i. (See Exercise 9.)

The following theorem gives conditions under which the QR algorithm will converge, and the proof gives some insight into possible difficulties and how to overcome them.

THEOREM 4.5

Let A have eigenvalues $\lambda_1, \lambda_2, \ldots, \lambda_n$ such that $|\lambda_1| > |\lambda_2| > \cdots > |\lambda_n| > 0$, and let $X = [X_1, X_2, \ldots, X_n]$ be the matrix of right eigenvectors of A. If the matrix of left eigenvectors, $Y = X^{-1}$, has an LU factorization, then the sequence $\{A_i\}$ generated by Algorithm (4.3) converges to an upper triangular matrix.

PROOF We have $X^{-1}A_1X = D = \mathrm{Dg}\{\lambda_1, \lambda_2, \ldots, \lambda_n\}$ so $A_1 = XDY = Y^{-1}DY$. The hypotheses of the theorem guarantee the existence of a factorization $Y = L_YR_Y$ where L_Y is unit lower triangular and R_Y is upper triangular. We also have the QR factorization of the eigenvector matrix $X = Q_XR_X$ where R_X (but not R_Y) can be assumed to have positive diagonal entries. We now define

$$Q^{(s)} = Q_1Q_2\cdots Q_s \quad \text{and} \quad R^{(s)} = R_sR_{s-1}\cdots R_2R_1$$

and note that

$$A_{s+1} = (Q^{(s)})^TA_1Q^{(s)} \quad \text{and that} \quad A_1^s = Q^{(s)}R^{(s)}.$$

To verify the last equality for $s = 4$ we note from Algorithm (4.3) that $A_4 = R_3Q_3 = Q_4R_4$; $A_3 = R_2Q_2 = Q_3R_3$; $A_2 = R_1Q_1 = Q_2R_2$; and $A_1 = Q_1R_1$. Then,

$$\begin{aligned} Q^{(4)}R^{(4)} &= Q_1Q_2Q_3(Q_4R_4)R_3R_2R_1 = Q_1Q_2Q_3(R_3Q_3)R_3R_2R_1 \\ &= Q_1Q_2(Q_3R_3)(Q_3R_3)R_2R_1 = Q_1Q_2(R_2Q_2)(R_2Q_2)R_2R_1 \\ &= Q_1(Q_2R_2)(Q_2R_2)(Q_2R_2)R_1 = Q_1(R_1Q_1)(R_1Q_1)(R_1Q_1)R_1 \\ &= (Q_1R_1)(Q_1R_1)(Q_1R_1)(Q_1R_1) = A_1^4. \end{aligned}$$

The general result follows in the same way. Thus we have a QR factorization of A_1^s where $R^{(s)}$ has positive diagonal entries. We will now obtain this factorization in a different way and then use the uniqueness part of Theorem 4.2 to compare the two factorizations.

Since $XY = I$ and Y has an LU factorization, we have

$$\begin{aligned} A_1^s &= (XDY)^s = XD^sY = XD^sL_YR_Y = X[D^sL_YD^{-s}]D^sR_Y \\ &= X[I + E_s]D^sR_Y. \end{aligned}$$

Note that

$$\text{ent}_{ij}(D^s L_Y D^{-s}) = \begin{cases} 0 & \text{if } i < j \\ 1 & \text{if } i = j \\ l_{ij}\left(\dfrac{\lambda_i}{\lambda_j}\right)^s & \text{if } i > j. \end{cases}$$

Thus $D^s L_Y D^{-s} = I + E_s$, where, as a consequence of the assumptions on the λ_i, E_s is a strictly lower triangular matrix whose elements approach zero as s gets large. Thus, as s gets large

$$A_1^s = X[I + E_s]D^s R_Y = Q_X R_X [I + E_s] D^s R_Y \to Q_X R_X D^s R_Y$$

where Q_X is orthogonal and $R_X D^s R_Y$ is upper triangular with diagonal entries that may not be positive. Now let F_s be an orthogonal diagonal matrix (diagonal entries are ± 1) such that $F_s R_X D^s R_Y$ has positive diagonal entries. Then

$$A_1^s \to (Q_X F_s)(F_s R_X D^s R_Y),$$

a QR factorization in which the upper triangular factor has positive diagonal entries. Comparing this with the factorization obtained earlier, we see that

$$(Q^{(s)} - Q_X F_s) \to 0 \text{ as } s \to \infty \quad \text{and}$$
$$(R^{(s)} - F_s R_X D^s R_Y) \to 0 \text{ as } s \to \infty.$$

Thus as s gets large we have, because $X = Q_X R_X$ implies that $Q_X^T X = R_X$ and $X^{-1} Q_X = R_X^{-1}$,

$$\begin{aligned} A_{s+1} &= (Q^{(s)})^T A_1 Q^{(s)} \to (Q_X F_s)^T A_1 (Q_X F_s) \\ &= F_s (Q_X^T X) D (X^{-1} Q_X) F_s = F_s (R_X D R_X^{-1}) F_s, \end{aligned}$$

which is upper triangular with diagonal entries $\lambda_1, \lambda_2, \ldots, \lambda_n$ in that order. This completes the proof. ■

Theorem 4.5 is a bit disappointing at first glance since it appears that the convergence is no better than the power method and that if there are eigenvalues of equal modulus, then the sequence $\{A_i\}$ will only converge to a block triangular matrix. This is indeed what happens as illustrated in Exercise 3. If $|\lambda_1| = |\lambda_2| = |\lambda_3|$, than the sequence will converge to a matrix with a 3×3 diagonal block

$$A_i \to \left[\begin{array}{ccc|ccc} y & y & y & x & \cdots & x \\ y & y & y & x & \cdots & x \\ y & y & y & x & \cdots & x \\ \hline & & & \lambda_4 & \cdots & x \\ & 0 & & & \ddots & \\ & & & 0 & \cdots & \lambda_n \end{array}\right]$$

where the 3×3 matrix whose entries are labeled y would have eigenvalues λ_1, λ_2, and λ_3.

If $\lambda_3 = \overline{\lambda_4}$, then

$$A_i \to \left[\begin{array}{cc:cc:ccc} \lambda_1 & x & x & x & x & \cdots & x \\ 0 & \lambda_2 & x & x & x & \cdots & x \\ \hdashline 0 & 0 & y & y & x & \cdots & x \\ 0 & 0 & y & y & x & \cdots & x \\ \hdashline 0 & 0 & 0 & 0 & \lambda_5 & \cdots & x \\ & & & & & \ddots & \\ 0 & 0 & 0 & 0 & 0 & \cdots & \lambda_n \end{array}\right]$$

where the 2×2 block whose entries are labeled y would have complex eigenvalues λ_3 and λ_4.

You can significantly reduce the amount of work per iteration if you first reduce the given matrix A to an appropriate compact form (tridiagonal if A is symmetric and upper Hessenberg if A is not symmetric). An upper Hessenberg matrix is said to be **unreduced** if all the subdiagonal entries are nonzero. If a subdiagonal element of an upper Hessenberg matrix is zero, then the matrix is block diagonal and the eigenvalue problem for that matrix splits into two smaller problems. An efficient implementation of the QR algorithm should test for this splitting of the problem into smaller problems. As pointed out in Exercise 5, the reduction to upper Hessenberg form also has the advantage that it tends to separate any repeated eigenvalues into separate diagonal blocks.

To really accelerate the convergence, as in the case of the power method, we need to include origin shifts. That is, we first choose a shift k_i, then we factor $A_i - k_iI = Q_iR_i$ and compute $A_{i+1} = k_iI + R_iQ_i$.

With this change in the algorithm note that

$$A_{i+1} = k_iI + R_iQ_i = k_iI + Q_i^T(A_i - k_iI)Q_i = Q_i^TA_iQ_i,$$

so including the origin shifts does not change the fact that we are generating a sequence of matrices orthogonally similar to A.

Initially we choose the shift as close as possible to λ_n to accelerate the convergence of the entry in the $(n, n - 1)$ position to zero. The Gerschgorin theorem tells us that there is an eigenvalue in a circle with center $a_{nn}^{(i)}$. Thus a very simple shift strategy would be to pick $k_i = a_{nn}^{(i)} = \text{ent}_{nn}(A_i)$. A more complicated shift strategy is to take k_i equal to the eigenvalue of the 2×2 lower-right-hand corner that is nearest $a_{nn}^{(i)}$. When the off-diagonal element in the last row becomes sufficiently small we accept the diagonal element in the last row as λ_n and then repeat the algorithm on the $n - 1 \times n - 1$ upper-left corner of A. While we are driving $a_{n,n-1}^{(i)}$ to zero some of the other subdiagonal entries of A may also be getting smaller.

The following version of the QR algorithm will converge to an upper triangular matrix if none of the eigenvalues of A are of equal modulus.

QR ALGORITHM WITH SHIFTS

1. INPUT: $A(n, n)$, TOL, Max
2. Reduce A to an upper Hessenberg matrix A_1.
3. For $j = 0, 1, 2, \ldots, n - 2$
4. $\quad m = n - j$
5. $\quad$ For $i = 1, 2, \ldots,$ Max
6. $\quad\quad$ Choose a shift $k_i(= a^i_{mm})$
7. $\quad\quad$ Factor $A_i - k_i I = Q_i R_i$ (4.4)
8. $\quad\quad$ Compute $A_{i+1} = k_i I + R_i Q_i$
9. $\quad\quad$ If $|a^{(i+1)}_{m,m-1}| < \text{TOL}$ accept $\lambda_m = a^{(i+1)}_{mm}$
10. $\quad$ Delete the last row and last column of A_i. (Equivalently, set $a^{(i+1)}_{m,m-1} = 0$.)
11. OUTPUT: An upper Hessenberg matrix similar to A with most subdiagonal entries zero

For Algorithm (4.4) there are stronger convergence results than Theorem 4.5. For the symmetric case, Wilkinson (1965) has proven Theorem 4.6.

THEOREM 4.6

With either of the shift strategies described above, the sequence generated by Algorithm (4.4), for a symmetric tridiagonal matrix A_1, converges to a block triangular matrix whose blocks are either 1×1 or 2×2. ■

Theorem 4.6 suggests that the convergence test in Algorithm (4.4) is not adequate since the entry in the $(n, n - 1)$ position may not approach 0. A better approach is to test the product of the last two entries on the subdiagonal to see if it is small enough to be considered zero. If the next-to-the-last subdiagonal entry goes to zero, then two eigenvalues are isolated in the 2×2 diagonal block in the lower-right-hand corner of the matrix.

Another convergence theorem proved by B. Parlett (1967) is Theorem 4.7.

THEOREM 4.7

For an unreduced upper Hessenberg matrix, the unshifted QR algorithm converges to a block triangular matrix whose blocks are either 1×1 or 2×2, if and only if in each set of all the distinct eigenvalues that have a common modulus there are not more than two of even or two of odd multiplicity. ■

■ EXAMPLE 6

Use Algorithm (4.4) to find the eigenvalues of $A_1 = \begin{bmatrix} 3 & 5 & 2 & 1 \\ 4 & -2 & 3 & 2 \\ 0 & 7 & 3 & 5 \\ 0 & 0 & 8 & 2 \end{bmatrix}$.

SOLUTION You can do all the calculations using either MATLAB or MATALG. You should attempt to reproduce the following MATLAB calculations. The MATALG results, although slightly different, are equivalent.

We will use the first shift strategy described above. Thus we choose $k_1 = a_{44}^{(1)} = 2$, compute the QR factorization of $B = A_1 - k_1 I = Q_1 R_1$, and then compute

$$A_2 = R_1 Q_1 + k_1 I = \begin{bmatrix} 0.4118 & -5.5820 & -2.2327 & 1.3213 \\ -8.8322 & 1.7756 & 4.0535 & 2.8354 \\ 0 & 6.1553 & 1.7093 & 5.1112 \\ 0 & 0 & 2.8201 & 2.1033 \end{bmatrix}.$$

We next choose $k_2 = a_{44}^{(2)} = 2.1033$ and repeat the above steps to obtain

$$A_3 = R_2 Q_2 + k_2 I = \begin{bmatrix} -0.9356 & 8.3393 & -4.9724 & 1.0632 \\ 8.0555 & 4.3669 & -2.5431 & -2.1518 \\ 0 & -2.8175 & -1.4940 & 0.1419 \\ 0 & 0 & 2.2297 & 4.0627 \end{bmatrix}.$$

Repeating this process two more times we find

$$A_5 = R_4 Q_4 + k_4 I = \begin{bmatrix} -5.1625 & 4.1410 & -5.1116 & 0.7857 \\ 4.6978 & 9.1836 & 3.6312 & -2.1044 \\ 0 & -2.3512 & -2.0984 & -1.3701 \\ 0 & 0 & 0.5550 \times 10^{-6} & 4.0774 \end{bmatrix}.$$

At this point the (4, 3) entry is close enough to zero that we accept 4.07735 as an adequate approximation to λ_4. Note that the first subdiagonal entry has grown while the second subdiagonal entry has descreased. We can now accomplish the deflation either by setting the (4, 3) entry of A_5 to zero or by deleting the last row and last column of A_5; the first method is easiest with MATALG, while the second method would normally be used in a computer program. We choose the second method and then proceed by taking

$k_5 = a_{33}^{(5)} - 2.0986$, computing the factorization $A_5 - k_5 I = Q_5 R_5$ and then

$$A_6 = R_5 Q_5 + k_5 I = \begin{bmatrix} 0.8575 & -8.4373 & 1.8033 \\ -8.3044 & 4.6069 & 4.6283 \\ 0 & -0.3522 & -3.5418 \end{bmatrix}.$$

Repeating this procedure four times we obtain

$$A_{10} = R_9 Q_9 + k_9 I = \begin{bmatrix} 11.2639 & -0.5401 & 2.6763 \\ 0.253404 \times 10^{-2} & -5.3291 & -4.5415 \\ 0 & 0.4710 \times 10^{-9} & -4.0122 \end{bmatrix}.$$

At this point we accept -4.0122 as a good approximation to λ_3 and deflate by deleting the last row and last column of the matrix A_{10} (equivalently we could set the $(3, 2)$ entry to zero). We now take $k_{10} = -5.3291$, compute the factorization $A_{10} - k_{10} I = Q_{10} R_{10}$, and then

$$A_{11} = R_{10} Q_{10} + k_{10} I = \begin{bmatrix} 11.2640 & 0.5376 \\ -0.6913 \times 10^{-9} & -5.3291 \end{bmatrix}.$$

Since the $(2, 1)$ entry is "small enough," we accept $\lambda_1 = 11.2640$ and $\lambda_2 = -5.3291$ as good approximations for the other two eigenvalues of the original matrix A. ■

■ SOFTWARE NOTES ■

Algorithm (4.3) can be implemented with the following MATLAB script (m-file):

```
for j=1:10
    [Q,R]=qr(A);
    A=R*Q
    pause(5)
end
```

MATLAB and MATALG may not produce the same matrices when implementing Algorithm (4.4) because MATALG always computes a QR factorization in which the diagonal elements of R are positive.

The EISPACK subroutine **ORTHES** will reduce A to upper Hessenberg form.

4.4 ■ Exercise Set

1. Show that the eigenvalues of an orthogonal matrix are all of modulus 1.

2. Compute five terms of the unshifted QR algorithm for

$$A = \begin{bmatrix} 2 & -1 & 0 & 0 \\ -1 & 2 & -1 & 0 \\ 0 & -1 & 2 & -1 \\ 0 & 0 & -1 & 2 \end{bmatrix}.$$

3. Compute seven terms of the unshifted QR algorithm for

$$A = \begin{bmatrix} 2 & 1 & -2 & -1 & 2 & 1 \\ -1 & 2 & 1 & -2 & -1 & 2 \\ 8 & 4 & 12 & 6 & -2 & -1 \\ -4 & 8 & -6 & 12 & 1 & -2 \\ 8 & 4 & 8 & 4 & 2 & 1 \\ -4 & 8 & -4 & 8 & -1 & 2 \end{bmatrix}.$$

Estimate the eigenvalues of A.

4. Compute 10 terms of the sequence generated by Algorithm (4.4), using the first shift strategy, for the matrix

$$A = \begin{bmatrix} 5 & 2 & 0 & 0 & 0 \\ 2 & -4 & 3 & 0 & 0 \\ 0 & 3 & 0 & 6 & 0 \\ 0 & 0 & 6 & 4 & 1 \\ 0 & 0 & 0 & 1 & 5 \end{bmatrix}.$$

Estimate the eigenvalues of A.

5. **a.** Show that the last $n - 1$ rows of an $n \times n$ unreduced Hessenberg matrix A are linearly independent.

   b. Let λ be an eigenvalue of an $n \times n$ unreduced Hessenberg matrix A. Show that rank $(A - \lambda I) = n - 1$ and hence that the dimension of the associated eigenspace is 1.

   c. Show that the eigenvalues of an unreduced, diagonalizable, Hessenberg matrix are simple.

   d. Show that a symmetric, unreduced, tridiagonal matrix has only simple eigenvalues.

6. The following 12×12 matrix has eigenvalues $\pm(3 \pm 4i)$, $\pm(4 \pm 3i)$, $\pm 5i$, and ± 5:

$$A = \begin{bmatrix} 24 & 3 & -20 & 10 & -1 & 5 & 24 & -11 & 16 & 40 & -22 & -29 \\ 3 & 13 & -17 & 6 & 1 & -1 & 12 & 4 & 12 & 18 & -11 & -3 \\ 4 & -5 & 7 & 2 & 12 & 5 & 9 & 4 & -7 & -7 & 1 & -4 \\ 5 & 8 & 6 & -3 & 4 & 6 & -5 & -2 & 3 & -1 & -13 & -5 \\ -23 & 40 & -18 & 26 & 13 & 15 & -3 & -17 & -26 & 18 & -17 & 23 \\ 8 & -9 & 1 & 10 & -4 & -3 & 0 & 4 & -10 & -1 & 1 & -8 \\ -19 & 10 & 13 & 2 & 1 & 0 & -4 & 2 & -2 & -13 & 9 & 19 \\ -16 & 21 & -4 & 17 & 16 & 11 & -3 & -20 & -17 & 4 & -5 & 16 \\ 3 & 5 & 9 & -7 & 4 & 6 & -5 & -2 & 7 & -4 & -8 & -3 \\ -2 & 2 & 3 & -1 & 12 & 5 & 9 & 4 & -4 & -3 & 0 & 2 \\ 7 & 6 & -17 & 6 & 1 & -1 & 12 & 4 & 12 & 18 & -8 & -7 \\ 18 & -11 & -3 & 4 & -2 & 6 & 12 & -15 & 4 & 22 & -7 & -23 \end{bmatrix}$$

a. Use Theorem 4.7 to predict what Algorithm (4.3) will produce for this matrix.

b. Use MATLAB, or MATALG and the Householder reduction, to find an upper Hessenberg matrix B that is orthogonally similar to A.

c. Run 40 iterations to Algorithm (4.4) for the matrix B.

7. Let A be an upper Hessenberg matrix. Show that if k_1 is an exact eigenvalue of A, then Algorithm (4.4) will deflate after one step.

8. Consider the companion matrices

$$C_1 = \begin{bmatrix} 0 & 0 & 0 & -1 \\ 1 & 0 & 0 & 0 \\ 0 & 1 & 0 & 0 \\ 0 & 0 & 1 & 0 \end{bmatrix} \quad \text{and}$$

$$C_2 = \begin{bmatrix} 0 & 0 & 0 & 0 & -1 \\ 1 & 0 & 0 & 0 & 0 \\ 0 & 1 & 0 & 0 & -1 \\ 0 & 0 & 1 & 0 & -1 \\ 0 & 0 & 0 & 1 & 0 \end{bmatrix}.$$

a. Show that C_1 is an orthogonal matrix and hence that Algorithm (4.3) goes nowhere.

b. Apply Algorithm (4.3) to the matrix $C_1 + 2I$. Explain what is happening to this sequence and why.

c. Show that the characteristic polynomial of C_2 is $(\lambda^3 + 1)(\lambda^2 + 1)$ and find the eigenvalues.

d. Predict the behavior of Algorithm (4.3) for the matrix C_2. Will any shift strategy help?

9. Let A_i be the sequence of matrices generated by Algorithm (4.3).

a. Show that if A is symmetric, then so is each A_i.

b. Show that if A is upper Hessenberg, then so is each A_i.

c. Show that if A is symmetric and tridiagonal, then so is each A_i.

d. Show that if A is tridiagonal (but not symmetric), then the A_i need not be tridiagonal.

4.5 ■ THE DOUBLE QR ALGORITHM

The QR algorithm, with the shift strategies described in the last section, may not converge if A has several complex eigenvalues of the same modulus (Exercise 6, Section 4.4). If the eigenvalues of the 2×2 lower-right corner of A were complex, then the second shift strategy of the last section would suggest a complex origin shift; this would be computationally inconvenient. Instead we consider two steps of the QR algorithm with shifts that are complex conjugates. We will show that the end result of these two steps can be computed using only real arithmetic. Thus suppose that k_1 and k_2 are the eigenvalues of the 2×2 lower-right corner of A_{2s} so that $k_1 + k_2 = \operatorname{tr}\begin{bmatrix} a^{(2s)}_{n-1,n-1} & a^{(2s)}_{n-1,n} \\ a^{(2s)}_{n,n-1} & a^{(2s)}_{nn} \end{bmatrix}$ and $k_1 k_2 = \det\begin{bmatrix} a^{(2s)}_{n-1,n-1} & a^{(2s)}_{n-1,n} \\ a^{(2s)}_{n,n-1} & a^{(2s)}_{nn} \end{bmatrix}$. The iterates would then satisfy

$$\begin{aligned}
A_{2s} - k_1 I &= Q_{2s} R_{2s} \\
A_{2s+1} &= R_{2s} Q_{2s} + k_1 I = Q_{2s}^H A_{2s} Q_{2s} \\
A_{2s+1} - k_2 I &= Q_{2s+1} R_{2s+1} \\
A_{2s+2} &= R_{2s+1} Q_{2s+1} + k_2 I = Q_{2s+1}^H A_{2s+1} Q_{2s+1} \\
&= Q_{2s+1}^H Q_{2s}^H A_{2s} Q_{2s} Q_{2s+1} \\
&= (Q_{2s} Q_{2s+1})^H A_{2s} (Q_{2s} Q_{2s+1})
\end{aligned}$$

where the Q_i are complex unitary matrices if the k_i are complex. Now

$$\begin{aligned}
(Q_{2s} Q_{2s+1})(R_{2s+1} R_{2s}) &= Q_{2s}(Q_{2s+1} R_{2s+1}) R_{2s} \\
&= Q_{2s}(A_{2s+1} - k_2 I) R_{2s} = (Q_{2s} A_{2s+1}) R_{2s} \\
&\quad - k_2 Q_{2s} R_{2s} = (A_{2s} Q_{2s}) R_{2s} - k_2 Q_{2s} R_{2s} \\
&= (A_{2s} - k_2 I) Q_{2s} R_{2s} = (A_{2s} - k_2 I)(A_{2s} - k_1 I) \\
&= A_{2s}^2 - (k_1 + k_2) A_{2s} + k_1 k_2 I
\end{aligned}$$

is a real matrix and hence has a real QR factorization. Thus $Q = Q_{2s} Q_{2s+1}$ is a real unitary matrix, and it is evident that we may go from A_{2s} to A_{2s+2} using real arithmetic. In fact we can compute Q by finding the QR factorization of the real matrix $A_{2s}^2 - (k_1 + k_2)A_{2s} + k_1k_2I$. We could then compute $A_{2s+2} = Q^TA_{2s}Q$.

EXAMPLE 7

Consider the matrix $A = \begin{bmatrix} 0 & 0 & 0 & -625 \\ 1 & 0 & 0 & 350 \\ 0 & 1 & 0 & -98 \\ 0 & 0 & 1 & 14 \end{bmatrix}$, which has four complex eigenvalues of absolute value $5(3 \pm 4i, 4 \pm 3i)$.

SOLUTION For the first double step we have $k_1 + k_2 = \text{tr}\begin{bmatrix} 0 & -98 \\ 1 & 14 \end{bmatrix} = 14$ and $k_1k_2 = \det\begin{bmatrix} 0 & -98 \\ 1 & 14 \end{bmatrix} = 98$. Thus $(A - k_1I)(A - k_2I)$

$$= A^2 - (k_1 + k_2)A + k_1k_2I = \begin{bmatrix} 98 & 0 & -625 & 0 \\ -14 & 98 & 350 & -625 \\ 1 & -14 & 0 & 350 \\ 0 & 1 & 0 & 0 \end{bmatrix}.$$

Note that this last matrix is one subdiagonal away from being upper Hessenberg.

The QR factorization of this matrix, as computed by MATALG, is then

$$\begin{bmatrix} 0.989899 & 0.141407 & 0.0101694 & -0.833871 \times 10^{-3} \\ -0.141414 & 0.979747 & 0.141759 & -0.148906 \times 10^{-2} \\ 0.010101 & -0.141407 & 0.988027 & 0.608726 \times 10^{-1} \\ 0 & -0.0102036 & -0.0600356 & 0.988144 \end{bmatrix}$$

$$\times \begin{bmatrix} 99 & -14 & -0.668182 & 91.9192 \\ 0 & 98.0051 & 254.532 & -661.834 \\ 0 & 0 & 43.2599 & 257.210 \\ 0 & 0 & 0 & 22.2361 \end{bmatrix}$$

and the matrix

$$A_2 = Q^TAQ = \begin{bmatrix} -0.141414 & -0.683803 & 40.1741 & -667.930 \\ 0.989950 & 2.738550 & -16.1215 & 268.035 \\ 0 & 0.441404 & 3.34856 & -54.3115 \\ 0 & 0 & 0.514012 & 8.05431 \end{bmatrix}.$$

Note that this matrix is still upper Hessenberg and that the smallest subdiagonal entry is in position (3, 2). This is the entry that the algorithm will make progressively smaller. Repeating the preceding calculations yields

$$k_1 + k_2 = 11.4029, \quad k_1 k_2 = 54.8871,$$

$$A_2^2 - (k_1 + k_2)A_2 + k_1 k_2 I$$

$$= \begin{bmatrix} 49.7503 & 77.9469 & -562.340 & -1683.69 \\ -8.71724 & 17.2740 & 263.242 & 50.8635 \\ 0.436968 & -2.34640 & -7.11611 & 118.312 \\ 0 & 0.226887 & 0 & 0 \end{bmatrix} = QR,$$

and

$$A_4 = Q^T A_2 Q = \begin{bmatrix} 1.30446 & -16.4681 & 101.833 & -698.341 \\ 0.600013 & 3.62334 & -22.0837 & 151.422 \\ 0 & 0.201096 & 5.00136 & -33.6549 \\ 0 & 0 & 0.412594 & 4.07084 \end{bmatrix}.$$

Note that the entry in the (3, 2) position is getting smaller as the two complex pairs of eigenvalues begin to uncouple. Repeating the calculation yields

$$a_{32}^{(6)} = 0.336322 \times 10^{-1}, \quad a_{32}^{(8)} = 0.917264 \times 10^{-2}, \quad a_{32}^{(10)} = 0.272661 \times 10^{-3},$$

and

$$A_{12} = \begin{bmatrix} 5.20596 & -28.4322 & 23.6299 & -721.548 \\ 0.733892 & 0.794174 & -0.546916 & 16.5963 \\ 0 & 0.107533 \times 10^{-4} & 1.06244 & -24.0040 \\ 0 & 0 & 0.734402 & 6.93743 \end{bmatrix}.$$

The algorithm has isolated the complex eigenvalues in 2×2 blocks containing conjugate pairs of eigenvalues. ■

The calculation illustrated in Example 7 is quite expensive as it requires computation of the matrix product A_{2s}^2. There is however a much more efficient way to find Q using some clever analyses developed by J. G. F. Francis, the details of which follow.

The key step is to observe that if $Q^T A Q = H$, where H is upper Hessenberg with positive subdiagonal and Q is orthogonal, then Q and H are completely determined by A and the first column of Q. We state this basic result as a uniqueness theorem for the reduction to upper Hessenberg form.

THEOREM 4.8

Let Q and V be orthogonal matrices and let $H = Q^TAQ$ and $G = V^TAV$ be upper Hessenberg matrices with positive subdiagonal entries. If Q and V have the same first column, then $Q = V$ and $G = H$.

PROOF We partition Q into its columns, $Q = [q_1, q_2, q_3 \cdots q_n]$, and observe that the matrix equality $AQ = QH$ implies the equality of the first columns, that is

$$Aq_1 = h_{11}q_1 + h_{21}q_2, \quad \text{which implies} \quad q_1^TAq_1 = h_{11}q_1^Tq_1 + h_{21}q_1^Tq_2 = h_{11}$$

since the columns of Q are orthonormal. Thus h_{11} is determined by A and q_1. Now,

$$\begin{aligned} h_{21}q_2 &= Aq_1 - h_{11}q_1 \Rightarrow |h_{21}| = \|Aq_1 - h_{11}q_1\|_2 \\ &\Rightarrow h_{21} = \|Aq_1 - h_{11}q_1\|_2 \end{aligned}$$

since h_{21} is to be positive. Finally, the second column of Q is $q_2 = (Aq_1 - h_{11}q_1)/h_{21}$ and we have determined the first column of H (and of G) and the second column of Q (and also of V). Now by comparing the second columns of AQ and QH we see that $Aq_2 = h_{12}q_1 + h_{22}q_2 + h_{32}q_3$ with $q_1^TAq_2 = h_{12}$ and $q_2^TAq_2 = h_{22}$. Then, since q_3 is a unit vector, $h_{32} = \|Aq_2 - h_{12}q_1 - h_{22}q_2\|_2$ and $q_3 = (Aq_2 - h_{12}q_1 - h_{22}q_2)/h_{32}$. Thus we have determined the second column of H (and of G) and the third column of Q (and of V). Continuing in this way we determine all of H and all of Q. Thus $H = G$ and $Q = V$. ■

Returning now to the problem of determining an orthogonal matrix Q such that $A_{2s+2} = Q^TA_{2s}Q$, we observe that if we could find an orthogonal P such that $P^TA_{2s}P = H$ was upper Hessenberg and $\mathrm{Col}_1(P) = \mathrm{Col}_1(Q)$, or equivalently $\mathrm{Row}_1(P^T) = \mathrm{Row}_1(Q^T)$, then it would follow that $P = Q$. Now the matrix $A_{2s}^2 - (k_1 + k_2)A_{2s} + k_1k_2I$ fails to be in upper Hessenberg form only because it has one extra nonzero subdiagonal. The Given's method for computing the QR factorization of this matrix would start with plane rotations R_{12} and R_{13}, and these are the only two rotations that would involve the first row. Thus the first row of Q^T is the first row of the product $R_{13}R_{12}$. The matrix $G = R_{13}R_{12}A_{2s}R_{12}^TR_{13}^T$ will be non-Hessenberg only in the first four rows, so it is easy to find an orthogonal matrix S such that $\mathrm{Col}_1(S) = \mathbf{e}_1$ and

$$S^TGS = S^TR_{13}R_{12}A_{2s}R_{12}^TR_{13}^TS = P^TA_{2s}P = H$$

is upper Hessenberg. Because $\mathrm{Col}_1(S) = \mathbf{e}_1$, $P^T = S^TR_{13}R_{12}$ is an orthogonal matrix whose first row is the same as the first row of $R_{13}R_{12}$, which is the same as the first row of Q^T. It follows that $P = Q$ and $H = A_{2s+2}$.

The computation of A_{2s+2} from A_{2s} thus requires the following steps:

1. Find the first three entries of the first column of $A' = A_{2s}^2 - (k_1 + k_2)A_{2s} + k_1k_2I$.
2. Determine rotations R_{12} and R_{13} so that $R_{13}R_{12}A'$ has a reduced first column.
3. Form $G = R_{13}R_{12}A_{2s}R_{12}^TR_{13}^T$.
4. Find an orthogonal S such that $S^TGS = H = A_{2s+2}$ is upper Hessenberg.

■ EXAMPLE 8

Use the Francis method to carry out one double QR step for the matrix A of Example 7.

SOLUTION In Example 7, we found that the first column of $A^2 - (k_1 + k_2)A + k_1k_2I$ was $C_1 = [98\ -14\ 1\ 0]^T$. The first step in the Francis calculation is to determine a rotation $R_{12}(\theta_1)$ so that $R_{12}(\theta_1)C_1$ has a zero in the second position. By Equation (4.1), $\tan\theta_1 = -14/98$ and

$$R_{12}(\theta_1)C_1 = \begin{bmatrix} 0.989949 & -0.0141421 & 0 & 0 \\ 0.141421 & 0.989949 & 0 & 0 \\ 0 & 0 & 1 & 0 \\ 0 & 0 & 0 & 1 \end{bmatrix}\begin{bmatrix} 98 \\ -14 \\ 1 \\ 0 \end{bmatrix} = \begin{bmatrix} 98.9949 \\ 0 \\ 1 \\ 0 \end{bmatrix}.$$

Next we determine θ_2 so that $R_{13}(\theta_2)(R_{12}(\theta_1)C_1)$ has a zero in the third position. From Equation (4.1), the proper angle is $\theta_2 = \tan^{-1}(1/98.9949)$. It follows that

$$R_{13}(\theta_2) = \begin{bmatrix} 0.999949 & 0 & 0.0101010 & 0 \\ 0 & 1 & 0 & 0 \\ -0.0101010 & 0 & 0.999949 & 0 \\ 0 & 0 & 0 & 1 \end{bmatrix}$$

and that

$$R_{13}(\theta_2)R_{12}(\theta_1)C_1 = \begin{bmatrix} 99 \\ 0 \\ 0 \\ 0 \end{bmatrix}.$$

Now,

$$G = R_{13}R_{12}AR_{12}^TR_{13}^T$$

$$= \begin{bmatrix} -0.141414 & -0.00999949 & 0.00142850 & -669.172 \\ 0.979950 & 0.140000 & -0.00989899 & 258.094 \\ -0.139993 & 0.990101 & 0.00141414 & -91.2453 \\ 0.0101010 & 0 & 0.999949 & 14 \end{bmatrix}$$

is non-Hessenberg in its first four rows. This matrix can be returned to upper

Hessenberg form by either two Householder transformations or three plane rotations. The result is

$$S^TGS = \begin{bmatrix} -0.141414 & -0.683803 & 40.1741 & -667.930 \\ 0.989950 & 2.73855 & -16.1215 & 268.035 \\ 0 & 0.441404 & 3.34856 & -54.3115 \\ 0 & 0 & 0.514012 & 8.05431 \end{bmatrix} = A_2,$$

which agrees with the first step in Example 7. ■

In general the technique illustrated in Example 8 is much more efficient than the technique of Example 7. (See Exercise 2.) A finely tuned implementation of this algorithm is contained in the EISPACK subroutine **HQR2**.

4.5 ■ Exercise Set

1. Use the procedure described in the proof of Theorem 4.8 to find an orthogonal Q such that Q^TAQ is upper Hessenberg if $q_1 = \mathbf{e}_1$ and

$$A = \begin{bmatrix} 4 & 3 & 5 & 7 \\ 2 & -4 & 0 & 1 \\ 2 & 2 & 3 & 8 \\ 1 & 3 & 6 & 2 \end{bmatrix}.$$

2. Count the number of multiplications required to compute A_{2s+2} from A_{2s} using the Francis method. Compare with the number required to compute $A' = A_{2s}^2 - (k_1 + k_2)A_{2s} + k_1k_2I$ and its QR factorization, as in Example 7.

3. Write a FORTRAN driver that uses appropriate EISPACK subroutines to find all eigenvalues of any $n \times n$ nonsymmetric matrix A with $n \leq 50$. See Appendix F. Test your program on the matrix of Exercise 1 and then on the matrix of Exercise 6 in Section 4.4.

4. Write a FORTRAN driver that uses appropriate EISPACK subroutines to find all the eigenvalues and eigenvectors of any $n \times n$ symmetric matrix A with $n \leq 50$. Test your program on the matrix of Exercise 4 of Section 4.3.

5

The Linear Least Squares Problem

5.1 ■ THE SINGULAR VALUE DECOMPOSITION

In this section we will present an important theorem that will be useful in the rest of this chapter. This theorem also sheds light on several important problems of numerical linear algebra that we considered earlier.

THEOREM 5.1

Singular value decomposition: If A is any $n \times n$ real matrix, then there exist orthogonal matrices U and V such that $U^TAV = \Sigma$ is a diagonal matrix. We may choose U and V so that the diagonal elements of Σ are $\sigma_1 \geq \sigma_2 \geq \cdots \geq \sigma_r > \sigma_{r+1} = \cdots = \sigma_n = 0$ where r is the rank of A.

PROOF The matrix $B = AA^T$ is a symmetric, positive semidefinite matrix of rank r. By Theorem 3.5, there exists an orthogonal matrix U such that

$$U^TAA^TU = U^TBU = \Sigma^2 = \text{Dg}\{\sigma_1^2, \sigma_2^2, \ldots, \sigma_n^2\}$$

where, without loss of generality, we may assume that the nonnegative diagonal elements of Σ^2 are arranged in descending order of magnitude; $\sigma_1 \geq \sigma_2 \geq \cdots \geq \sigma_r > \sigma_{r+1} = \cdots = \sigma_n = 0$.

Now consider the matrix $F = U^TA$. Note that $FF^T = U^TAA^TU = \Sigma^2$, which says that the nonzero rows of F (the first r rows) are mutually orthogonal (but not orthonormal) vectors with $\|\text{Row}_i(F)\|_2^2 = \sigma_i^2$. Now let V^T be an orthogonal matrix such that $\text{Row}_i(V^T) = \dfrac{1}{\sigma_i}\text{Row}_i(F)$, $i = 1, 2, \ldots, r = \text{rank}(A)$, and the remaining rows form an orthonormal basis for the orthogonal complement of the row space of F. Thus $\Sigma V^T = F$, which is equivalent to saying that $V\Sigma$ is a QR factorization of F^T. Now $F = \Sigma V^T = U^TA \Rightarrow U^TAV = \Sigma$ or $A = U\Sigma V^T$. ■

If the matrix A is not square, we can still construct a decomposition similar to that given in Theorem 5.1. The more general result is given in the next theorem; a proof that does not depend on Theorem 3.5 is outlined in Exercise 11.

THEOREM 5.2

Let the matrix A be $n \times k$ of rank r. Then there exists an $n \times n$ orthogonal matrix U and a $k \times k$ orthogonal matrix V such that

$$U^T A V = \left[\begin{array}{cccc:c} \sigma_1 & & & & \\ & \sigma_2 & & & 0 \\ & & \ddots & & \\ & & & \sigma_r & \\ \hdashline & & 0 & & 0 \end{array}\right] = \underset{n \times k}{\begin{bmatrix} \Sigma & 0 \\ 0 & 0 \end{bmatrix}} \}r .$$

PROOF As in the proof of Theorem 5.1, there exists an $n \times n$ orthogonal matrix U such that $U^T(AA^T)U = \begin{bmatrix} \Sigma^2 & 0 \\ 0 & 0 \end{bmatrix}$ where Σ^2 is $r \times r$ and nonsingular. As in the proof of Theorem 5.1, let $U^T A = F = \begin{bmatrix} F_1 \\ F_2 \end{bmatrix}$ where F_1 is $r \times k$. Then $U^T AA^T U = \begin{bmatrix} F_1 \\ F_2 \end{bmatrix} [F_1^T \;\; F_2^T] = \begin{bmatrix} F_1F_1^T & F_1F_2^T \\ F_2F_1^T & F_2F_2^T \end{bmatrix} = \begin{bmatrix} \Sigma^2 & 0 \\ 0 & 0 \end{bmatrix}$. It follows that $F_2 = 0$. Since $F_1F_1^T = \Sigma^2$, it follows that the rows of F_1 are mutually orthogonal vectors. Thus the rows of the $r \times k$ matrix $V_1 = \Sigma^{-1}F_1$ are orthonormal vectors. Let the rows of the $(k - r) \times k$ matrix V_2 be an orthonormal basis for the orthogonal complement of the row space of V_1 so that the $k \times k$ orthogonal matrix

$$V^T = \begin{bmatrix} V_1 \\ V_2 \end{bmatrix} \quad \text{is such that} \quad U^T A = \begin{bmatrix} F_1 \\ 0 \end{bmatrix} = \begin{bmatrix} \Sigma V_1 \\ 0 \end{bmatrix}$$

$$= \begin{bmatrix} \Sigma & 0 \\ 0 & 0 \end{bmatrix} \begin{bmatrix} V_1 \\ V_2 \end{bmatrix} = \begin{bmatrix} \Sigma & 0 \\ 0 & 0 \end{bmatrix} V^T.$$

Note that we could obtain V by finding the QR factorization of $F_1^T(V = Q)$. It follows that $U^T AV = \begin{bmatrix} \Sigma & 0 \\ 0 & 0 \end{bmatrix}$ as claimed. ■

Note that if we partition $V = [V_1 \;\; V_2]$ and $U = [U_1 \;\; U_2]$, where both V_1 and U_1 have r columns, we have, from Theorem 5.2,

$$AV = A[V_1 \;\; V_2] = [AV_1 \;\; AV_2] = [U_1 \;\; U_2] \begin{bmatrix} \Sigma & 0 \\ 0 & 0 \end{bmatrix} = [U_1\Sigma \;\; 0].$$

From this last equation it follows that $AV_1 = U_1\Sigma$ or, since the columns of V_1

are orthonormal, $A = U_1 \Sigma V_1^T$. This condensed form of the factorization is also called the singular value decomposition of the matrix A.

Note that $AV_2 = 0$, so the columns of V_2 form an orthonormal basis for the nullspace of A. The columns of V_1 thus form an orthonormal basis for the orthogonal complement of $\mathcal{NS}(A)$. Note also that the columns of U_1 form an orthonormal basis for the column space of AV_1 and that $\mathcal{CS}(AV_1) = \mathcal{CS}(A)$. Moreover, the columns of U_2 form an orthonormal basis for the orthogonal complement of $\mathcal{CS}(A)$.

In either of the above decompositions the non-negative diagonal elements of Σ are called the **singular values** of A. The columns of V are called **right singular vectors** of A and the rows of U are called **left singular vectors** of A. The number of nonzero singular values of A is the rank of the matrix A.

The rank of a matrix is a discontinuous function of the elements of the matrix and as such is very difficult to compute because it is difficult for the machine to distinguish between zero and very small numbers generated by round off. The singular value decomposition provides the best computational means of investigating the rank of a singular matrix. In practice the user will need to decide whether certain small singular values are to be treated as zero.

■ EXAMPLE *1*

Compute the singular value decomposition of the 5×5 matrix

$$A = \begin{bmatrix} 1 & 6 & 11 \\ 2 & 7 & 12 \\ 3 & 8 & 13 \\ 4 & 9 & 14 \\ 5 & 10 & 15 \end{bmatrix}.$$

SOLUTION We will follow the proof of Theorem 5.2, using MATLAB to make the calculations; the MATLAB commands will be listed inside brackets. We can also do the calculations by using MATALG although some of the intermediate steps will be different. We first compute the matrix

$$B = AA^T = \begin{bmatrix} 158 & 176 & 194 & 212 & 230 \\ 176 & 197 & 218 & 239 & 260 \\ 194 & 218 & 242 & 266 & 290 \\ 212 & 239 & 266 & 293 & 320 \\ 230 & 260 & 290 & 320 & 350 \end{bmatrix} \quad \{\text{B=A*A'}\}.$$

The next step is to compute the eigenvectors and eigenvalues of

B**{[U,D1]=eig(B)}**. The results are

$$U = \begin{bmatrix} -0.3784 & 0.4967 & -0.1006 & -0.6887 & 0.3546 \\ 0.7995 & -0.2191 & -0.1131 & -0.03756 & 0.3987 \\ -0.4644 & -0.7636 & -0.0350 & -0.0624 & 0.4428 \\ 0.0440 & 0.1979 & 0.8117 & 0.2507 & 0.4870 \\ -0.0007 & 0.2881 & -0.5630 & 0.5638 & 0.5311 \end{bmatrix},$$

$$D1 = \text{Dg}\{\mathbf{0, 0, 0, 0.0061, 1.2339}\} \times 10^3.$$

Note that MATLAB has computed the eigenvalues in increasing order of magnitude rather than in decreasing order as specified in the proof of Theorem 5.2. To rearrange the eigenvalues we need to rearrange the columns of U. We can do this by multiplying U on the right by the matrix

$$R = \begin{bmatrix} 0 & 0 & 0 & 0 & 1 \\ 0 & 0 & 0 & 1 & 0 \\ 0 & 0 & 1 & 0 & 0 \\ 0 & 1 & 0 & 0 & 0 \\ 1 & 0 & 0 & 0 & 0 \end{bmatrix};$$

$$U = UR = \begin{bmatrix} 0.3546 & -0.6887 & -0.1006 & 0.4967 & -0.3784 \\ 0.3987 & -0.3756 & -0.1131 & -0.2191 & 0.7995 \\ 0.4428 & -0.0624 & -0.0350 & -0.7636 & -0.4644 \\ 0.4870 & 0.2507 & 0.8117 & 0.1979 & 0.0440 \\ 0.5311 & 0.5638 & -0.5630 & 0.2881 & -0.0007 \end{bmatrix}.$$

We can use the software to verify that this new U is in fact orthogonal **{U'*U}**. Thus we now have an orthogonal matrix U such that $U^TBU = \text{Dg}\{1233.9, 6.07818, 0, 0, 0\}$. We can see at this point that B, and hence A, is of rank 2. Next we form the matrix $F = U^TA$**{F=U'*A}**

$$F = U^TA = \begin{bmatrix} F_1 \\ F_2 \end{bmatrix} = \begin{bmatrix} 7.0839 & 18.1548 & 29.2257 \\ 2.1950 & 0.6344 & -0.9261 \\ \hdashline 0 & 0 & 0 \\ 0 & 0 & 0 \\ 0 & 0 & 0 \end{bmatrix},$$

and note that $F_2 = 0$. The matrix V can now be found by computing the QR factorization of F^T {**[V,D2]=qr(F')**}:

$$V = \begin{bmatrix} -0.2017 & 0.8903 & 0.04082 \\ -0.5168 & 0.2573 & -0.8165 \\ -0.8320 & -0.3757 & 0.4082 \end{bmatrix},$$

$$D2 = \begin{bmatrix} -35.1272 & 0 & 0 & 0 & 0 \\ 0 & 2.4654 & 0 & 0 & 0 \\ 0 & 0 & 0 & 0 & 0 \end{bmatrix}.$$

Note that the first diagonal entry of $D2$ is negative. If we multiply the first column of V by -1 {**V(:,1)=-V(:,1)**}, we obtain the required orthogonal matrix V

$$V = \begin{bmatrix} 0.2017 & 0.8903 & 0.04082 \\ 0.5168 & 0.2573 & -0.8165 \\ 0.8320 & -0.3757 & 0.4082 \end{bmatrix},$$

and

$$U^TAV = \left[\begin{array}{cc|c} 35.1272 & 0 & 0 \\ 0 & 2.4654 & 0 \\ \hline 0 & 0 & 0 \\ 0 & 0 & 0 \\ 0 & 0 & 0 \end{array}\right] = \begin{bmatrix} \Sigma & 0 \\ 0 & 0 \end{bmatrix}$$

is the decomposition of Theorem 5.2. The nonzero singular values of A are $\sigma_1 = 35.1272$ and $\sigma_2 = 2.4655$. If we write $U = [U_1, U_2]$ and $V = [V_1, V_2]$ where U_1 and V_1 have two columns, then we have the alternate decomposition $A = U_1 \Sigma V_1^T$; {**U1=U(:,1:2),V1=V(:,1:2)**}

$$A = \begin{bmatrix} 1 & 6 & 11 \\ 2 & 7 & 12 \\ 3 & 8 & 13 \\ 4 & 9 & 14 \\ 5 & 10 & 15 \end{bmatrix} = \begin{bmatrix} 0.3546 & -0.6887 \\ 0.3987 & -0.3756 \\ 0.4428 & -0.0624 \\ 0.4870 & 0.2507 \\ 0.5311 & 0.5638 \end{bmatrix} \begin{bmatrix} 35.1272 & 0 \\ 0 & 2.4654 \end{bmatrix}$$

$$\times \begin{bmatrix} 0.2017 & 0.5168 & 0.8320 \\ 0.8903 & 0.2573 & -0.3757 \end{bmatrix}.$$

■

In practice the singular value decomposition is not computed as described in the preceding proofs. Good subroutines for implementing the decomposition are **SVD** in EISPACK or **SSVDC** in LINPACK. These subroutines return square matrices U and V rather than just the matrices U_1 and V_1 described earlier. The reduction technique used in these subroutines involves first reducing A to upper bidiagonal form, $B = U^TAV$, by either Householder transformations or plane rotations, and then proceeds with an iterative reduction of B to diagonal form using a variant of the QR algorithm applied to B^TB (B^TB is not actually formed). For further details the reader is referred to the documentation for either EISPACK (Smith et al., 1972) or LINPACK (Dongarra et al., 1979).

■ SOFTWARE NOTES ■

The singular value decomposition of A can be computed using the MATLAB command **[U,S,V]=svd(A)** or by the MATALG command **M SVD A**. MATLAB returns the factorization described in Theorem 5.2, while MATALG returns the condensed form of the factorization $A = U_1\Sigma V_1^T$.

For a symmetric matrix, the MATLAB **eig** function will compute an orthogonal matrix of eigenvectors. The MATALG **evec** function does not compute an orthogonal matrix; the result of Exercise 1 can be used to find an orthogonal matrix of eigenvectors.

5.1 ■ Exercise Set

1. **a.** What are the singular values of a symmetric matrix?

   b. Let B be a symmetric matrix of rank r and let P be any matrix whose columns are an independent set of eigenvectors of B with $P^{-1}BP = \text{Dg}\{\lambda_1, \lambda_2, \ldots, \lambda_r, 0, \ldots, 0\}$. Let $P = QR$ be the QR factorization of P. Show that $Q^TBQ = P^{-1}BP = \text{Dg}\{\lambda_1, \lambda_2, \ldots, \lambda_r, 0, \ldots, 0\}$.

2. Find the singular value decomposition of $A = \begin{bmatrix} 1 & 2 & 3 \\ 4 & 5 & 6 \\ 7 & 8 & 9 \end{bmatrix}$ by following the proof of Theorem 5.2. Use the software to help with the calculations.

3. Find the singular value decomposition of $A = \begin{bmatrix} 1 & 6 & 11 & 19 \\ 2 & 7 & 12 & 20 \\ 3 & 8 & 13 & 24 \\ 4 & 9 & 14 & 29 \\ 5 & 10 & 14 & 30 \end{bmatrix}$ using the EISPACK subroutine **SVD**.

4. Show, as a consequence of the singular value decomposition, that

$$\|A\|_2 = \sigma_1 = \sqrt{\rho(AA^T)} \text{ for any matrix } A.$$

5. Show that for any $n \times n$ nonsingular matrix A, $\text{cond}_2(A) = \sigma_1/\sigma_n$.

6. For the matrix A of Exercise 3 show that $\text{cond}_2(A^TA) = \sigma_1/\sigma_n = \infty$. How does this condition number change if a_{53} is changed to 16?

7. From the result of Exercise 2 find orthonormal bases for $\mathcal{NS}(A)$, $\mathcal{CS}(A)$, $\mathcal{NS}(A)^\perp$, $\mathcal{CS}(A)^\perp$.

8. Find orthogonal U and V so that U^TAV is bidiagonal where A is the matrix of Exercise 3.

9. Show that if P and Q are orthogonal matrices, then A and PAQ have the same singular values. Are the singular vectors the same?

10. Show that if the matrix B is upper bidiagonal, then B^TB is a symmetric tridiagonal matrix.

11. An inductive proof of the singular value decomposition for an $m \times n$ matrix A can be constructed by filling in the details below.
    - **a.** Show that there exist unit vectors X in $\mathcal{R}^n$ and Y in $\mathcal{R}^m$ such that $AX = \|A\|_2\, Y$.
    - **b.** Let $V = [X\ V_1]$ in $\mathcal{R}^{n\times n}$ and $U = [Y\ U_1]$ in $\mathcal{R}^{m\times m}$ be orthogonal matrices (Exercise 4, Section 3.2). Show that $U^TAV = \begin{bmatrix} Y^T \\ U_1^T \end{bmatrix}[AX\ AV_1]$ $= \begin{bmatrix} \sigma_1 & \mathbf{w}^T \\ 0 & B \end{bmatrix} = A_1.$
    - **c.** Verify that
    $$\left\|A_1\begin{bmatrix} \sigma_1 \\ \mathbf{w} \end{bmatrix}\right\|_2^2 = \left\|\begin{bmatrix} \sigma_1^2 + \mathbf{w}^T\mathbf{w} \\ B\mathbf{w} \end{bmatrix}\right\|_2^2 \geq \left\|\begin{bmatrix} \sigma_1^2 + \mathbf{w}^T\mathbf{w} \\ B\mathbf{w} \end{bmatrix}\right\|_\infty^2 \geq (\sigma_1^2 + \mathbf{w}^T\mathbf{w})^2.$$
    Hence show that $\|A_1\|_2 \geq \sigma_1^2 + \mathbf{w}^T\mathbf{w}$ and, since $\sigma_1^2 = \|A\|_2^2 = \|A_1\|_2^2$, that $\mathbf{w} = 0$.
    - **d.** Now complete an inductive proof of Theorem 5.2.

12. Let A be an $n \times r$ matrix whose columns are orthonormal. Show that A^T is a left inverse of A but not a right inverse.

5.2 ■ LEAST SQUARES PROBLEMS

We now consider the problem of finding a functional relationship $y = f(t)$ that, in some sense, agrees with a given set of data points

$$(t_i, y_i);\ i = 1, 2, \ldots, m. \tag{5.1}$$

We seek an approximating function in some fixed, finite dimensional subspace $\mathscr{W}$ of the infinite dimensional vector space of all functions with the appropriate domain and range. Let $\{b_1(t), b_2(t), b_3(t), \ldots, b_n(t)\}$ be a basis for $\mathscr{W}$. We thus seek n scalars x_i, such that

$$y(t) = x_1b_1(t) + x_2b_2(t) + x_3b_3(t) + \cdots + x_nb_n(t) \tag{5.2}$$

"best fits" the data in (5.1). An exact fit with the data given in (5.1) would require that

$$\sum_{j=1}^{n} x_jb_j(t_i) = y_i; \quad i = 1, 2, \ldots, m; \quad \text{or} \quad MX = Y \quad \text{where} \quad m_{ij} = b_j(t_i).$$

In most cases of interest, $m > n$, so the system $MX = Y$ has more equations than unknowns. In this case we say that the system is **overdetermined**. Such a system cannot be expected to have an exact solution.

For the approximate solution given by (5.2), we define the **residual** at the point (t_i, y_i) to be $r_i = \sum_{j=1}^{n} x_jb_j(t_i) - y_i = \sum_{j=1}^{n} m_{ij}x_j - y_i$. The vector $R = MX - Y$ is called the residual vector for this approximate solution. For our purposes, "best fit" will mean that $\|R\|_2^2 = \sum_{i=1}^{m} r_i^2$ is minimum. If the system has an exact solution, then the residual vector is zero.

■ DEFINITION 5.1 ■

The vector X_0 is said to be a least squares solution of the system $MX = Y$ if the 2-norm of the residual vector $R_0 = MX_0 - Y$ is minimum.

For any vector X, MX is an element of $\mathscr{CS}(M)$, the column space of M. Thus, we wish to choose X_0 so that MX_0 is the element of $\mathscr{CS}(M)$ that is closest to Y. If $n = 3$, it is clear from Figure 5.1 that we need to choose X_0 so

FIGURE 5.1 ■ THE LEAST SQUARES PROBLEM IN $\mathscr{R}^3$

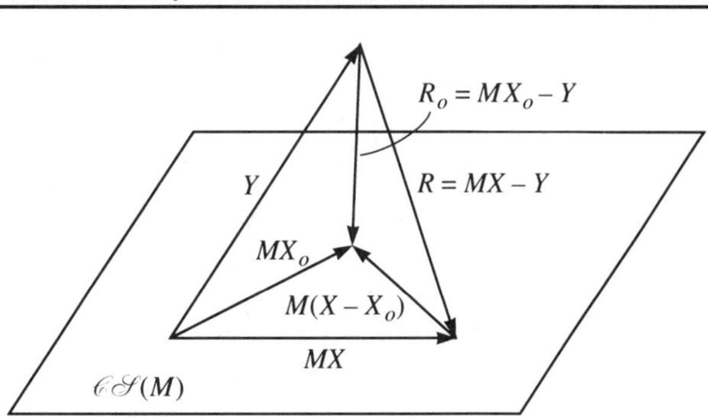

that MX_0 is the orthogonal projection of Y on $\mathscr{CS}(M)$ and hence that the residual vector $R_0 = MX_0 - Y$ is orthogonal to $\mathscr{CS}(M)$.

In the general case, where the geometry is no longer convincing, we write

$$R = MX - Y = MX - MX_0 + MX_0 - Y = (MX_0 - Y) + M(X - X_0)$$

and use the generalization of the Pythagorean theorem given in Exercise 1 to note that if $MX_0 - Y$ is orthogonal to $\mathscr{CS}(M)$, and hence to $M(X - X_0)$, then

$$\|R\|_2^2 = \|MX_0 - Y\|_2^2 + \|M(X - X_0)\|_2^2 = \|R_0\|_2^2 + \|M(X - X_0)\|_2^2.$$

Thus $\|R_0\|_2 \leq \|R\|_2$ for every choice of the vector X. It follows that we must choose the optimal X so that $MX - Y$ is orthogonal to $\mathscr{CS}(M)$. Thus every column of M must be orthogonal to $MX - Y$, and the optimal X must satisfy the **normal equations** $M^T(MX - Y) = 0$ or $M^TMX = M^TY$. The matrix M^TM is $n \times n$ and, if the rank of M is n, the matrix MM^T will also have rank n (Theorem 1.1) and hence be invertible. In this case the normal equations will have a unique solution. We have proven the following theorem.

THEOREM 5.3

Let M be an $m \times n$ matrix with $m > n$. The least squares solution of the overdetermined system $MX = Y$ is a solution of the normal equations

$$M^TMX = M^TY. \tag{5.3}$$

If M has rank n, then M^TM is nonsingular and the unique solution of Equation (5.3) is given by

$$X = (M^TM)^{-1}M^TY.$$

■

■ **EXAMPLE 2**

Find the parabola $y = a + bx + cx^2$, which best fits the five data points

$$(0, 4), (1, 1), (2, 4), (3, 6), (4, 8).$$

SOLUTION Attempting to fit $y = a + bx + cx^2$ to the given data leads to the overdetermined system

$$MX = \begin{bmatrix} 1 & 0 & 0 \\ 1 & 1 & 1 \\ 1 & 2 & 4 \\ 1 & 3 & 9 \\ 1 & 4 & 16 \end{bmatrix} \begin{bmatrix} a \\ b \\ c \end{bmatrix} = \begin{bmatrix} 4 \\ 1 \\ 4 \\ 6 \\ 8 \end{bmatrix} = Y$$

FIGURE 5.2 ■ THE BEST-FITTING PARABOLA FOR FIVE DATA POINTS

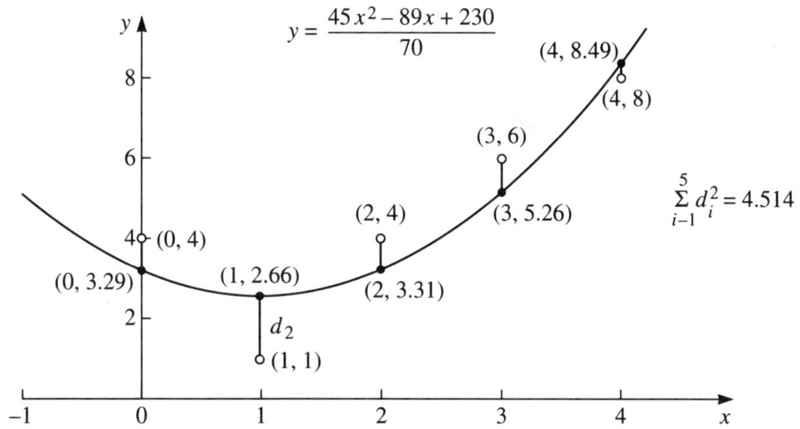

for which the normal equations are

$$M^TMX = \begin{bmatrix} 5 & 10 & 30 \\ 10 & 30 & 100 \\ 30 & 100 & 354 \end{bmatrix} \begin{bmatrix} a \\ b \\ c \end{bmatrix} = \begin{bmatrix} 23 \\ 59 \\ 199 \end{bmatrix} = M^TY.$$

The straightforward solution of this linear system leads to

$$\begin{bmatrix} a \\ b \\ c \end{bmatrix} = \frac{1}{70} \begin{bmatrix} 230 \\ -89 \\ 45 \end{bmatrix},$$

so the best-fitting parabola is

$$y = \frac{1}{70}(230 - 89x + 45x^2).$$

This curve and the given data points are shown in Figure 5.2. ■

The normal equations always have a coefficient matrix that is symmetric and positive semidefinite—positive definite if the coefficient matrix M has full rank. If you wish to solve the normal equations directly, it is best to use procedures that take advantage of the structure of the coefficient matrix. For example, as shown in Section 2.5, a solution algorithm based on the Cholesky factorization of M^TM would be roughly twice as efficient as one based on the LU factorization of M^TM.

In general, for machine calculations, it is best to avoid the direct construction and solution of the normal equations. This is because in many cases the coefficient matrix M^TM has a very high condition number, which means that the reliability of the computed solution must be questioned. The condition number of the matrix M^TM of Example 1 is about 27, which is quite reasonable. However, for the matrix A of Exercise 3 in Section 5.1, $\text{cond}_2(A^TA) = 78087$. Changing the $(5, 3)$ entry from 14 to 15 would make $\text{cond}_2(A^TA) = \infty$.

A more satisfactory approach to finding a solution of the linear least squares problem is to use the QR factorization of the coefficient matrix M. If M is $m \times n$ of rank n, then we can find $M = QR$ where Q is $m \times n$ with orthonormal columns ($Q^TQ = I$) and R is $n \times n$, upper triangular, and nonsingular. In this case the normal equations reduce to

$$\begin{aligned} M^TMX = M^TY &\Rightarrow (QR)^T(QR)X = (QR)^TY \\ &\Rightarrow R^T(Q^TQ)RX = R^TQ^TY \\ &\Rightarrow R^TRX = R^TQ^TY \end{aligned}$$

Since R^T is nonsingular we obtain the triangular system

$$RX = Q^TY, \tag{5.4}$$

which is equivalent to the normal equations. Note that we can obtain Equation (5.4) by substituting $M = QR$ into the "equation" $MX = Y$ and multiplying by Q^T. In general the condition number of R will be significantly smaller than the condition number of M^TM. For the matrix of Example 2,

$$M = QR = \begin{bmatrix} 0.447214 & -0.632456 & 0.534522 \\ 0.447214 & -0.316228 & -0.267261 \\ 0.447214 & 0 & -0.534523 \\ 0.447214 & 0.316228 & -0.267261 \\ 0.447214 & 0.632456 & 0.534522 \end{bmatrix}$$

$$\times \begin{bmatrix} 2.23607 & 4.47214 & 13.4164 \\ 0 & 3.16228 & 12.6491 \\ 0 & 0 & 3.74166 \end{bmatrix},$$

and the solution of the system $RX = Q^TY$ will present no difficulties.

A still better way of solving the problem is to use the singular value decomposition of the matrix M. This method does not make any assumption on the rank of the coefficient matrix M. If we have the singular value decomposition $M = U\Sigma V^T$, where U is $m \times m$ and orthogonal and V is

$n \times n$ and orthogonal, then the normal equations $M^TMX = M^TY$ reduce to

$$\begin{aligned} M^TMX = M^TY &\Rightarrow V\Sigma^TU^TU\Sigma V^TX = V\Sigma^TU^TY \\ &\Rightarrow \Sigma^T\Sigma(V^TX) = \Sigma^T(U^TY) \\ &\Rightarrow \Sigma^T\Sigma X' = \Sigma^TY' \end{aligned}$$

where $X' = V^TX$ and $Y' = U^TY$. These last equations are of the form

$$\sigma_i^2 x_i' = \sigma_i y_i', \quad i = 1, 2, \ldots, n.$$

If Σ is of rank r, then the general solution is $X = VX'$ where

$$\begin{aligned} x_i' &= y_i'/\sigma_i, i = 1, 2, \ldots, r \\ x_i' &\text{ arbitrary for } i = r + 1, \ldots, n. \end{aligned}$$

It is normally desirable to have $\|X\|_2$ as small as possible, so we set the arbitrary variables to zero and compute $X = VX'$.

For the matrix of Example 2, the singular value decomposition, computed by MATALG, is

$$M = U\Sigma V^T$$

where

$$U = \begin{bmatrix} -0.422219 \times 10^{-2} & 0.423731 & -0.840327 \\ -0.669587 \times 10^{-1} & 0.605760 & -0.215273 \times 10^{-3} \\ -0.227406 & 0.540992 & 0.375937 \\ -0.485565 & 0.229427 & 0.288129 \\ -0.841435 & -0.328934 & -0.263637 \end{bmatrix},$$

$$\Sigma = \begin{bmatrix} 19.6215 & 0 & 0 \\ 0 & 1.86319 & 0 \\ 0 & 0 & 0.723699 \end{bmatrix},$$

and

$$V^T = \begin{bmatrix} -0.828470 \times 10^{-1} & -0.272364 & -0.958621 \\ 0.789492 & 0.569070 & -0.229915 \\ -0.608144 & 0.775872 & -0.167884 \end{bmatrix}.$$

Then

$$Y' = U^TY = \begin{bmatrix} -10.6383 \\ 3.20974 \\ -2.23810 \end{bmatrix},$$

$$X' = V^TX = \Sigma^{-1}Y' = \begin{bmatrix} -0.542177 \\ 1.72271 \\ -3.09258 \end{bmatrix},$$

and

$$X = VX' = \begin{bmatrix} 3.2857 \\ -1.2714 \\ 0.6429 \end{bmatrix}.$$

The same solution is obtained using the MATLAB factorization, although the two factorizations are not the same. (See the first software note in Section 5.1.)

■ **EXAMPLE 3**

An exact solution of $\begin{bmatrix} 1 & 6 & 11 & 19 \\ 2 & 7 & 12 & 20 \\ 3 & 8 & 13 & 24 \\ 4 & 9 & 14 & 29 \\ 5 & 10 & 15 & 30 \end{bmatrix} X = \begin{bmatrix} 103 \\ 108 \\ 131 \\ 160 \\ 165 \end{bmatrix}$ is $X = \begin{bmatrix} -2 \\ 4 \\ -3 \\ 6 \end{bmatrix}$. Use the singular value decomposition to find a least squares solution of the system

$$\begin{bmatrix} 1 & 6 & 11 & 19 \\ 2 & 7 & 12 & 20 \\ 3 & 8 & 13 & 24 \\ 4 & 9 & 14 & 29 \\ 5 & 10 & 15 & 30 \end{bmatrix} X = \begin{bmatrix} 100 \\ 105 \\ 135 \\ 160 \\ 160 \end{bmatrix}.$$

SOLUTION We will use MATLAB to do the calculations. The singular-value decomposition of the coefficient matrix $A = USV^T$, computed by the command **[USV]=svd(A)**, is

$$A = USV^T = \begin{bmatrix} 0.3460 & 0.6083 & -0.4855 & 0.5232 & 0.0278 \\ 0.3715 & 0.5026 & 0.4906 & -0.3990 & 0.4576 \\ 0.4357 & 0.1132 & 0.1696 & -0.2173 & 0.8494 \\ 0.5127 & -0.3708 & -0.5838 & -0.4611 & 0.2149 \\ 0.5382 & -0.4765 & 0.3923 & 0.5542 & 0.1491 \end{bmatrix}$$

$$\times \begin{bmatrix} 65.6446 & 0 & 0 & 0 \\ 0 & 2.7608 & 0 & 0 \\ 0 & 0 & 1.0791 & 0 \\ 0 & 0 & 0 & 0 \\ 0 & 0 & 0 & 0 \end{bmatrix}$$

$$\times \begin{bmatrix} 0.1087 & -0.6927 & 0.5845 & -0.4082 \\ 0.2766 & -0.0103 & 0.5067 & 0.8165 \\ 0.4445 & 0.6722 & 0.4288 & -0.4082 \\ 0.8450 & -0.2611 & -0.4666 & -0.0000 \end{bmatrix}^T.$$

Note that the coefficient matrix has only three nonzero singular values. Thus the rank of the coefficient matrix A is only 3 and the coefficient matrix in the normal equations ($B = A^TA$) would be 4×4 of rank 3. The solution of the reduced system $\Sigma^T\Sigma X' = \Sigma^T Y'$, where $Y' = U^TY = \begin{bmatrix} 300.5579 \\ -6.6782 \\ -4.7826 \\ -4.0128 \\ -5.5990 \end{bmatrix}$, is obtained by the MATLAB commands **for i=1:3, Xp(i) = Yp(i)/S(i,i), end; Xp(4)=0**; $X' = \begin{bmatrix} y_1'/\sigma_1 \\ y_2'/\sigma_2 \\ y_3'/\sigma_3 \\ 0 \end{bmatrix} = \begin{bmatrix} 4.5786 \\ -2.4189 \\ -4.4319 \\ 0 \end{bmatrix}$. A least squares solution of the original problem is $X = VX' = \begin{bmatrix} -0.4170 \\ -0.9542 \\ -1.4915 \\ 6.5686 \end{bmatrix}$. This is the solution with the least residual and the least norm. The residual for this solution is $R = AX - Y = \begin{bmatrix} 2.2549 \\ 0.9608 \\ -5.6275 \\ -0.6471 \\ 3.0588 \end{bmatrix}$, for which the 2-norm is 6.8885.

If we compare the relative change in the solution with the relative change in the data, we see that the solution process has multiplied the relative change in the data by about 27:

$$\frac{\|\Delta B\|_2}{\|B\|_2} = \frac{\sqrt{59}}{\sqrt{92259}} = 0.025288$$

and

$$\frac{\|\Delta X\|_2}{\|X\|_2} = \frac{\sqrt{29.6494}}{\sqrt{65}} = 0.67538.$$

This calculation was done on a Macintosh. Note that MATLAB will produce different intermediate results on different machines. In particular the last two columns of U are very different on a Macintosh than on an IBM PC, and both are different from the VAX result. The final solution is the same in each case.

5.2 ▪ Exercise Set

1. Show that vectors X and Y in $\mathscr{R}^n$ are orthogonal if and only if

$$\|X\|_2^2 + \|Y\|_2^2 = \|X - Y\|_2^2.$$

2. Find the straight line that best fits the points $(1,4)$, $(2,9)$, $(3,10)$, $(4,15)$, $(5,16)$, and $(6,21)$.

3. Find a plane $a + bx + cy = z$ that best fits the points $(1,2,-1)$, $(2,3,-3)$, $(3,2,5)$, $(2,1,8)$, and $(2,2,3)$.

4. **a.** If $M^T = M$ and $q = X^TMX$ is the associated quadratic form, show that

$$\frac{\partial q}{\partial x_i} = 2m_{i1}x_1 + 2m_{i2}x_2 + \cdots + 2m_{in}x_n = 2\,\text{Row}_i(M)X.$$

b. Show that $\displaystyle\frac{\partial q}{\partial X} = \frac{\partial(X^TMX)}{\partial X} \equiv \begin{bmatrix} \frac{\partial q}{\partial x_1} \\ \frac{\partial q}{\partial x_2} \\ \vdots \\ \frac{\partial q}{\partial x_n} \end{bmatrix} = 2\begin{bmatrix} \text{Row}_1(M) \\ \text{Row}_2(M) \\ \vdots \\ \text{Row}_n(M) \end{bmatrix} X = 2MX$

c. Show that $\displaystyle\frac{\partial(\|R\|_2^2)}{\partial X} = \frac{\partial(\|MX - Y\|_2^2)}{\partial X} = 2(M^TM)X - 2M^TX$ and hence that $\|R\|_2$ is minimum when $M^TMX = M^TY$.

5. Use the QR factorization to solve Exercise 3.

6. Use the singular value decomposition to solve $\begin{bmatrix} 1 & 6 & 11 \\ 2 & 7 & 12 \\ 3 & 8 & 13 \\ 4 & 9 & 14 \\ 5 & 10 & 15 \end{bmatrix} X = \begin{bmatrix} 1 \\ 2 \\ 3 \\ 4 \\ 5 \end{bmatrix}$.

7. Solve the least squares problem of Example 3 by forming the normal equations and solving them by Gaussian elimination. Compare the relative changes in the data and the solution.

8. Determine scalars x_i such that $p(t) = x_1 \sin t + x_2 \cos t + x_3 \sin 2t + x_4 \cos 2t$ best fits the data

$$(-2, -4),\ (-1, -4),\ (0, -2),\ (1, 2),\ (2, 8),\ (3, 16).$$

Plot the resulting curve, using MATLAB, showing the data points on the same graph.

5.3 ■ APPLICATIONS

In Section 3.4 we discussed the power method for finding the dominant eigenvalue of an $n \times n$ real matrix A. In the case where the dominant eigenvalue was complex, we observed that the sequence generated by Algorithm (3.10) would not converge and indeed could be expected to oscillate wildly. This case is thus detectable and we now consider the question of how to determine the eigenvalues from the terms of the iteration.

Suppose that the dominant eigenvalue is $\lambda_1 = \overline{\lambda_2}$ and that the corresponding eigenvectors are X_1 and X_2; $AX_1 = \lambda_1 X_1$ and $AX_2 = \lambda_2 X_2$. Now λ_1 and λ_2 are the roots of a real quadratic equation $x^2 + bc + c = 0$, and we will be able to determine the eigenvalues if we can determine the scalars b and c. Suppose that the power method has proceeded for enough iterations so that the current vector $Y_k \approx a_1 X_1 + a_2 X_2$. Then for the three vectors $Y_k \approx a_1 X_1 + a_2 X_2$, $AY_k \approx a_1 \lambda_1 X_1 + a_2 \lambda_2 X_2$, and $A^2 Y_k \approx a_1 \lambda_1^2 X_1 + a_2 \lambda_2^2 X_2$, we have

$$\begin{aligned} &A^2 Y_k + bAY_k + cY_k \\ &\approx a_1 \lambda_1^2 X_1 + a_2 \lambda_2^2 X_2 + b(a_1 \lambda_1 X_1 + a_2 \lambda_2 X_2) + c(a_1 X_1 + a_2 X_2) \\ &= a_1(\lambda_1^2 + b\lambda_1 + c)X_1 + a_2(\lambda_2^2 + b\lambda_2 + c)X_2 = 0. \end{aligned}$$

Thus we could attempt to find the monic quadratic polynomial whose zeros are λ_1 and λ_2 by looking for scalars b and c so that

$$bAY_k + cY_k = -A^2Y_k. \tag{5.5}$$

Equation (5.5) is a system of n equations in two unknowns and thus is strongly overdetermined. Since Equation (5.5) is most likely inconsistent, the best we can do is to find a least squares solution of the system. This will provide us with an approximate solution of Equation (5.5) and thus with estimates of the complex dominant eigenvalues λ_1 and λ_2. We can repeat this procedure for another k and compare the results to the first calculation. If the results agree, it is safe to accept the current eigenvalue estimates.

■ EXAMPLE 4

Use the power method to find the dominant eigenvalue of

$$A = \begin{bmatrix} 2 & 3 & 4 & 6 \\ -3 & 2 & -6 & 4 \\ -3 & -6 & 4 & 6 \\ -6 & -6 & -6 & 4 \end{bmatrix}.$$

SOLUTION The power method, with infinity norm scaling and initial vector $Y_1 = [1\ 1\ 1\ 1]^T$, yields $Y_{10} = [0.5108\ -0.4494\ -0.2419\ -1]^T$, and the sequence of Y_k seems to be oscillating wildly. Thus we suspect that the dominant eigenvalue is complex. Following the preceding discussion we now compute AY_{10} and A^2Y_{10}. The system $bAY_{10} + cY_{10} = -A^2Y_{10}$, in matrix form, is

$$\begin{bmatrix} -7.2942 & 0.5108 \\ -4.9799 & -0.4494 \\ -5.8033 & -0.2419 \\ -2.9168 & -1.000 \end{bmatrix} \begin{bmatrix} b \\ c \end{bmatrix} = \begin{bmatrix} 70.2422 \\ -35.0752 \\ -11.0484 \\ -96.7973 \end{bmatrix}.$$

Using the methods of Section 5.2 we find that the least squares solution of this system is $[b\ c]^T = [-2.3723\ 103.7423]^T$. Thus the roots of the quadratic $x^2 - 2.3723x + 103.7423 = 0$ should be approximate eigenvalues of A. These roots are $1.186 \pm 10.116i$, which in this case are excellent approximations of the dominant eigenvalues of A, as we can verify using the software. ■

Another application of least squares results to eigenvalue calculations is to the estimation of the eigenvalue associated with an approximate eigenvector. We can view the equation $\lambda X = AX$ as a system of n equations in the single unknown λ; the $n \times 1$ coefficient matrix is the approximate eigenvector X. The normal equation for this overdetermined system is $X^TX\lambda = X^TAX$, so the best estimate (in the least squares sense) of the eigenvalue λ is given by the Rayleigh quotient $\lambda = X^TAX/X^TX$.

Let us now use the least squares techniques to estimate the population of the United States in 1990 from the U.S. Census Bureau data provided in Table 5.1 (the figures are rounded to the nearest thousand).

Let us try to fit a quadratic $p(t) = a + bt + ct^2$ to the data in Table 5.1 and then use the quadratic we find to estimate $p(1990)$, the population of the United States in 1990. If the quadratic were to fit the data exactly, then the overdetermined linear system

$$AX = \begin{bmatrix} 1 & 1900 & 3610000 \\ 1 & 1910 & 3648100 \\ 1 & 1920 & 3686400 \\ 1 & 1930 & 3724900 \\ 1 & 1940 & 3763600 \\ 1 & 1950 & 3802500 \\ 1 & 1960 & 3841600 \\ 1 & 1970 & 3880900 \\ 1 & 1980 & 3920400 \end{bmatrix} \begin{bmatrix} a \\ b \\ c \end{bmatrix} = \begin{bmatrix} 75{,}995{,}000 \\ 91{,}972{,}000 \\ 105{,}711{,}000 \\ 122{,}775{,}000 \\ 131{,}669{,}000 \\ 150{,}697{,}000 \\ 179{,}323{,}000 \\ 203{,}235{,}000 \\ 226{,}546{,}000 \end{bmatrix} = Y$$

would have an exact solution. It is unlikely that this is the case, so we would expect to have to resort to a least squares solution. The solution of the normal equations $A^TAX = A^TY$, by Gaussian elimination (single precision) on a Macintosh, is

$$a = 0.390766 \times 10^{10}, \quad b = -0.566429 \times 10^7, \quad c = 1935.13,$$

which leads to the population prediction

$$p(1990) = a + b1990 + c1990^2 = 299{,}031{,}000.$$

If we compute the least squares solution using the QR factorization of the coefficient matrix A (MATALG on Mac), we obtain, from the triangular

TABLE 5.1 ■ U.S. POPULATION

Year	Population
1900	75,995,000
1910	91,972,000
1920	105,711,000
1930	122,775,000
1940	131,669,000
1950	150,697,000
1960	179,323,000
1970	203,235,000
1980	226,546,000

system $RX = Q^TY$,

$$a = 0.362254 \times 10^{11}, \quad b = -0.390576 \times 10^8, \quad c = 10543.7,$$

which leads to the population prediction

$$p(1990) = a + b1990 + c1990^2 = 255{,}025{,}000.$$

The wide disparity between these two estimates, which theoretically should be the same, indicates that this problem is poorly conditioned. In fact, the eigenvalues of A^TA (the squares of the singular values of A), computed using the software, are 0, 0.127×10^{15}, and 4546.7. Thus the condition number of A^TA is essentially infinite and we should not put much faith in the solution computed by Gaussian elimination. Note that the matrix A is close to being of rank 2 since the third column is close to a scalar multiple of the second column.

The condition of the problem can be improved by shifting the origin so that the numbers stay smaller. Thus, if we assume that the approximating quadratic is

$$p(t) = a + b(t - 1900) + c(t - 1900)^2,$$

we are led to the equations

$$BX = \begin{bmatrix} 1 & 0 & 0 \\ 1 & 10 & 100 \\ 1 & 20 & 400 \\ 1 & 30 & 900 \\ 1 & 40 & 1600 \\ 1 & 50 & 2500 \\ 1 & 60 & 3600 \\ 1 & 70 & 4900 \\ 1 & 80 & 6400 \end{bmatrix} \begin{bmatrix} a \\ b \\ c \end{bmatrix} = \begin{bmatrix} 75{,}995{,}000 \\ 91{,}972{,}000 \\ 105{,}711{,}000 \\ 122{,}775{,}000 \\ 131{,}669{,}000 \\ 150{,}697{,}000 \\ 179{,}323{,}000 \\ 203{,}235{,}000 \\ 226{,}546{,}000 \end{bmatrix} = Y.$$

For the matrix B, the singular values are 9366.9, 35.425, and 1.2290, so the coefficient matrix is better conditioned than the original matrix A. Solving the normal equations $B^TBX = B^TY$ by Gaussian elimination yields

$$a = 0.788738 \times 10^8, \quad b = 0.100781 \times 10^7, \quad c = 10551.2,$$

and the predicted 1990 population is

$$p(1990) = a + b90 + c90^2 = 255{,}041{,}000.$$

If we compute the least squares solution of $BX = Y$ using the QR factorization we obtain essentially the same results. Note that the shift of origin has removed the disparity between the results computed by the two

FIGURE 5.3 ■ POPULATION OF THE UNITED STATES

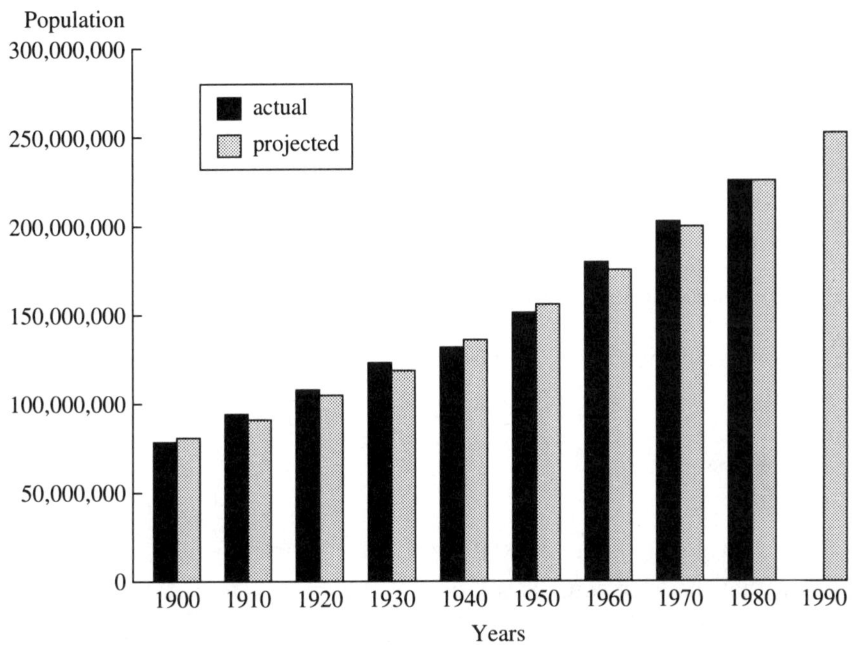

different methods. We could expect still better results if the origin were shifted to 1940 (Exercise 2).

Because this problem is quite ill-conditioned, it is also likely that we will obtain different results on different computers. For example, the population predictions we obtain from the unshifted model on a VAX 750 is 379,630,000 using Gaussian elimination, and 255,062,000 using the QR factorization. MATLAB, which does all calculations in double precision, will produce more consistent results on all machines.

The graph in Figure 5.3 shows the data in Table 5.1 and the projections using the preceding solution.

5.3 ■ Exercise Set

1. Estimate the complex dominant eigenvalue of the matrix $B = \begin{bmatrix} -10 & 5 & 12 & -8 \\ -8 & 4 & 12 & -6 \\ -3 & 0 & 5 & -3 \\ 9 & -5 & -6 & 7 \end{bmatrix}$ using 10 iterations of the power method and a least squares fit of X_{10}, AX_{10}, and A^2X_{10}.

2. Solve the population-projection problem by assuming that

$$p(t) = a + b(t - 1940) + c(t - 1940)^2.$$

3. Use the singular value decomposition to solve the population problem in the text.

4. Find the straight line that best fits the data in Table 5.1 and use this line to predict the 1990 population.

5. Redo the calculations used to solve the population-projection problem using the double precision arithmetic of MATLAB.

6. For an $m \times n$ matrix A whose singular value decomposition is $A = U\begin{bmatrix} \Sigma & 0 \\ 0 & 0 \end{bmatrix}V^T$, we define the **generalized inverse** of A to be $A^\dagger = V\begin{bmatrix} \Sigma^{-1} & 0 \\ 0 & 0 \end{bmatrix}U^T$.

a. Show that if $m = n$ and $\text{rank}(A) = n$, then $A^\dagger = A^{-1}$.

b. Show that the least squares solution of $AX = Y$ is $X = A^\dagger Y$.

c. Show that if $\text{rank}(A) = n$, then $A^\dagger = (A^TA)^{-1}A^T$.

d. Prove that $A^\dagger$ satisfies the following properties:

$$A^\dagger A A^\dagger = A^\dagger$$
$$AA^\dagger A = A$$
$$(AA^\dagger)^T = AA^\dagger$$
$$(A^\dagger A)^T = A^\dagger A$$

e. Show that $A^\dagger$ is the unique matrix satisfying the four properties in part d.

7. The 1990 census data have been released; the official 1990 population is 248,709,873. Compare it with the projections obtained in this section. Estimate the 1991 and 2000 populations using the data in Table 5.1; compare it with a projection obtained using the data in Table 5.1 and the actual 1990 population. Then estimate today's population.

6 Iterative Methods

6.1 INTRODUCTION

In this chapter we will again consider the most common problem of computational linear algebra: the solution of a system of linear equations $AX = B$ where the coefficient matrix A is square and nonsingular. The methods we studied in Chapter 2 for solving this problem are called direct methods; they would produce the exact solution of the system, in a finite number of steps, if all the arithmetic could be done exactly. In this chapter we will take a brief look at iterative methods for solving the system $AX = B$. These methods produce a sequence of approximate solutions converging to the unique solution of the system. These methods cannot be expected to produce the exact solution, even if the arithmetic were done exactly; in that sense they are like the power method for finding the dominant eigenvalue of a matrix or Newton's method for finding the roots of the equation $f(x) = 0$.

The basic approach to the problem is like Picard's method for solving the equation $x = f(x)$ (x_0 arbitrary, $x_k = f(x_{k-1})$ for $k = 1, 2, \ldots$). (See Exercise 8.) We begin with a useful theorem about matrix norms and then proceed to the basic theorems about the convergence of such methods.

Although the spectral radius of A, $\rho(A) \equiv \max\{|\lambda| : \lambda$ is an eigenvalue of $A\}$, is not a matrix norm, we have observed, in Section 3.3, that for any induced matrix norm $\|\cdot\|$, $\|A\| \geq \rho(A)$. We will not prove the useful fact that there is always a consistent matrix norm such that $\|A\|$ is arbitrarily close to $\rho(A)$. Recall that a matrix norm $\|\cdot\|$ is said to be **consistent** if $\|AB\| \leq \|A\| \, \|B\|$.

THEOREM 6.1

For any $n \times n$ real matrix A and any $\varepsilon > 0$, there is a consistent matrix norm (depending on A and ε) such that $\|A\| \leq \rho(A) + \varepsilon$.

PROOF By Schur's theorem (Theorem 3.7), there exists a nonsingular matrix P so that $P^{-1}AP = T$ is upper triangular with diagonal elements $\lambda_1, \lambda_2, \ldots, \lambda_n$. Now, set $b = \max\{|t_{ij}| : i \neq j\}$, choose $0 < r < 1$ such that $(n - 1)br \leq \varepsilon$, and then construct the diagonal matrix

$$D = \mathrm{Dg}\{1, r, r^2, r^3, \ldots, r^{n-1}\}.$$

The matrix $D^{-1}TD$ is still upper triangular and has the form

$$D^{-1}TD = \begin{bmatrix} \lambda_1 & t_{12}r & t_{13}r^2 & \cdots & t_{1n}r^{n-1} \\ 0 & \lambda_2 & t_{23}r & \cdots & t_{2n}r^{n-2} \\ 0 & 0 & \lambda_3 & \cdots & t_{3n}r^{n-3} \\ \vdots & \vdots & \vdots & \ddots & \vdots \\ 0 & 0 & 0 & \cdots & \lambda_n \end{bmatrix}.$$

It follows, from the way we have chosen r, that

$$\begin{aligned} \|D^{-1}TD\|_\infty = \|D^{-1}P^{-1}APD\|_\infty &\le \max\{|\lambda_i|\} + (n-1)\max\{|t_{ij}|\, r\} \\ &\le \rho(A) + \varepsilon. \end{aligned}$$

We now can define the norm $\|\cdot\|$ by $\|M\| = \|D^{-1}P^{-1}MPD\|_\infty$, and from the above calculation we have $\|A\| \le \rho(A) + \varepsilon$. This new norm is consistent because the infinity norm is consistent. ■

Our next theorem is an easy corollary of Theorem 6.1. This theorem provides a sharper result than the simple result that $\|A\| < 1 \Rightarrow A^n \to 0$.

THEOREM 6.2

For a square matrix A, $\lim_{n\to\infty} A^n = 0$ if and only if $\rho(A) < 1$.

PROOF From Schur's theorem we know that $A = PTP^{-1}$ where the diagonal entries of the upper triangular matrix T are the eigenvalues of A. Since $T^n = P^{-1}A^nP$ it follows that

$$\begin{aligned} \lim_{n\to\infty} A^n = 0 \Rightarrow \lim_{n\to\infty} T^n = 0 \Rightarrow \lim_{n\to\infty} \lambda_i^n = 0 \quad \text{for} \quad i = 1, 2, \ldots, n \\ \Rightarrow |\lambda_i| < 1 \Rightarrow \rho(A) < 1. \end{aligned}$$

Conversely, suppose that $\rho(A) < 1$. Then, by Theorem 6.1, there exists a consistent matrix norm such that $\|A\| < 1$. Since, by consistency, $\|A^n\| \le \|A\|^n$, it follows that $\|A^n\| \to 0$ as $n \to \infty$ and hence that $A^n \to 0$ as $n \to \infty$. ■

Theorem 6.2 allows us to establish the basic convergence theorem for iterative methods.

THEOREM 6.3

Consider the linear system $X = GX + B$, where $I - G$ is nonsingular. Let X_0 be arbitrary and define the sequence $\{X_k\}$ by $X_k = GX_{k-1} + B$ for $k = 1, 2, \ldots$. This sequence converges to the unique solution of the system $X = GX + B$ if and only if the spectral radius of G is less than 1.

PROOF Let X_t be the exact solution of $X = GX + B$ and define the error vector at step k by $E_k = X_t - X_k$. It then follows that

$$\begin{aligned} E_k &= X_t - X_k = (GX_t + B) - (GX_{k-1} + B) = G(X_t - X_{k-1}) = GE_{k-1} \\ &= G^2E_{k-2} = G^3E_{k-3} = \cdots = G^kE_0. \end{aligned}$$

The sequence $\{X_k\}$ will converge to the true solution X_t if and only if the sequence of error vectors, $\{E_k\}$, converges to 0. Certainly $E_k \to 0$, for any choice of E_0, if and only if $G^k \to 0$. By Theorem 6.2, $G^k \to 0$ if and only if $\rho(G) < 1$.

■

Solving the linear system $AX = B$ by an iterative method first requires that we rewrite the system in the form $X = GX + B'$. It is essential that the two systems be consistent in the sense that they have precisely the same solutions. The easiest way to construct an iterative method is to choose a suitable nonsingular matrix Q (a **splitting matrix**) and write $A = Q - (Q - A)$. Then

$$AX = B \Leftrightarrow QX = (Q - A)X + B \Leftrightarrow X = (I - Q^{-1}A)X + Q^{-1}B.$$

Thus the iteration matrix G, associated with the splitting matrix Q, is $G = I - Q^{-1}A$. In the actual implementation of the iterative method based on this splitting we will not need to form Q^{-1} explicitly. The important thing is that we choose Q so that linear systems $QX = Y$ are easy to solve. In most cases Q will be either diagonal or triangular.

6.1 ■ Exercise Set

1. Given $A = \begin{bmatrix} 2 & -1 & 0 \\ -1 & 3 & -1 \\ 0 & -1 & 2 \end{bmatrix}$, for each of the following choices of splitting matrix Q, find the iteration matrix M and its spectral radius:

a. $Q = I$

b. $Q = \begin{bmatrix} 2 & 0 & 0 \\ 0 & 3 & 0 \\ 0 & 0 & 2 \end{bmatrix}$

c. $Q = \begin{bmatrix} 2 & 0 & 0 \\ -1 & 3 & 0 \\ 0 & -1 & 2 \end{bmatrix}$

d. $Q = \begin{bmatrix} 4 & 0 & 0 \\ -1 & 6 & 0 \\ 0 & -1 & 4 \end{bmatrix}$

2. Given $A = \begin{bmatrix} 4 & -1 & -1 & 0 \\ -1 & 4 & 0 & -1 \\ -1 & 0 & 4 & -1 \\ 0 & -1 & -1 & 4 \end{bmatrix}$, for each of the following choices of splitting matrix Q, find the iteration matrix M and its spectral radius:

a. $Q = I$

b. $Q = 4I$

c. $Q = \begin{bmatrix} 4 & 0 & 0 & 0 \\ -1 & 4 & 0 & 0 \\ -1 & 0 & 4 & 0 \\ 0 & -1 & -1 & 4 \end{bmatrix}$

d. $Q = \begin{bmatrix} \frac{2}{3} & 0 & 0 & 0 \\ -1 & \frac{2}{3} & 0 & 0 \\ -1 & 0 & \frac{2}{3} & 0 \\ 0 & -1 & -1 & \frac{2}{3} \end{bmatrix}$

3. Let B be any normal matrix ($B^H B = BB^H$).

a. Show that $\|B\|_2 = \rho(B)$.

b. Show that $\|B^k \mathbf{e_0}\|_2 \le \|B^k\|_2 \, \|\mathbf{e_0}\|_2 = \rho(B)^k \, \|\mathbf{e_0}\|_2$ and hence that a smaller spectral radius means a faster rate of convergence for the iterative method $X_{k+1} = BX_{k-1} + C$.

4. Suppose the eigenvalues of the iteration matrix G are real and satisfy $m(G) < \lambda < M(G) < 1$. Let $G_\gamma = \gamma G + (1 - \gamma)I$. Show that the spectral radius of G_γ is minimized if we choose $\gamma = \bar{\gamma} = 2/(2 - (m(G) + M(G)))$ and in that case $\rho(G_{\bar{\gamma}}) = (M(G) - m(G))/(2 - (M(G) + m(G)))$.

5. For $T = \begin{bmatrix} 0.5 & 1 \\ 0 & 0.5 \end{bmatrix}$, compute T^k in closed form and show that $T^k \to 0$ despite the fact that $\|T\|_1 = \|T\|_\infty > 1$.

6. For the iterative method $X_k = GX_{k-1} + B$, define the average rate of convergence after n iterations to be $R_n(G) \equiv -(1/n)\ln\|G^n\|$.

- **a.** Show that if $\rho(G) < 1$, then $\lim_{n\to\infty} \sqrt[n]{\|G^n\|} = \rho(G)$.
- **b.** Define the asymptotic rate of convergence by $R(G) = \lim_{n\to\infty} R_n(G) = -\ln(\rho(G))$.
- **c.** Show that the number of iterations needed to reduce the error by a factor of q is $n = -(\ln q/R(G))$.

7. What happens to the iterative method $X_{n+1} = GX_n + B$ if $X_{k+1} = X_k$?

8. Picard's Method: Let $y = f(x)$ be a function, and define a sequence $\{x_k\}$ by $x_{k+1} = f(x_k)$ where x_0 is arbitrary. Show that the sequence converges to a solution of the equation $x = f(x)$ if $|f'(x)| < 1$ (f is a contraction) in some interval containing x_0. In this case the solution is called a fixed point of the function. Use Picard's method to find a solution of the equation $x = \cos x$.

9. Which of the defining properties of a vector norm (Definition 2.2) are satisfied by the function $\rho(A)$? Is the inequality $\rho(AB) \leq \rho(A)\rho(B)$ true for all matrices?

6.2 ■ BASIC METHODS

For the linear system $AX = B$ we will now consider two basic iterative methods. If we assume that the diagonal entries of A are all nonzero, we can write the coefficient matrix A as $A = D - L - R$, where D is a nonsingular diagonal matrix, L is lower triangular, and R is upper triangular. Then the equation $AX = B$ can be rewritten as

$$DX = (L + R)X + B \quad \text{or as} \quad X = D^{-1}(L + R)X + D^{-1}B.$$

The iterative method based on this last form of the equation is called the

Jacobi method. In matrix form the iteration is $X_k = D^{-1}(L + R)X_{k-1} + D^{-1}B$. In terms of the entries of A the iteration would take the form

JACOBI ITERATIVE METHOD

1. INPUT: $A(n, n)$, $B(n, 1)$, $X_0(n, 1)$ arbitrary
2. For $k = 1, 2, \ldots,$ convergence
3. For $i = 1, 2, \ldots, n$
4. $x_i^{(k)} = \dfrac{1}{a_{ii}}\left\{-\sum_{j=1}^{i-1} a_{ij}x_j^{(k-1)} - \sum_{j=i+1}^{n} a_{ij}x_j^{(k-1)} + b_i\right\}$ (6.1)
5. OUTPUT: $X(n, 1)$, an approximate solution of $AX = B$

Note that the inversion of the splitting matrix D involves only the reciprocals $1/a_{ii}$. Note also that Jacobi's method, as detailed in Algorithm (6.1), requires that we save all of X_{k-1} until X_k has been completely computed. At the time we are computing $x_i^{(k)}$ we have already computed the better (presumably) estimates $x_j^{(k)}$ for $j = 1, 2, \ldots, i - 1$. If we use these new values in Algorithm (6.1) we obtain another iterative method. This new method is called the **Gauss-Seidel iterative method.**

GAUSS-SEIDEL ITERATIVE METHOD

1. INPUT: $A(n, n)$, $B(n, 1)$, $X_0(n, 1)$ arbitrary
2. For $k = 1, 2, \ldots$ convergence
3. For $i = 1, 2, \ldots, n$
4. $x_i^{(k)} = \dfrac{1}{a_{ii}}\left\{-\sum_{j=1}^{i-1} a_{ij}x_j^{(k)} - \sum_{j=i+1}^{n} a_{ij}x_j^{(k-1)} + b_i\right\}$ (6.2)
5. OUTPUT: $X(n, 1)$, an approximate solution of $AX = B$

In matrix form Algorithm (6.2) is equivalent to

$$DX_k = LX_k + RX_{k-1} + B.$$

This last equation is equivalent to rewriting $AX = B$ as

$$(D - L)X = RX + B \quad \text{or} \quad X = (D - L)^{-1}RX + (D - L)^{-1}B.$$

The Gauss-Seidel iteration matrix is $(D - L)^{-1}R$. Note that the inverse of the splitting matrix $Q = D - L$ is never formed explicitly in Algorithm (6.2). Algorithm (6.2) requires exactly the same number of multiplications as

Algorithm (6.1) and uses less storage because the new iterate can overwrite the old iterate as it is formed.

■ EXAMPLE 1

Consider the linear system

$$\begin{bmatrix} 4 & -1 & -1 & 0 \\ -1 & 4 & 0 & -1 \\ -1 & 0 & 4 & -1 \\ 0 & -1 & -1 & 4 \end{bmatrix} X = \begin{bmatrix} 9 \\ 12 \\ 0 \\ -3 \end{bmatrix}$$

$$\text{for which } X = \begin{bmatrix} 3.5 \\ 4 \\ 1 \\ 0.5 \end{bmatrix} \text{ is the true solution.}$$

Find the Jacobi and Gauss-Seidel iteration matrices and use them to find approximate solutions of the given system.

SOLUTION We first write the coefficient matrix as

$$A = \begin{bmatrix} 4 & -1 & -1 & 0 \\ -1 & 4 & 0 & -1 \\ -1 & 0 & 4 & -1 \\ 0 & -1 & -1 & 4 \end{bmatrix} = D - L - R$$

$$= \begin{bmatrix} 4 & 0 & 0 & 0 \\ 0 & 4 & 0 & 0 \\ 0 & 0 & 4 & 0 \\ 0 & 0 & 0 & 4 \end{bmatrix} - \begin{bmatrix} 0 & 0 & 0 & 0 \\ 1 & 0 & 0 & 0 \\ 1 & 0 & 0 & 0 \\ 0 & 1 & 1 & 0 \end{bmatrix} - \begin{bmatrix} 0 & 1 & 1 & 0 \\ 0 & 0 & 0 & 1 \\ 0 & 0 & 0 & 1 \\ 0 & 0 & 0 & 0 \end{bmatrix}.$$

Then the Jacobi iteration matrix is $J = D^{-1}(L + R) = \dfrac{1}{4}\begin{bmatrix} 0 & 1 & 1 & 0 \\ 1 & 0 & 0 & 1 \\ 1 & 0 & 0 & 1 \\ 0 & 1 & 1 & 0 \end{bmatrix}$ and the given system is equivalent to $X = JX + D^{-1}B = JX + (1/4)B$. The Jacobi iteration is then defined by $X_{k+1} = JX_k + (1/4)B$ where X_0 is arbitrary. If we take

$$X_0 = \begin{bmatrix} 1 \\ 1 \\ 1 \\ 1 \end{bmatrix}, \quad \text{then } X_1 = JX_0 + \frac{1}{4}B = \begin{bmatrix} 2.75 \\ 3.50 \\ 0.50 \\ -0.25 \end{bmatrix},$$

$$X_2 = JX_1 + \frac{1}{4}B = \begin{bmatrix} 3.25 \\ 3.625 \\ 0.625 \\ 0.250 \end{bmatrix}, \quad X_3 = \begin{bmatrix} 3.3125 \\ 3.8750 \\ 0.8750 \\ 0.3125 \end{bmatrix},$$

$$X_4 = \begin{bmatrix} 3.4375 \\ 3.9063 \\ 0.9063 \\ 0.4375 \end{bmatrix}, \quad X_5 = \begin{bmatrix} 3.4531 \\ 3.9688 \\ 0.9688 \\ 0.4531 \end{bmatrix}, \quad X_6 = \begin{bmatrix} 3.4844 \\ 3.9766 \\ 0.9766 \\ 0.4844 \end{bmatrix},$$

$$X_7 = \begin{bmatrix} 3.4883 \\ 3.9922 \\ 0.9922 \\ 0.4883 \end{bmatrix}, \quad X_8 = \begin{bmatrix} 3.4961 \\ 3.9941 \\ 0.9941 \\ 0.4961 \end{bmatrix}, \ldots$$

This sequence of approximate solutions is clearly converging to the true solution of the system.

The Gauss-Seidel iteration matrix is

$$\mathscr{L} = (D - L)^{-1}R = \frac{1}{32}\begin{bmatrix} 0 & 8 & 8 & 0 \\ 0 & 2 & 2 & 8 \\ 0 & 2 & 2 & 8 \\ 0 & 1 & 1 & 4 \end{bmatrix}, \quad B' = (D - L)^{-1} = \frac{1}{32}\begin{bmatrix} 72 \\ 114 \\ 18 \\ 9 \end{bmatrix},$$

and the given system is equivalent to $(D - L)X = RX + B$, or to $X = \mathscr{L}X + B'$. The Gauss-Seidel iteration is then $X_{k+1} = \mathscr{L}X_k + B'$ with X_0 arbitrary. If we take

$$X_0 = \begin{bmatrix} 1 \\ 1 \\ 1 \\ 1 \end{bmatrix}, \quad \text{then } X_1 = \mathscr{L}X_0 + B' = \begin{bmatrix} 2.7500 \\ 3.9375 \\ -0.9375 \\ 0.4688 \end{bmatrix},$$

$$X_2 = \begin{bmatrix} 3.4688 \\ 3.9844 \\ 0.9844 \\ 0.4922 \end{bmatrix}, \quad X_3 = \begin{bmatrix} 3.4922 \\ 3.9961 \\ 0.9961 \\ 0.4980 \end{bmatrix}, \quad X_4 = \begin{bmatrix} 3.4980 \\ 3.9990 \\ 0.9990 \\ 0.4995 \end{bmatrix}$$

which is already very close to the true solution. For the given system both the

Jacobi method and the Gauss-Seidel method yield convergent sequences. The Gauss-Seidel method seems to be converging more rapidly. ■

Either of the two iterative methods just described is relatively unattractive if A is a full matrix. In that case each iteration would require about n^2 multiplications, so the iterative method would be more expensive than Gaussian elimination if more than $n/3$ iterations were required.

Iterative methods are frequently useful in solving systems whose coefficient matrix is large, sparse, and regularly structured. In these cases the matrix vector product is substantially less expensive and you do not need to actually form, or store, the coefficient matrix of the system. As an example of this situation we consider the solution of a boundary value problem for an elliptic partial differential equation.

■ EXAMPLE 2

Determine the internal temperature in a rectangular plate if one edge is immersed in live steam at 100°C while the other three edges are kept in a solution of ice water at 0°C. The top and bottom of the plate are assumed to be insulated. (See Figure 6.1.)

SOLUTION If $u(x, y)$ is the steady-state temperature at (x, y), then it is known from physics that $u(x, y)$ satisfies the Laplace equation

$$\frac{\partial^2 u(x, y)}{\partial x^2} + \frac{\partial^2 u(x, y)}{\partial y^2} = u_{xx}(x, y) + u_{yy}(x, y) = 0. \tag{6.3}$$

Instead of trying to find an analytic expression for $u(x, y)$, we will try to find a solution at the nine interior grid points shown in Figure 6.2. To find the relationship between the temperatures at the grid points we expand $u(x,y)$ in

FIGURE 6.1 ■ RECTANGULAR PLATE IMMERSED IN STEAM AND ICE WATER

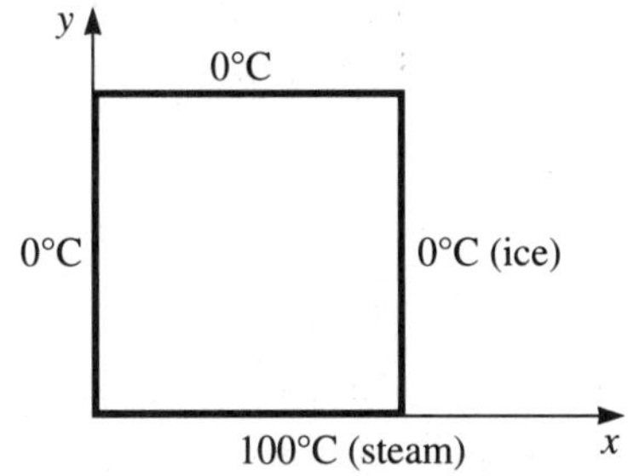

FIGURE 6.2 ■ PLATE WITH SUPERIMPOSED GRID

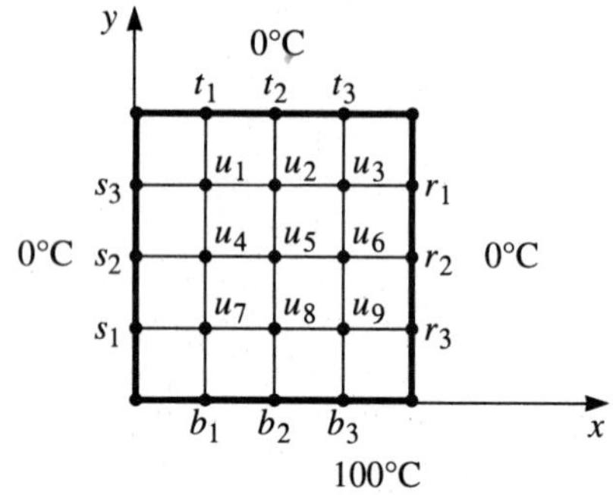

a Taylor series. If $u(x, y)$ is sufficiently differentiable, it is shown in calculus that for any interior grid point (x_0, y_0) of the rectangle

$$u(x_0 \pm h, y_0) = u(x_0, y_0) \pm hu_x(x_0, y_0) + \frac{h^2}{2}u_{xx}(x_0, y_0) \pm \frac{h^3}{3!}u_{xxx}(x_0, y_0) + \frac{h^4}{4!}u_{xxxx}(x_0, y_0) + \cdots$$

and

$$u(x_0, y_0 \pm h) = u(x_0, y_0) \pm hu_y(x_0, y_0) + \frac{h^2}{2}u_{yy}(x_0, y_0) \pm \frac{h^3}{3!}u_{yyy}(x_0, y_0) + \frac{h^4}{4!}u_{yyyy}(x_0, y_0) + \cdots.$$

Using these expansions we find that, for small h,

$$\begin{aligned}
&\frac{1}{h^2}\{u(x_0 + h, y_0) + u(x_0 - h, y_0) + u(x_0, y_0 + h) \\
&\qquad + u(x_0, y_0 - h) - 4u(x_0, y_0)\} \\
&= \frac{u(x_0 + h, y_0) + u(x_0 - h, y_0) - 2u(x_0, y_0)}{h^2} \\
&\quad + \frac{u(x_0, y_0 + h) + u(x_0, y_0 - h) - 2u(x_0, y_0)}{h^2} \\
&= \{u_{xx}(x_0, y_0) + u_{yy}(x_0, y_0)\} + \frac{h^2}{12}\{u_{xxxx}(x_0, y_0) + u_{yyyy}(x_0, y_0) + \cdots\} \\
&= 0 + \frac{h^2}{12}\{u_{xxxx}(x_0, y_0) + u_{yyyy}(x_0, y_0) + \cdots\} \approx 0.
\end{aligned}$$

Note that

$$\frac{u(x_0 + h, y_0) + u(x_0 - h, y_0) - 2u(x_0, y_0)}{h^2}$$

is the standard finite-difference approximation for $u_{xx}(x_0, y_0)$ and that

$$\frac{u(x_0, y_0 + h) + u(x_0, y_0 - h) - 2u(x_0, y_0)}{h^2}$$

is the corresponding approximation for $u_{yy}(x_0, y_0)$. Thus, if we ignore terms of order h^2 and higher, we have, as a finite difference approximation to Equation (6.3),

$$\begin{aligned} u(x_0, y_0) \approx \frac{1}{4}\{&u(x_0 + h, y_0) + u(x_0 - h, y_0) \\ &+ u(x_0, y_0 + h) + u(x_0, y_0 - h)\}. \end{aligned} \tag{6.4}$$

Equation (6.4) says an intuitively appealing thing: The steady-state temperature at each interior grid point is approximately equal to the average of the temperatures at the four neighboring grid points. Thus we have the following nine equations:

$$u_1 = \frac{1}{4}(u_2 + u_4 + s_3 + t_1), \quad u_2 = \frac{1}{4}(u_1 + u_3 + u_5 + t_2),$$

$$u_3 = \frac{1}{4}(u_2 + u_6 + r_1 + t_3), \quad u_4 = \frac{1}{4}(u_1 + u_5 + u_7 + s_2),$$

$$u_5 = \frac{1}{4}(u_2 + u_4 + u_6 + u_8), \quad u_6 = \frac{1}{4}(u_3 + u_5 + u_9 + r_2),$$

$$u_7 = \frac{1}{4}(u_4 + u_8 + b_1 + s_1), \quad u_8 = \frac{1}{4}(u_5 + u_7 + u_9 + b_2),$$

$$u_9 = \frac{1}{4}(u_6 + u_8 + b_3 + r_3),$$

which in matrix form can be written as

$$U = \frac{1}{4}\left[\begin{array}{ccc:ccc:ccc} 0 & 1 & 0 & 1 & 0 & 0 & 0 & 0 & 0 \\ 1 & 0 & 1 & 0 & 1 & 0 & 0 & 0 & 0 \\ 0 & 1 & 0 & 0 & 0 & 1 & 0 & 0 & 0 \\ \hdashline 1 & 0 & 0 & 0 & 1 & 0 & 1 & 0 & 0 \\ 0 & 1 & 0 & 1 & 0 & 1 & 0 & 1 & 0 \\ 0 & 0 & 1 & 0 & 1 & 0 & 0 & 0 & 1 \\ \hdashline 0 & 0 & 0 & 1 & 0 & 0 & 0 & 1 & 0 \\ 0 & 0 & 0 & 0 & 1 & 0 & 1 & 0 & 1 \\ 0 & 0 & 0 & 0 & 0 & 1 & 0 & 1 & 0 \end{array}\right] U + \frac{1}{4}\begin{bmatrix} s_3 + t_1 \\ t_2 \\ r_1 + t_3 \\ s_2 \\ 0 \\ r_2 \\ b_1 + s_1 \\ b_2 \\ b_3 + r_3 \end{bmatrix} = JU + B.$$

In this case, the system is already written in a form suitable for solution by an iterative method. The coefficient matrix J is the Jacobi iteration matrix, for which we find, using the software, that $\rho(J) = 0.707107 < 1$. Thus the sequence generated by Algorithm (6.1) will converge by Theorem 6.3. We can obtain Gauss-Seidel matrix in this case by writing $J = L + R$ and then $\mathscr{L} = (I - L)^{-1}R$. The Gauss-Seidel iteration matrix is

$$\mathscr{L} = \frac{1}{512}\begin{bmatrix} 0 & 128 & 0 & 128 & 0 & 0 & 0 & 0 & 0 \\ 0 & 32 & 128 & 32 & 128 & 0 & 0 & 0 & 0 \\ 0 & 8 & 32 & 8 & 32 & 128 & 0 & 0 & 0 \\ 0 & 32 & 0 & 32 & 128 & 0 & 128 & 0 & 0 \\ 0 & 16 & 32 & 16 & 64 & 128 & 32 & 128 & 0 \\ 0 & 6 & 16 & 6 & 24 & 64 & 8 & 32 & 128 \\ 0 & 8 & 0 & 8 & 32 & 0 & 32 & 128 & 0 \\ 0 & 6 & 8 & 6 & 24 & 32 & 16 & 64 & 128 \\ 0 & 3 & 6 & 3 & 12 & 24 & 6 & 24 & 64 \end{bmatrix}$$

and the spectral radius is $\rho(\mathscr{L}) = 0.50$, which is less than the spectral radius of the Jacobi matrix computed earlier. Note that J is much sparser than $\mathscr{L}$.

A more efficient way of dealing with this system is to treat the temperatures at the grid points as the entries of a 5×5 matrix V. The association of the temperatures and the $v_{ij} = v(i, j)$ is indicated in Figure 6.3. With this notation Equation (6.4) becomes

FIGURE 6.3 ■ ASSOCIATION OF TEMPERATURES AND v_{ij}

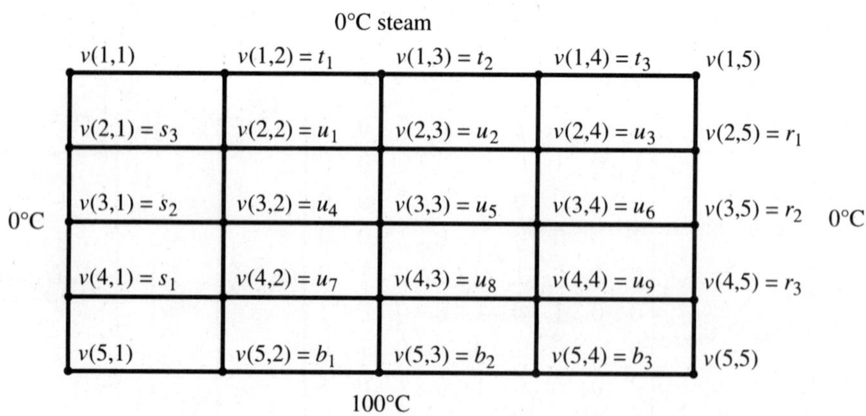

$$v(i, j) = \frac{1}{4}\{v(i - 1, j) + v(i + 1, j) + v(i, j - 1) + v(i, j + 1)\}$$

for all interior grid points, that is, for $i, j = 2, 3, 4$.

Thus the Jacobi iterative method of Algorithm (6.1), for this problem, would amount to

1. Set $v(1, j)$, $v(5, j)$, $v(i, 1)$, and $v(i, 5)$ equal to the given boundary values.
2. Set the other $v(i, j)$ to some arbitrary starting values.
3. For $k = 1, 2, \ldots$ convergence
4. For $i = 2, 3, 4$
5. For $j = 2, 3, 4$ (6.5)
6. $v_{\text{new}}(i, j) = \frac{1}{4}\{v(i - 1, j) + v(i + 1, j) + v(i, j - 1) + v(i, j + 1)\}$
7. $V = V_{\text{new}}$

Note that this implementation of the Jacobi method does not require the explicit formation or storage of the coefficient matrix of the linear system; the only storage required is two 5×5 matrices.

After 20 iterations of Algorithm (6.5), with starting values $v_{ij} = 1$, we find that the matrix V is

$$V = \left[\begin{array}{c|ccc|c} 0 & 0 & 0 & 0 & 0 \\ \hline 0 & 7.12528 & 9.79799 & 7.12528 & 0 \\ 0 & 18.7266 & 24.9648 & 18.7266 & 0 \\ 0 & 42.8396 & 52.6551 & 42.8396 & 0 \\ \hline 100 & 100 & 100 & 100 & 100 \end{array}\right],$$

from which we conclude that

$$u_1 \approx v(2, 2) = 7.12528, \quad u_2 \approx v(2, 3) = 9.79799, \quad u_3 \approx v(2, 4) = 7.12528,$$
$$u_4 \approx v(3, 2) = 18.7266, \quad u_5 \approx v(3, 3) = 24.9648, \quad u_6 \approx v(3, 4) = 18.7266,$$
$$u_7 \approx v(4, 2) = 42.8396, \quad u_8 \approx v(4, 3) = 52.6551, \quad u_9 = v(4, 4) = 42.8396.$$

The Gauss-Seidel iteration, in this matrix format, would take the form

1. Set $v(1, j)$, $v(5, j)$, $v(i, 1)$, and $v(i, 5)$ equal to the given boundary values.
2. Set the other $v(i, j)$ to some arbitrary starting values.
3. For $k = 1, 2, \ldots$ convergence
4.   For $i = 2, 3, 4$ (6.6)
5.   For $j = 2, 3, 4$
6. $$v(i, j) = \frac{1}{4}\{v(i-1, j) + v(i+1, j) + v(i, j-1) + v(i, j+1)\}$$

We obtain numerical results consistent with the Jacobi values after only 10 iterations. At least in this case the Gauss-Seidel iteration appears to be considerably more efficient than the Jacobi iteration. Note that for this problem $\rho(\mathscr{L})^{10} = (0.5)^{10} \approx \rho(J)^{20} = (0.707107)^{20}$.

Note that the corner values of the matrix are never encountered in the above algorithms. This is a good thing since these corner values are not clearly defined.

6.2 ■ Exercise Set

1. What does the Gerschgorin theorem tell you about the eigenvalues of the 9×9 matrices J and $\mathscr{L}$ of Example 2? Use the software to find the eigenvalues of these matrices.

2. Use the software to solve the system from Example 2 using Gaussian elimination (MATMAN with rational arithmetic will produce the exact solution). Compare this solution with the approximate solution obtained using 20 iterations of the

a. Jacobi method

b. Gauss-Seidel method

3. For the matrix $C = \begin{bmatrix} 2 & -1 & 0 & 0 & 0 \\ -1 & 2 & -1 & 0 & 0 \\ 0 & -1 & 2 & -1 & 0 \\ 0 & 0 & -1 & 2 & -1 \\ 0 & 0 & 0 & -1 & 2 \end{bmatrix}$ find the associated Jacobi and

Gauss-Seidel iteration matrices and find their spectral radii.

4. Solve $AX = \begin{bmatrix} 1 & 0 & -\frac{1}{4} & -\frac{1}{4} \\ 0 & 1 & -\frac{1}{4} & -\frac{1}{4} \\ -\frac{1}{4} & -\frac{1}{4} & 1 & 0 \\ -\frac{1}{4} & -\frac{1}{4} & 0 & 1 \end{bmatrix} X = \begin{bmatrix} 1 \\ 2 \\ 3 \\ 4 \end{bmatrix}$ using

a. Gaussian elimination

b. Jacobi iteration

c. Gauss-Seidel iteration.

5. Write a computer program or a MATLAB m-file that will solve the temperature-distribution problem of Example 2 on a $2n + 1 \times 2n + 1$ grid using the Gauss-Seidel iterative method. Stop the iteration when the middle element $v(n, n)$ changes less than 10^{-4}. Test your program for $n = 3, 5$, and 11.

6. Consider the problem of Example 2 with the mesh points relabeled (red-black ordering) as shown below. Find the Jacobi matrix associated with this ordering of the nodes and compare its spectral radius with that obtained with the natural ordering in Example 2. Find the Gauss-Seidel matrix associated with this ordering and compare its spectral radius with that obtained in Example 2.

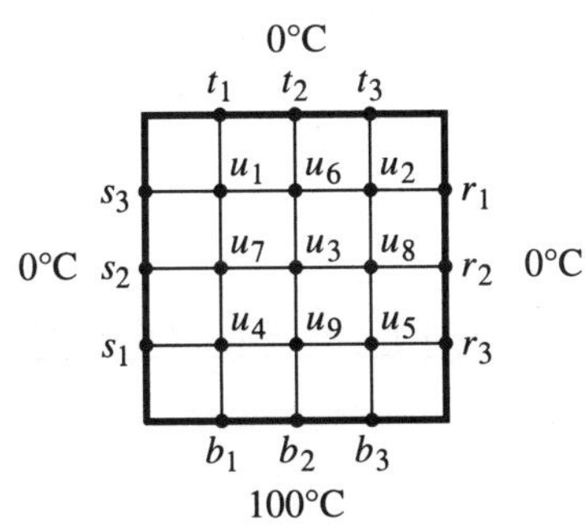

7. Let X_0 be an approximate solution of the linear system $AX = B$. Define the residual $R = B - AX_0$ and then show that $R = AE$, where $E = X - X_0$ is the error in the approximate solution. This is the basis for the following iterative refinement algorithm:

For $i = 0, 1, 2, \ldots, nc$

Compute the residual $R = B - AX_i$ (in double precision)

Solve $AE = R$ (in single precision)

$X_{i+1} = X_i + E$ (in double precision)

Show that if we replace A by D (the diagonal of A) in the second step, we generate the same sequence as in the Jacobi iteration.

8. Use the Gauss-Seidel method to solve the linear system

$$\frac{1}{4}\begin{bmatrix} 4 & -1 & -1 & 0 & 0 & 0 \\ -1 & 4 & -1 & -1 & 0 & 0 \\ -1 & -1 & 4 & -1 & -1 & 0 \\ 0 & -1 & -1 & 4 & -1 & -1 \\ 0 & 0 & -1 & -1 & 4 & -1 \\ 0 & 0 & 0 & -1 & -1 & 4 \end{bmatrix} X = \begin{bmatrix} 1 \\ 0 \\ 0 \\ 0 \\ 1 \\ 1 \end{bmatrix}.$$

6.3 ■ CONVERGENCE THEOREMS

The Gerschgorin theorem falls just short of establishing the convergence of the Jacobi method for the problem of Example 2 in Section 6.2. In this case the Gerschgorin region contains some points of absolute value 1 (see Figure 6.4), so all we can conclude from the Gerschgorin analysis is that $\rho(A) \leq 1$. For the Gauss-Seidel method, the Gerschgorin theorem shows that the spectral radius is less than 1 and hence that the iteration will converge. (See Figure 6.5.)

In this section we will sharpen the Gerschgorin theorem and then mention a few of the many theorems about the convergence of the basic iterative methods.

FIGURE 6.4 ■ GERSCHGORIN REGION FOR THE JACOBI METHOD

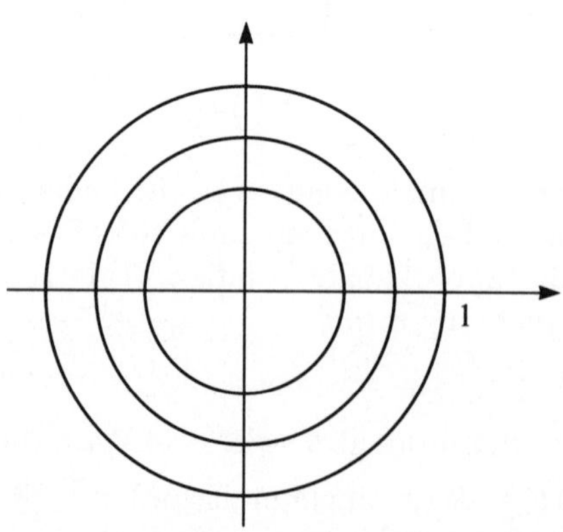

FIGURE 6.5 ■ GERSCHGORIN REGION FOR THE GAUSS-SEIDEL METHOD

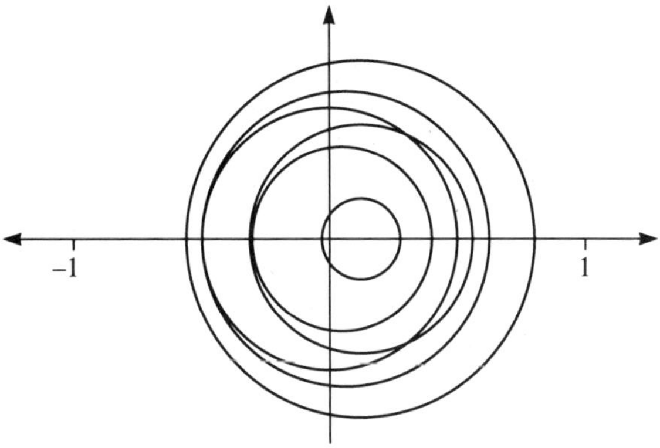

■ DEFINITION 6.1 ■

An $n \times n$ matrix A is said to be **reducible** if there exists a permutation matrix P such that $P^TAP = \left[\begin{array}{c|c} A_{11} & A_{12} \\ \hline 0 & A_{22} \end{array}\right]$ is block upper triangular with square diagonal blocks. If A is not reducible it is said to be **irreducible**.

To understand the notion of irreducibility we will find it helpful to examine the directed graph associated with a matrix. A graph consists of a set of nodes and a set of directed edges connecting some or all of the nodes. For an $n \times n$ matrix A, the graph has n nodes numbered from 1 to n and there is a directed edge from node i to node j if and only if $a_{ij} \neq 0$.

■ EXAMPLE 3

The graphs for the matrices $A = \begin{bmatrix} 0 & 0 & 0 & 1 \\ 1 & 0 & 0 & 0 \\ 0 & 1 & 0 & 5 \\ 0 & 0 & 1 & 2 \end{bmatrix}$ and

$B = \begin{bmatrix} 4 & -1 & 0 & -1 & 0 & 0 \\ -1 & 4 & -1 & 0 & -5 & 0 \\ 0 & -1 & 0 & 0 & 0 & -1 \\ -1 & 0 & 0 & 4 & -1 & 0 \\ 0 & -1 & 0 & -1 & 4 & -1 \\ 0 & 0 & -1 & 0 & -1 & 4 \end{bmatrix}$, are as shown in Figure 6.6 and Figure 6.7.

■

FIGURE 6.6 ■ GRAPH OF MATRIX A IN EXAMPLE 3

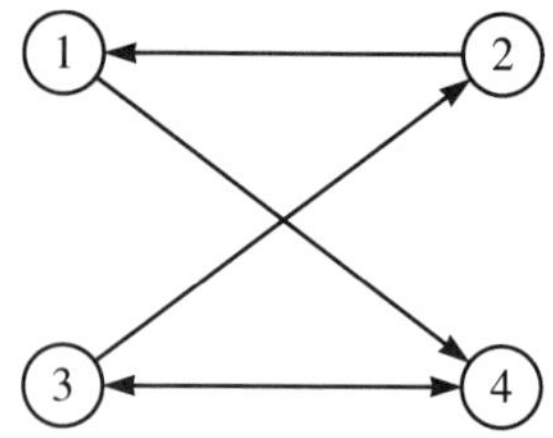

We say that node i is connected to node j if there is a directed path from node i to node j. In the graph of the matrix A of Example 2, node 1 is connected to node 3 by means of the path from 1 to 4 to 3; node 4 is connected to node 1 by the path from 4 to 3 to 2 to 1. In the graph of the matrix B, node 5 is connected to node 1 by means of the path from 5 to 6 to 3 to 2 to 1, or by the path from 5 to 4 to 1. In terms of the entries of the matrix A, node i is connected to node j if and only if there exists integers $k_1, k_2, k_3, \ldots, k_r$ such that the product $a_{ik_1}a_{k_1k_2}a_{k_2k_3}\cdots a_{k_rj} \neq 0$.

■ **DEFINITION 6.2** ■

The matrix A is said to be **strongly connected** if, in the graph of A, every node is connected to every other node.

Both of the matrices of Example 3 are strongly connected.

The following theorem shows that a matrix's being irreducible is

FIGURE 6.7 ■ GRAPH OF MATRIX B IN EXAMPLE 3

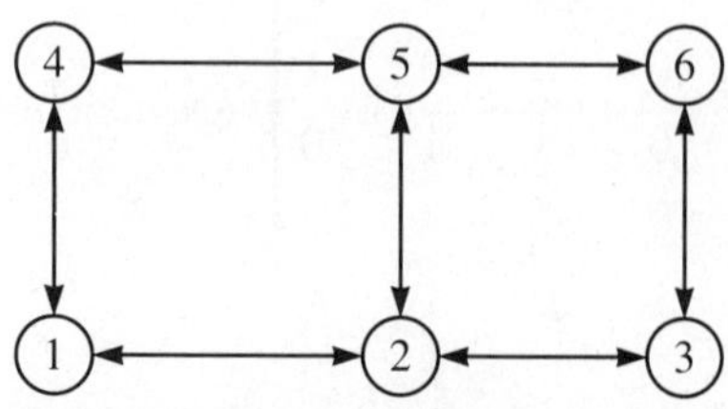

equivalent to its graph being strongly connected. Note that for any permutation matrix P, the graph of A and the graph of P^TAP are the same except for a relabeling of the nodes. (See Exercise 8.)

THEOREM 6.4

A matrix A is irreducible if and only if it is strongly connected.

PROOF If A is reducible, then it follows that not every node is connected to every other node; in particular, if $P^TAP = \left[\begin{array}{c|c} A_{11} & A_{12} \\ \hline 0 & A_{22} \end{array}\right]$ and A_{11} is $r \times r$, then none of the last $n - r$ nodes is connected to node 1. Thus A is not strongly connected. The contrapositive is that if A is strongly connected, then A is irreducible.

Suppose now that the matrix A is not strongly connected and let

$$\mathcal{Y} = \{i \mid \text{node } i \text{ is connected to every other node}\} \quad \text{and}$$
$$\mathcal{N} = \{1, 2, \ldots, n\} \backslash \mathcal{Y}.$$

Note that node $k \varepsilon \mathcal{N}$ is not connected to any node in $\mathcal{Y}$ because otherwise node k would be connected to every node. Now let σ be a permutation of $\{1, 2, \ldots, n\}$ such that if $i \varepsilon \mathcal{Y}$ and $j \varepsilon \mathcal{N}$, then $\sigma(i) < \sigma(j)$. If P^T is the permutation matrix we obtain by applying the permutation σ to the rows of I, then A is reducible because

$$P^TAP = \left[\begin{array}{c|c} A_{11} & A_{12} \\ \hline 0 & A_{22} \end{array}\right]$$

where A_{11} has as many rows as the number of nodes in $\mathcal{Y}$. The $(2, 1)$ block is 0 because, as observed before, no node in $\mathcal{N}$ is connected to any node of $\mathcal{Y}$. ■

The next theorem shows that it is quite difficult for an irreducible matrix to have an eigenvalue that lies on the boundary of the Gerschgorin region.

THEOREM 6.5

If A is an $n \times n$ irreducible matrix and λ is an eigenvalue of A that is on the boundary of the Gerschgorin region of A, then all the Gerschgorin circles of A must pass through λ.

PROOF Let X be an eigenvector for λ, $AX = \lambda X$, and assume, without loss of generality, that $\|X\|_\infty = 1$ with $|x_r| = 1 \geq |x_i|$ for all i. Then, as in the proof of the Gerschgorin theorem,

$$(\lambda - a_{rr})x_r = \sum_{j=1,\neq r}^{n} a_{rj}x_j \Rightarrow |\lambda - a_{rr}| \leq \sum_{j=1,\neq r}^{n} |a_{rj}|\,|x_j| \leq \sum_{j=1,\neq r}^{n} |a_{rj}| \equiv \Lambda_r.$$

Since λ is a boundary point of the Gerschgorin region, we must have $|\lambda - a_{rr}| = \Lambda_r$, and it follows that

$$|\lambda - a_{rr}| = \sum_{j=1,\neq r}^{n} |a_{rj}|\,|x_j| = \sum_{j=1,\neq r}^{n} |a_{rj}| \equiv \Lambda_r \quad \text{so that} \quad a_{rl} \neq 0 \Rightarrow |x_l| = 1.$$

Since A is irreducible, there exists an index p such that $a_{rp} \neq 0$. It now follows, from the last remark, that $|x_p| = 1$. Repeating the preceding argument we can now conclude that λ must lie on the boundary of the pth Gerschgorin circle, that is, $|\lambda - a_{pp}| = \Lambda_p$. Since A is irreducible we can find, for any index j, a sequence of nonzero entries of A, $a_{rp}, a_{pp_2}, a_{p_2p_3}, \ldots, a_{p_kj}$. Since $a_{p_kj} \neq 0$ it follows, as above, that λ must lie on the boundary of the jth Gerschgorin circle. ■

Two examples of Gerschgorin regions for matrices, which could have an eigenvalue on the boundary, are given in Figure 6.8.

The Gerschgorin region for the 9×9 Jacobi matrix J in Example 1 is shown in Figure 6.4. Since the matrix J is irreducible, Theorem 6.5 tells us that it is not possible for J to have an eigenvalue λ of absolute value 1. Thus $\rho(J) < 1$ and the Jacobi method for this problem will converge. The Gauss-Seidel iteration matrix for this problem is reducible since its first column is zero. However, the 8×8 lower-right-hand corner of G is irreducible. Since the Gerschgorin circles (see Figure 6.5) have no point in common, it follows from Theorem 6.5 that no eigenvalue of G can be on the boundary of the Gerschgorin region.

One of the most useful corollaries of the Gerschgorin theorem is the fact that a diagonally dominant matrix must be nonsingular (Theorem 3.11). This is one of the few results that allow you to make important conclusions about

FIGURE *6.8* ■ GERSCHGORIN REGIONS DESCRIBED IN THEOREM 6.5

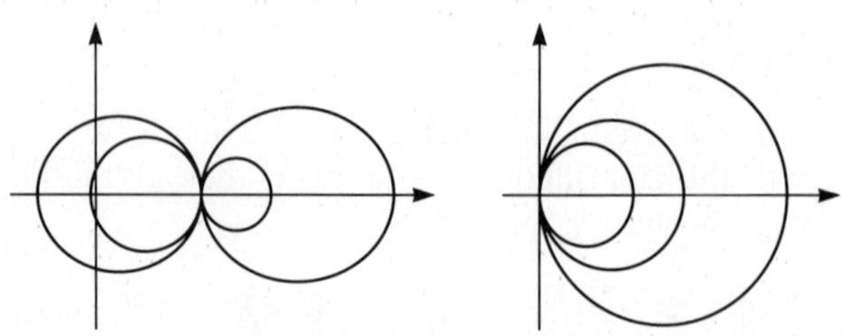

a matrix by "eyeballing" it. This result falls just short of establishing the nonsingularity of the coefficient matrix for the systems arising from the finite difference approximations of the partial differential equation in Example 2. We will now show that, for an irreducible matrix, strict diagonal dominance is not required to establish nonsingularity.

■ DEFINITION 6.3 ■

The $n \times n$ matrix A is **weakly diagonally dominant** if

$$|a_{ii}| \geq \sum_{\substack{j=1 \\ j \neq i}}^{n} |a_{ij}| = \Lambda_i; \quad i = 1, 2, \ldots, n$$

with strict inequality for at least one index i.

For the finite-difference matrices in Example 2, we will have weak diagonal dominance, with strict inequality for all those nodes that are next to the boundary and equality for all the other nodes. For the 3×3 grid shown in Figure 6.2 there is only one interior node (u_5), and hence there will be a strict inequality in eight of the nine rows of the matrix. In this case the coefficient matrix is

$$A = I - J = \frac{1}{4}\begin{bmatrix} 4 & -1 & 0 & -1 & 0 & 0 & 0 & 0 & 0 \\ -1 & 4 & -1 & 0 & -1 & 0 & 0 & 0 & 0 \\ 0 & -1 & 4 & 0 & 0 & -1 & 0 & 0 & 0 \\ -1 & 0 & 0 & 4 & -1 & 0 & -1 & 0 & 0 \\ 0 & -1 & 0 & -1 & 4 & -1 & 0 & -1 & 0 \\ 0 & 0 & -1 & 0 & -1 & 4 & 0 & 0 & -1 \\ 0 & 0 & 0 & -1 & 0 & 0 & 4 & -1 & 0 \\ 0 & 0 & 0 & 0 & -1 & 0 & -1 & 4 & -1 \\ 0 & 0 & 0 & 0 & 0 & -1 & 0 & -1 & 4 \end{bmatrix}.$$

Note that we do not have strict diagonal dominance in the fifth row. The Gerschgorin region is like that shown in Figure 6.4, with the origin shifted one unit to the left. Since A is irreducible and 0 is on the boundary of the Gerschgorin region, it follows from Theorem 6.5 that 0 cannot be an eigenvalue of A. Hence A is nonsingular. The same argument applies in general so we have the following theorem.

THEOREM 6.6

If a matrix A is both irreducible and weakly diagonally dominant, then A is also nonsingular. ■

There are many theorems about the convergence of iterative methods. Among them are the next two theorems.

THEOREM 6.7

Ostrowski-Reich: If A is positive definite, then the Gauss-Seidel method for $AX = B$ is convergent.

PROOF We write $A = D - L - R$, where D is diagonal, L is strictly lower triangular, and R is strictly upper triangular. Since A is positive definite, the diagonal matrix D is nonsingular. The Gauss-Seidel matrix is $\mathscr{L} = (D - L)^{-1}R = (D - L)^{-1}(D - L - A) = I - (D - L)^{-1}A$. Now let $A^{1/2}$ be the unique positive definite square root of A and define the $A^{1/2}$ norm of B by $\|B\|_{A^{1/2}} = \|A^{1/2}BA^{-1/2}\|_2$. This is a consistent matrix norm and the matrix $\mathscr{L}$ is convergent if $\|\mathscr{L}\|_{A^{1/2}} = \|A^{1/2}\mathscr{L}A^{-1/2}\|_2 = \|\mathscr{L}'\|_2 < 1$. Since $\|\mathscr{L}'\|_2^2 = \rho(\mathscr{L}'\mathscr{L}'^T)$, we compute

$$\begin{aligned}
\mathscr{L}'\mathscr{L}'^T &= [I - A^{1/2}(D - L)^{-1}A^{1/2}][I - A^{1/2}(D - R)^{-1}A^{1/2}] \\
&= I - A^{1/2}(D - L)^{-1}A^{1/2} - A^{1/2}(D - R)^{-1}A^{1/2} \\
&\quad + A^{1/2}(D - L)^{-1}A(D - R)^{-1}A^{1/2} \\
&= I - A^{1/2}(D - L)^{-1}[(D - R) + (D - L) - A](D - R)^{-1}A^{1/2} \\
&= I - A^{1/2}(D - L)^{-1}D(D - R)^{-1}A^{1/2} \\
&= I - [A^{1/2}(D - L)^{-1}D^{1/2}][A^{1/2}(D - L)^{-1}D^{1/2}]^T.
\end{aligned}$$

Since $\mathscr{L}'\mathscr{L}'^T$ is I minus a positive definite matrix, it follows that $\rho(\mathscr{L}'\mathscr{L}'^T) < 1$ and hence that $\|\mathscr{L}'\|_2 = \|\mathscr{L}\|_{A^{1/2}} < 1$. Thus, $\mathscr{L}$ is a convergent matrix. ■

The next theorem confirms our intuitive feeling that the Gauss-Seidel method should converge faster than the Jacobi method. The proof depends on the theory of nonnegative matrices and is omitted (Varga, 1962).

THEOREM 6.8

Stein-Rosenberg: If the Jacobi matrix $J = D^{-1}(L + R)$ is nonnegative and $\mathscr{L} = (D - L)^{-1}R$ is the Gauss-Seidel matrix, then one and only one of the following mutually exclusive relations holds:

$$\rho(J) = \rho(\mathscr{L}) = 0$$
$$0 < \rho(\mathscr{L}) < \rho(J) < 1$$
$$1 = \rho(J) = \rho(\mathscr{L})$$
$$1 < \rho(J) < \rho(\mathscr{L})$$

6.3 ■ Exercise Set

1. Find the graphs of the following matrices and decide whether they are irreducible:

$$A = \begin{bmatrix} 1 & -7 & 2 & 5 \\ 0 & 5 & 0 & -2 \\ -3 & 1 & 4 & 5 \\ 0 & 4 & 0 & 3 \end{bmatrix}, \quad B = \begin{bmatrix} 0 & 0 & 4 & 2 \\ 0 & 0 & -2 & 5 \\ 3 & 2 & 0 & 0 \\ -8 & 6 & 0 & 0 \end{bmatrix}, \quad C = \begin{bmatrix} 0 & 0 & 0 & 1 \\ 1 & 0 & 0 & 0 \\ 0 & 1 & 0 & 0 \\ 0 & 0 & 1 & 0 \end{bmatrix}.$$

2. Compare the Gerschgorin regions for the matrices B and $X^{-1}BX$ where

$$B = \begin{bmatrix} 1 & \frac{1}{2} & \frac{1}{2} \\ \frac{1}{2} & 4 & \frac{1}{2} \\ \frac{3}{4} & \frac{1}{2} & 6 \end{bmatrix} \quad \text{and} \quad X = \begin{bmatrix} 1 & 0 & 0 \\ 0 & \frac{5}{3} & 0 \\ 0 & 0 & \frac{15}{7} \end{bmatrix}.$$

3. For the Gauss-Seidel matrix $\mathscr{L}$ of Example 2, find $\|\mathscr{L}\|_1$, $\|\mathscr{L}\|_\infty$, and $\|\mathscr{L}\|_{A^{1/2}}$.

4. Verify, as a consequence of Theorem 3.5, that any symmetric positive definite matrix has a unique positive definite square root. Find the positive definite square root of

$$B = \begin{bmatrix} 2 & -1 & 0 & 0 & 0 \\ -1 & 2 & -1 & 0 & 0 \\ 0 & -1 & 2 & -1 & 0 \\ 0 & 0 & -1 & 2 & -1 \\ 0 & 0 & 0 & -1 & 2 \end{bmatrix}.$$

5. Verify that if A is positive definite, then $\|B\|_{A^{1/2}} \equiv \|A^{1/2}BA^{-1/2}\|_2$ defines a consistent matrix norm.

6. Given the system $X = MX + B$, where $M = \begin{bmatrix} 1.2 & -0.6 & 0.2 & 0 \\ 0.2 & 0.7 & -0.6 & 0.5 \\ 0.2 & 0.2 & -0.4 & 0.8 \\ 0.2 & 0.2 & -1.2 & 1.6 \end{bmatrix}$, find the Jacobi and Gauss-Seidel matrices and their spectral radii.

7. Consider the insulated metal rod pictured below. The t_i represent the temperatures at the indicated points. Assume that the temperature at each node is the average of the temperatures at the two neighboring nodes. Find a system of linear equations that could be used to determine the t_i. Solve this system using the Gauss-Seidel method.

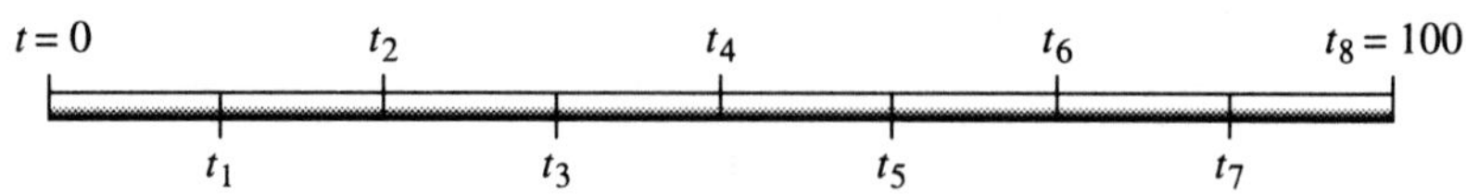

8. Let A be any $n \times n$ matrix and let P' be an $n \times n$ permutation matrix. Show that the graphs of A and P^TAP are the same except for a renumbering of the nodes.

6.4 ■ THE SOR METHOD

For the system $X = BX + C$, with $B = L + R$, the Gauss-Seidel iteration is

$$x_i^{(n+1)} = \sum_{j=1}^{i-1} b_{ij} x_j^{(n+1)} + \sum_{j=i+1}^{n} b_{ij} x_j^{(n)} + c_i; \quad i = 1, 2, \ldots, n \tag{6.7}$$

or, in matrix form,

$$X_{n+1} = LX_{n+1} + RX_n + C.$$

Let us attempt to improve the rate of convergence of the sequence generated by Equation (6.7) by taking, as the next iterate, a weighted average of X_n and the new Gauss-Seidel iterate from Equation (6.7). Specifically, let us consider a new iterative method, called the SOR (successive overrelaxation) method (Young, 1971) defined by

$$X_{n+1} = \omega \hat{X}_{n+1} + (1 - \omega)X_n = \omega[LX_{n+1} + RX_n + C] + (1 - \omega)X_n. \tag{6.8}$$

TABLE 6.1 ■ SEARCH FOR THE OPTIMUM ω

ω	$\rho(\mathcal{L}_\omega)$	ω	$\rho(\mathcal{L}_\omega)$	ω	$\rho(\mathcal{L}_\omega)$	ω	$\rho(\mathcal{L}_\omega)$
0	1			1.160	0.250676	1.1700	0.199820
0.2	0.936886			1.162	0.243220	1.1702	0.197927
0.4	0.862710			1.164	0.234994	1.1704	0.196930
0.6	0.773019			1.166	0.225671	**1.1706**	**0.193689**
0.8	0.659333			1.168	0.214582	**1.1708**	**0.1708**
1.0	**0.5**	1.10	0.378586	**1.170**	**0.199821**	1.1710	0.1710
1.2	**0.20**	1.12	0.345524	**1.172**	**0.172000**	1.1712	0.1712
1.4	0.40	1.14	0.305681	1.174	0.174	1.1714	0.1714
1.6	0.60	**1.16**	**0.250676**	1.176	0.176	1.1716	0.1716
1.8	0.80	**1.18**	**0.18**	1.178	0.178	1.1718	0.1718
2.0	1.0	1.20	0.20	1.180	0.180	1.1720	0.1720

Equivalently,

$$(I - \omega L)X_{n+1} = [\omega R + (1 - \omega)I]X_n + \omega C \Rightarrow$$
$$X_{n+1} = (I - \omega L)^{-1}[\omega R + (1 - \omega)I]X_n + \omega(I - \omega L)^{-1}C.$$

Thus the SOR iteration matrix is

$$\mathcal{L}_\omega = (I - \omega L)^{-1}[\omega R + (1 - \omega)I].$$

Note that the Gauss-Seidel iteration matrix $\mathcal{L}$ is the special case $\omega = 1$ of the above matrix.

For the temperature-distribution problem of Example 2 in Section 6.2, we can experimentally determine the optimum ω using the software. After a brief search, during which we collect the data in Table 6.1, we find that the optimum value of ω is about 1.1707.

From the data in Table 6.1 we can plot the graph of $\rho(\mathcal{L}_\omega)$ versus ω shown in Figure 6.9. Note that because of the steepness of the curve to the

FIGURE 6.9 ■ GRAPH OF DATA FROM TABLE 6.1

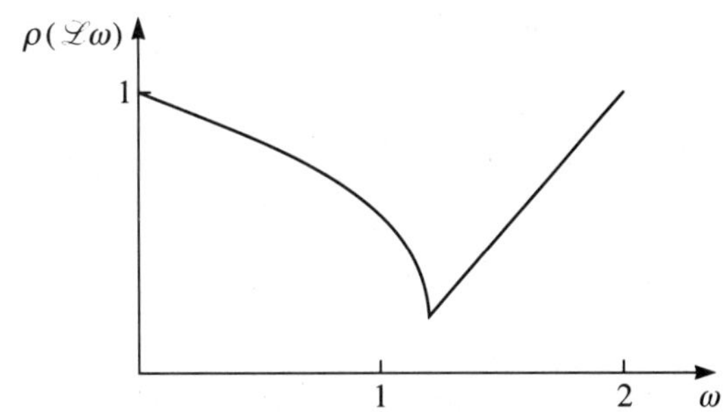

left of the minimum it would be better to overestimate the optimum ω than to underestimate it.

The next theorem shows that we did the above search in the right regions.

THEOREM 6.9

Consider the linear system $X = BX + K$ where $B = L + R$ is an $n \times n$ matrix with zero diagonal entries, L is strictly lower triangular, and R is strictly upper triangular. Then the SOR iteration matrix $\mathcal{L}_\omega = (I - \omega L)^{-1} \times [\omega R + (1 - \omega)I]$ is such that $\rho(\mathcal{L}_\omega) \geq |\omega - 1|$. Thus $\mathcal{L}_\omega$ can be convergent only for $0 < \omega < 2$.

PROOF Let $c(x) = \det(xI - \mathcal{L}_\omega) = \prod_{i=1}^{n} [x - \lambda_i(\omega)]$ be the characteristic polynomial of $\mathcal{L}_\omega$. The constant term of this polynomial is $\prod_{i=1}^{n} [-\lambda_i(\omega)] = c(0) = \det(-\mathcal{L}_\omega)$. Since $(I - \omega L)^{-1}$ is unit lower triangular and R is strictly upper triangular we have

$$\begin{aligned}\det(-\mathcal{L}_\omega) &= \det(I - \omega L)^{-1} \det[-\omega R - (1 - \omega)I] \\ &= \det[-(1 - \omega)I] = (\omega - 1)^n.\end{aligned}$$

It follows that $\rho(\mathcal{L}_\omega) \geq |\omega - 1|$. It is immediate that $\rho(\mathcal{L}_\omega) < 1$ only if $0 < \omega < 2$. ■

For certain special kinds of matrices it is possible to theoretically determine the optimum relaxation factor ω_b such that $\min_\omega \rho(\mathcal{L}_\omega) = \rho(\mathcal{L}_{\omega_b})$.

For the model problem, introduced in Example 2 of Section 6.2, and for many other important problems, it can be shown that if all the eigenvalues of the Jacobi matrix are real, then the theoretically best ω is given by

$$\omega_b = 1 + \left(\frac{\rho(J)}{1 + \sqrt{1 + \rho(J)^2}}\right)^2.$$

A detailed discussion of the convergence of the SOR method is beyond the scope of this text. Details are given in the books by Varga (1962), Young (1971), and Young and Hageman (1981). The various versions of the ITPACK software package contain many finely tuned programs to implement these methods in a variety of hardware environments.

■ SOFTWARE NOTES ■

The MATALG **L** command can be used to generate the data in Table 6.1. This command lists the values of the last function between specified values of the variable. The commands

```
w=1
=EVAL(INV(1-w*L)*(W*trans(L)+(1-w)))
L w, 0, 2, 10
```

will generate the data in the first two columns of Table 6.1. The following MATLAB commands will generate the first two columns of Table 6.1:

```
I=eye(9)
for j=0:10
    w=(0.2)*j
    v=eig(inv(I-w*L)*(w*L'+(1-w)*I));
    sr=norm(v,inf); [w,sr]
end
```

6.4 ■ Exercise Set

1. Find the splitting matrix that determines the SOR method.
2. Find the optimum ω for the SOR method associated with the model problem on a 2×2 grid.
3. Find the optimum ω for the SOR method associated with the model problem on a 2×3 grid.
4. Find the SOR iteration matrix for $A = \begin{bmatrix} 2 & -1 & 0 & 0 & 0 \\ -1 & 2 & -1 & 0 & 0 \\ 0 & -1 & 2 & -1 & 0 \\ 0 & 0 & -1 & 2 & -1 \\ 0 & 0 & 0 & -1 & 2 \end{bmatrix}$.
5. Compute five iterations of SOR for the problem of Example 2 in Section 6.2 using $\omega = 1.1707$. Compare the results with those we obtained using the Jacobi and Gauss-Seidel methods.
6. For the problem of Example 1 in Section 6.2, use the software to estimate the value of ω that makes $\|\mathcal{L}_\omega\|_1$ as small as possible.

6.5 ■ THE CONJUGATE GRADIENT METHOD

In this section we will consider another type of iterative method for solving the linear system $AX = B$ in the important case where A is a symmetric positive definite matrix. The method we will consider, although iterative in nature, is quite different from the earlier iterative methods we have discussed in that it uses techniques from optimization theory. Throughout this section we will use the notation $\langle X, Y\rangle$ to denote the standard inner product on $\mathscr{R}^n$, that is $\langle X, Y\rangle = X^TY$.

THEOREM 6.10

If A is an $n \times n$ symmetric positive definite matrix, then the problem of solving the linear system $AX = B$ is equivalent to finding the minimum value of the quadratic function

$$q(X) = X^TAX - 2X^TB = \langle X, AX\rangle - 2\langle X, B\rangle.$$

Moreover, the minimum value of q, on the ray $X_0 + tV$, is

$$q(X_0 + \hat{t}V) = q(X_0) - \frac{\langle V, B - AX\rangle^2}{\langle V, AV\rangle} \quad \text{where} \quad \hat{t} = \frac{\langle V, B - AX_0\rangle}{\langle V, AV\rangle}.$$

PROOF We consider first the values of q on the ray $X_0 + tV$, determined by the initial approximation X_0 and the search direction V. Using elementary properties of the inner product we have

$$\begin{aligned} q(X_0 + tV) &= \langle X_0 + tV, A(X_0 + tV)\rangle - 2\langle X_0 + tV, B\rangle \\ &= \langle X_0, AX_0\rangle + t\langle V, AX_0\rangle + t\langle X_0, AV\rangle \\ &\quad + t^2\langle V, AV\rangle - 2\langle X_0, B\rangle - 2t\langle V, B\rangle \\ &= \langle X_0, AX_0\rangle - 2\langle X_0, B\rangle + 2t\langle V, AX_0\rangle \\ &\quad - 2t\langle V, B\rangle + t^2\langle V, AV\rangle \\ &= q(X_0) + 2t\langle V, AX_0 - B\rangle + t^2\langle V, AV\rangle. \end{aligned} \tag{6.9}$$

Since A is positive definite, the coefficient of t^2 is positive and this quadratic will have an absolute minimum when $dq/dt = 0$. Computing this derivative we have

$$\frac{d[q(X_0 + tV)]}{dt} = 2\langle V, AX_0 - B\rangle + 2t\langle V, AV\rangle = 0 \quad \text{only if}$$

$$t = \hat{t} = -\frac{\langle V, AX_0 - B\rangle}{\langle V, AV\rangle} = \frac{\langle V, B - AX_0\rangle}{\langle V, AV\rangle}.$$

From Equation (6.9) we see that the minimum value of q, on the ray $X_0 + tV$, is

$$\begin{aligned} q(X_0 + \hat{t}V) &= q(X_0) + 2\frac{\langle V, B - AX_0\rangle}{\langle V, AV\rangle}\langle V, AX_0 - B\rangle \\ &\quad + \frac{\langle V, B - AX_0\rangle^2}{\langle V, AV\rangle^2}\langle V, AV\rangle = q(x_0) - \frac{\langle V, B - AX_0\rangle^2}{\langle V, AV\rangle}. \end{aligned}$$

Thus, in moving from X_0 to $X_0 + \hat{t}V$, we obtain a reduction in the value of q

unless the search direction V is orthogonal to the residual $R_0 = B - AX_0$. It follows that if $B - AX_0 \neq 0$, then X_0 does not minimize q. If $B - AX_0 = 0$, then there is no ray $X_0 + tV$ on which q takes a smaller value than $q(X_0)$; that is, q is minimum at X_0. ■

Let us consider using Theorem 6.10 as the basis for an iterative method for solving the linear system $AX = B$. The plan is to start with an initial approximation X_0 and an initial search direction V_0. We then move to a new approximation $X_1 = X_0 + \hat{t}_0 V_0$ and repeat. Thus we propose to search for the minimum value of q by minimizing q along a sequence of rays.

Before we can implement this search we must decide how to choose the search directions. From calculus we know that the most rapid decrease in q is in the direction of the negative of the gradient of q. Since

$$\frac{\partial q(X)}{\partial x_i} = 2\sum_{j=1}^{n} x_{ij}x_j - 2b_i,$$

it follows that

$$\text{grad}(q) = \begin{bmatrix} \vdots \\ \dfrac{\partial q(X)}{\partial x_i} \\ \vdots \end{bmatrix} = 2AX - 2B = -2(B - AX).$$

Thus the optimum search direction, from X_0, is in the direction of the residual $R_0 = B - AX_0$.

These observations are the basis for an iterative method known as the **method of steepest descent**, described in Algorithm (6.10).

METHOD OF STEEPEST DESCENT

1. INPUT: X_0, A, B, m
2. For $k = 0, 1, 2, \ldots, m - 1$
3. $\quad V_k = B - AX_k$
4. $\quad t_k = \langle V_k, V_k\rangle / \langle V_k, AV_k\rangle$ (6.10)
5. $\quad X_{k+1} = X_k + t_k V_k$
6. OUTPUT: X_m

It can be shown (Exercise 2) that the method of steepest descent is globally convergent. Unfortunately, the convergence may be very slow. For example, suppose that the level curves of $q(X)$ are long, thin ellipses and the minimum is at the low part of a relatively flat valley with very steep sides. The search directions used in Algorithm (6.10) would then take you back and

FIGURE *6.10* ■ THE METHOD OF STEEPEST DESCENT

forth across the valley with little progress down the valley, as shown in Figure 6.10.

As we noted in the proof of Theorem 6.10, we are free to choose the search directions in any way we like as long as they are not orthogonal to the residuals. Let us consider the possibility of choosing independent search directions $V_1, V_2, \ldots, V_n$ so that

$$
\begin{aligned}
&X_1 = \min\{q(X): X \in \operatorname{Span}\{V_1\}\}, \quad X_2 = \min\{q(X): X \in \operatorname{Span}\{V_1, V_2\}\},\\
&X_3 = \min\{q(X): X \in \operatorname{Span}\{V_1, V_2, V_3\}\}, \quad \ldots,\\
&X_n = \min\{q(X): X \in \operatorname{Span}\{V_1, V_2, \ldots, V_n\}\}.
\end{aligned}
$$

If we could do this, with the search directions $\{V_1, V_2, \ldots, V_n\}$ forming a basis for $\mathscr{R}^n$, then we could be certain that $q(X_n)$ was the global minimum of $q(X)$ and, by Theorem 6.10, the solution of the linear system $AX = B$. In this case the iterative method would converge after only n steps.

Let $P_k = [V_1, V_2, \ldots, V_k] \varepsilon \mathscr{R}^{n\times k}$ be the matrix of the first k search directions. If X_{k-1} is in $\operatorname{Span}\{V_1, V_2, \ldots, V_{k-1}\}$, then $X_{k-1} = P_{k-1}Y$ for some Y in $\mathscr{R}^{k-1}$. If we set $X = X_{k-1} + tV_k = P_{k-1}Y + tV_k$, then, from Equation (6.9), we have

$$
\begin{aligned}
q(X) &= q(P_{k-1}Y) + 2t\langle V_k, AP_{k-1}Y - B\rangle - t^2\langle V_k, AV_k\rangle\\
&= q(P_{k-1}Y) + 2t\langle V_k, AP_{k-1}Y\rangle + t^2\langle V_k, AV_k\rangle - 2t\langle V_k, B\rangle,
\end{aligned}
$$

and we want to choose t to minimize $q(X)$ on the ray $X_{k-1} + tV_k$. The calculation will be significantly simplified if we choose the V_i so that the second term in the last equation is zero, that is, if $\langle V_k, AP_{k-1}Y\rangle = 0$ for $k = 1, 2, \ldots, n$. In that case the minimum would be obtained for

$$
t_k = \frac{\langle V_k, B\rangle}{\langle V_k, AV_k\rangle} = \frac{\langle V_k, B\rangle}{\langle V_k, AV_k\rangle} - \frac{\langle V_k, AP_{k-1}Y\rangle}{\langle V_k, AV_k\rangle} = \frac{\langle V_k, R_{k-1}\rangle}{\langle V_k, AV_k\rangle},
$$

which is consistent with the result of Theorem 6.10.

■ DEFINITION *6.4* ■

Let A be a positive definite matrix. The vectors $\{V_1, V_2, \ldots, V_n\}$ are said to be **A-conjugate**, or **A-orthogonal**, if $\langle V_i, AV_j\rangle = 0$ for $i \neq j$. If the vectors are scaled so that $\langle V_i, AV_j\rangle = \delta_{ij}$, then they are said to be **A-orthonormal**.

The next theorem summarizes the discussion preceding Definition 6.4. Note that it asserts that a certain iterative method converges in a finite number of steps.

THEOREM 6.11

Let A be a positive definite symmetric matrix and assume that the vectors $\{V_1, V_2, \ldots, V_n\}$ form an A-conjugate set of search directions. Then, for arbitrary X_0 and

$$X_i = X_{i-1} + \frac{\langle V_k, B - AX_{i-1}\rangle}{\langle V_k, AV_k\rangle} V_k, \quad i = 1, 2, \ldots, n,$$

it follows that X_n is the global minimum of the quadratic function $q(X) = \langle X, AX\rangle - 2\langle X, B\rangle$ and hence that the solution of the system $AX = B$ is $X = X_n$. ■

The conjugate gradient method of Hestenes and Stiefel is a particular conjugate direction method in which the search direction V_k is chosen to be as close as possible to the residual vector R_{k-1}, subject to the constraint that it be A-conjugate to $\{V_1, V_2, \ldots, V_{k-1}\}$. That is,

$$\|V_k - R_{k-1}\|_2 = \min\{\|V - R_{k-1}\|_2 : V \in \text{Span}\{AV_1, AV_2, \ldots, AV_{k-1}\}^{\perp}\}.$$

In this case it turns out that the residuals R_k form an orthogonal set of vectors in the ordinary sense. That is, $\langle R_j, R_i\rangle = 0$ for $i \neq j$.

It can be shown (see Theorem 6.12) that the proper choice of V_k is

$$V_k = R_{k-1} + b_k V_{k-1} \quad \text{where} \quad b_k = -\frac{\langle V_{k-1}, AR_{k-1}\rangle}{\langle V_{k-1}, AV_{k-1}\rangle}.$$

The algorithm describing the conjugate gradient method follows. Note that in coding this algorithm it would be unnecessary to use subscripts; the new iterates can overwrite the previous ones.

CONJUGATE GRADIENT METHOD

1. INPUT: $X_0, A, B, m, \varepsilon$
2. $R_0 = B - AX_0$
3. $V_0 = R_0$
4. For $k = 0, 1, 2, \ldots, m-1$
5.   If $V_k = 0$, STOP
6. $t_k = \langle R_k, R_k\rangle / \langle V_k, AV_k\rangle$
7. $X_{k+1} = X_k + t_k V_k$

(6.11)

8. $R_{k+1} = R_k - t_k A V_k$
9. If $\|R_{k+1}\|_2 < \varepsilon$, STOP
10. $b_k = \langle R_{k+1}, R_{k+1}\rangle / \langle R_k, R_k\rangle$
11. $V_{k+1} = R_{k+1} + b_k V_k$
12. OUTPUT: $k + 1$, $X = X_{k+1}$, R_{k+1}.

■ **EXAMPLE 4**

Use the method of conjugate gradients to solve the system

$$\begin{bmatrix} 4 & -1 & 1 \\ -1 & 4 & -1 \\ 1 & -1 & 4 \end{bmatrix} X = \begin{bmatrix} 1 \\ 2 \\ 3 \end{bmatrix} \quad \text{for which the true solution is} \quad X_t = \frac{1}{9}\begin{bmatrix} 2 \\ 7 \\ 8 \end{bmatrix}.$$

SOLUTION The computations in the following solution were done with MATALG; you could also do them with MATLAB. The variables c, d, and Z are introduced to avoid recalculations. We take as our initial approximation $X_0 = \begin{bmatrix} 1 \\ 1 \\ 1 \end{bmatrix}$, for which $R_0 = B - AX_0 = \begin{bmatrix} -3 \\ 0 \\ -1 \end{bmatrix}$. We take as our first search direction $V_0 = R_0$, compute $c = R_0^T R_0 = 10$, and $Z = AV_0 = \begin{bmatrix} -13 \\ 4 \\ -7 \end{bmatrix}$. Next we compute $t_1 = c/(V_0^T Z) = 0.217391$ and the new approximation $X_1 = X_0 + t_1 V = \begin{bmatrix} 0.347826 \\ 1 \\ 0.782609 \end{bmatrix}$. The residual for this new approximation is $R_1 = B - AX_1 = B - A(X_0 + t_1 V_0) = R_0 - t_1 A V_0 = R_0 - t_1 Z = \begin{bmatrix} -0.173913 \\ -0.869565 \\ 0.521739 \end{bmatrix}$. We now compute $d = R_1^T R_1 = 1.05860$ and then the new search direction $V_1 = R_1 + \dfrac{d}{c} V_0 = \begin{bmatrix} -0.491494 \\ -0.869565 \\ 0.415879 \end{bmatrix}$. This completes one step of Algorithm (6.11).

To start the second iteration we set $c = d$ and compute $Z = AV_1 = \begin{bmatrix} -0.680530 \\ -3.40265 \\ 2.04159 \end{bmatrix}$, $t_2 = c/(V_1^T Z) = 0.255556$, and $X_2 = X_1 + t_2 V_1 =$

$\begin{bmatrix} 0.222222 \\ 0.777778 \\ 0.888889 \end{bmatrix}$. For this new approximation the residual is $R_2 = R_1 - t_2 Z =$ $\begin{bmatrix} -0.0298023 \times 10^{-7} \\ -0.0596046 \times 10^{-7} \\ 0 \end{bmatrix} \approx 0$. Thus the sequence has essentially converged after only two iterations.

This example illustrates a very important fact about the conjugate gradient method; frequently the iteration will converge in fewer than the n steps guaranteed by Theorem 6.11.

Our final theorem will establish that the residuals in the conjugate gradient Algorithm (6.11) are orthogonal in the usual sense. In addition the proof establishes some other useful properties of the conjugate gradient method and gives some insight into how you determine the constants b_k.

THEOREM 6.12

If the search directions $V_0, V_1, \ldots, V_n$ in Algorithm (6.11) are all nonzero, then the residuals $R_i = B - AX_i$, $i = 0, 1, 2, \ldots, n$ form an orthogonal set of nonzero vectors.

PROOF We shall now show, by induction on n, that

(a) $\langle R_n, V_i \rangle = 0$: $i = 0, 1, 2, \ldots, n-1$,

(b) $\langle R_i, R_i \rangle = \langle R_i, V_i \rangle$: $i = 0, 1, 2, \ldots, n$,

(c) $\langle V_n, AV_i \rangle = 0$: $i = 0, 1, 2, \ldots, n-1$,

(d) $R_i = B - AX_i$: $i = 0, 1, 2, \ldots, n$,

(e) $\langle R_n, R_i \rangle = 0$: $i = 0, 1, 2, \ldots, n-1$,

(f) $R_i \neq 0$: $i = 0, 1, 2, \ldots, n$.

For $m = 0$ we assume that $V_0 = R_0 \neq 0$ and note that assertions (a), (c), and (e) assert nothing. The remaining assertions follow at once from Algorithm (6.11).

Assume now that the six assertions are true for $n = m$. Then

$$\begin{aligned} \langle R_{m+1}, V_m \rangle &= \langle R_m - t_m A V_m, V_m \rangle = \langle R_m, V_m \rangle - t_m \langle A V_m, V_m \rangle \\ &= \langle R_m, V_m \rangle - \frac{\langle R_m, R_m \rangle}{\langle V_m, A V_m \rangle} \langle A V_m, V_m \rangle \\ &= \langle R_m, V_m \rangle - \langle R_m, R_m \rangle = 0, \end{aligned}$$

where the last equality follows from part (b) of the inductive hypothesis.

Now, for $i = 0, 1, 2, \ldots, m - 1$, we have

$$\begin{aligned}\langle R_{m+1}, V_i\rangle &= \langle R_m - t_m A V_m, V_i\rangle = \langle R_m, V_i\rangle - t_m\langle Av_m, V_i\rangle \\ &= \langle R_m, V_i\rangle - t_m\langle V_m, AV_i\rangle = 0 - 0 = 0,\end{aligned}$$

where the last equality uses parts (a) and (c) of the inductive hypothesis. Thus part (a) holds for $m + 1$.

Now, from assertion (a) for $m + 1$, we have

$$\begin{aligned}\langle R_{m+1}, V_{m+1}\rangle &= \langle R_{m+1}, R_{m+1} + b_m V_m\rangle = \langle R_{m+1}, R_{m+1}\rangle + b_m\langle R_{m+1}, V_m\rangle \\ &= \langle R_{m+1}, R_{m+1}\rangle + b_m 0 = \langle R_{m+1}, R_{m+1}\rangle,\end{aligned}$$

independent of the choice of b_m. This establishes assertion (b) for $m + 1$.

To establish part (c) for $m + 1$, we compute, for $i = 0, 1, 2, \ldots, m$, using steps 8 and 11 of Algorithm (6.11),

$$\begin{aligned}\langle V_{m+1}, AV_i\rangle &= \langle R_{m+1} + b_m V_m, AV_i\rangle = \langle R_{m+1}, AV_i\rangle + b_m\langle V_m, AV_i\rangle \\ &= t_i^{-1}\langle R_{m+1}, R_i - R_{i+1}\rangle + b_m\langle V_m, AV_i\rangle \\ &= t_i^{-1}\langle R_{m+1}, (V_i - b_{i-1}V_{i-1}) - (V_{i+1} - b_i V_i)\rangle + b_m\langle V_m, AV_i\rangle \\ &= t_i^{-1}[\langle R_{m+1}, V_i\rangle - b_{i-1}\langle R_{m+1}, V_{i-1}\rangle - \langle R_{m+1}, V_{i+1}\rangle \\ &\quad + b_i\langle R_{m+1}, V_i\rangle] + b_m\langle V_m, AV_i\rangle.\end{aligned}$$

If $i < m$, each term in the last expression is zero by parts (a) and (c) of the inductive hypothesis (independent of the choice of the b_i). If $i = m$, we have, as in the case $i < m$,

$$\begin{aligned}\langle V_{m+1}, AV_m\rangle &= t_m^{-1}[\langle R_{m+1}, V_m\rangle - b_{m-1}\langle R_{m+1}, V_{m-1}\rangle - \langle R_{m+1}, V_{m+1}\rangle \\ &\quad + b_m\langle R_{m+1}, V_m\rangle] + b_m\langle V_m, AV_m\rangle.\end{aligned}$$

By parts (a) and (b), for $m + 1$, this reduces to

$$\begin{aligned}\langle V_{m+1}, AV_m\rangle &= -t_m^{-1}\langle R_{m+1}, V_{m+1}\rangle + b_m\langle V_m, AV_m\rangle \\ &= -\frac{\langle V_m, AV_m\rangle}{\langle R_m, R_m\rangle}\langle R_{m+1}, V_{m+1}\rangle + \frac{\langle R_{m+1}, R_{m+1}\rangle}{\langle R_m, R_m\rangle}\langle V_m, AV_m\rangle = 0.\end{aligned}$$

Note that this last equation determines b_m. We have established part (c) for $m + 1$.

Now we have, from the inductive hypothesis and step 8 of Algorithm (6.11),

$$\begin{aligned}B - AX_{m+1} &= B - A(X_m + t_m V_m) = B - AX_m - t_m A V_m \\ &= R_m + (R_{m+1} - R_m) = R_{m+1},\end{aligned}$$

which establishes part (d) for $m + 1$.

For $i \leq m$, we then have

$$\langle R_{m+1}, R_i \rangle = \langle R_{m+1}, V_i - b_{i-1}V_{i-1} \rangle = \langle R_{m+1}, V_i \rangle - b_{i-1}\langle R_{m+1}, V_{i-1} \rangle = 0$$

by part (a) for $m + 1$. Thus the residuals are orthogonal in the usual sense.

Finally, since A is positive definite and $V_{m+1} \neq 0$,

$$0 < \langle V_{m+1}, AV_{m+1} \rangle = \langle R_{m+1} + b_m V_m, AV_{m+1} \rangle = \langle R_{m+1}, AV_{m+1} \rangle + b_m\langle V_m, AV_{m+1} \rangle = \langle R_{m+1}, AV_{m+1} \rangle$$

by part (c) of the inductive hypothesis. It follows that $R_{m+1} \neq 0$. ■

Note that Algorithm (6.11) involves only one matrix-vector multiplication per iteration. For a full $n \times n$ matrix there are thus about n^2 multiplications per step. If the matrix A is large, sparse, and regularly structured, it may be possible to compute this matrix-vector product much more efficiently. In this case the method of conjugate gradients can be quite efficient when compared to Gaussian elimination.

6.5 ■ Exercise Set

1. Show how you could use the Cholesky decomposition of A to determine a set of A-orthonormal directions to use with a conjugate direction method.

2. For the method of steepest descent, show that

$$q(X_k) + \frac{1}{2}B^TA^{-1}B \leq \left(1 - \frac{1}{\text{cond}_2(A)}\right)\left(q(X_{k-1}) + \frac{1}{2}B^TA^{-1}B\right)$$

and hence that the method is globally convergent.

3. Rewrite Algorithm (6.10) without subscripts and redundant calculations.

4. Write a FORTRAN subroutine or a MATLAB m-file to implement Algorithm (6.11).

5. Use the conjugate gradient method to solve the system of Example 2 in Section 6.2.

6. Use the software and the method of conjugate gradients to solve the system

$$\frac{1}{4}\begin{bmatrix} 4 & -1 & -1 & 0 & 0 & 0 \\ -1 & 4 & -1 & -1 & 0 & 0 \\ -1 & -1 & 4 & -1 & -1 & 0 \\ 0 & -1 & -1 & 4 & -1 & -1 \\ 0 & 0 & -1 & -1 & 4 & -1 \\ 0 & 0 & 0 & -1 & -1 & 4 \end{bmatrix} X = \begin{bmatrix} 1 \\ 0 \\ 0 \\ 0 \\ 1 \\ 1 \end{bmatrix}.$$

7. Count the number of multiplications required in one iteration of Algorithm (6.11).

Appendix A

The MATLAB Program

MATLAB, the MATrix LABoratory, is an interactive computing environment for scientific calculation. It supports matrix calculation, complex arithmetic, a programming command set, 2 and 3 dimensional graphics, and is expandable to meet the needs of various users. The basic data element in MATLAB is a matrix that does not require dimensioning. The program is easy to learn and to use and has extensive on line help and demonstrations. The professional version of the program is very powerful and quite expensive. It may be available to you on the main frame computer at your base of operations. The inexpensive student version [SEM] , which is available for both Macintosh and IBM PC's, is more than adequate for the needs of this text.

To see a list of MATLAB commands type "help" followed by a RETURN or ENTER."

Any MATLAB command must be followed by a RETURN or ENTER in order to activate it. Several commands can be typed on the same line separated by commas or semicolons.

To get help on a particular command type "help command".

To run a MATLAB demonstration type "demo".

To exit MATLAB type "exit" or "quit".

<u>DATA ENTRY and DISPLAY</u>:

The simplest way to enter a matrix is to type its entries inside of square brackets with the entries separated by spaces or commas and the rows separated by semicolons or returns. Note that MATLAB is case sensitive (A ≠ a), so you need to be consistent in the names you give to variables. Typing either

A = [1 2 3
4 5 6
7 8 9]   or A = [1 2 3; 4 5 6; 7 8 9] will enter $A = \begin{bmatrix} 1 & 2 & 3 \\ 4 & 5 & 6 \\ 7 & 8 & 9 \end{bmatrix}$.

Once the matrix *A* has been entered into the program it, or its parts, may be displayed:

> Type "A" to see the entire matrix.
> Type "A(i,j)" to see the entry in row *i* and column *j* .
> Type "A(i,:)" to see row *i* of *A*. "v = A(1,:)" would construct a row vector **v** from the first row of *A*.
> Type "A(:,j)" to see column *j* of *A*. "w = A(:,3)" constructs a column vector **w** from the third column of *A*.
> Type "A(2:3,[1 3])" to see the submatrix obtained by deleting row 1 and column 2 of *A*.

If the data to be entered contains some complex numbers, then they are to be entered in the form "a + b*i" where "i = sqrt(-1)". If complex numbers are involved, do not use i as an index in a loop.

If the column vectors **u**, **v**, and **w** are already defined, then the command "A = [u v w]" builds a matrix whose columns are the given vectors. The command "B = [u'; v'; w'] "builds a matrix whose rows are $\mathbf{u}' = \mathbf{u}^T$, $\mathbf{v}^T$, and $\mathbf{w}^T$. The augmented matrix of the system $AX = B$ is constructed by the command " M = [A, B]".

Some special matrices are easily constructed:

> "eye(n)" is the $n \times n$ identity matrix and eye(A) is an identity matrix of the same size as *A*.
> "zeros(n)" is the $n \times n$ zero matrix.
> "ones(n)" is the $n \times n$ matrix of all ones.
> "rand(n)" is an $n \times n$ matrix with random entries.
> "hilb(n)" is the $n \times n$ Hilbert matrix.
> "x = [a:h:b]" = [a , a+h, a+2h, ..., a+nh] where n is as large as possible so that $a + nh \le b$.

The display format can be changed to suit the user. The choice of display format does not effect the computations, all of which are done in double precision. The choices are:

"format short "(the default) displays 4 decimal places.
"format long" displays 16 decimal places.
"format short e" uses floating point notation with 5 significant figures.
"format long e" uses floating point notation with 16 significant digits.

If you do not wish to see the result of a computation displayed, then end the command line with a semicolon. This is frequently convenient when all you are interested in is the final result of a sequence of caculations.

If you wish to keep a record of your MATLAB session on your disk, then type "diary filename". Where "filename" is a name you choose for the diary file. This file can be edited, annotated, and printed after you exit MATLAB.

The currently active variables can be saved to a file, for latter use by MATLAB, using the command "save filename". The command "load filename" will enter the variables saved from an earlier session. Do not try to use the "save" command to construct a diary file.

MATRIX OPERATIONS:

MATLAB will perform all of the standard matrix operations. If a requested operation is undefined, because of size incompatibilities, the program will give appropriate explanations.

The basic matrix operations are:

+, -, *, ^, ' for addition, subtraction, multiplication, exponentiation, and transpose.

In addition the following element-wise operations are available:

$A .* B = (a_{ij}*b_{ij})$, $A ./ B = (a_{ij}/b_{ij})$, and A $.\wedge k = (a_{ij}^k)$.

These vectorized operations can be very useful in avoiding loops and in preparing data for graphs.

Note that the operation A + 2 is treated as A .+ 2, that is 2 is added to every element of the matrix A. In order to compute $3A^2 + 5A - 6I$ you should type "B = 3*A^2 + 5*A - 6*eye(A)".

MATRIX FUNCTIONS:

A full list of the available matrix functions can be obtained by typing "help" There follows a short list of those functions which will be useful in this text. For additional information type help followed by the name of the command you are interested in.

"inv(A)" = inverse of the matrix A.

"rref"(A) = row reduced echelon form of the matrix A.

"A\B" generates the solution of the linear system $AX = B$. The results may be inaccurate if A is singular.

"det(A)" = determinant of A.

"trace(A)" = trace of A.

"rank(A)" = rank of A.

"null(A)" generates a basis for the null space of A.

"norm(A)" = norm(A, 2) = $\|A\|_2$. (A may be a matrix or a vector)

"norm(A, 1)" = $\|A\|_1$. (A may be either a matrix or a vector)

"norm(A, inf)" = $\|A\|_\infty$. (A may be either a matrix or a vector)

"cond(A)" = $\|A\|_2 \, \|A^{-1}\|_2$ is the condition number of A using the 2-norm.

"lu(A)" generates the LU factorization of the matrix A.
"[L, U, P] = lu(A)" generates a permutation matrix P , an upper-triangular matrix U and a unit lower -triangular matrix L such that $PA = LU$.

"chol(A)" computes the Cholesky factorization of the positive definite matrix A. If A is positive definite, then "R = chol(A)" generates an upper triangular matrix R such that $A = R^T R$.

"v = poly(A)" computes the coefficients of the characteristic polynomial of A and stores them in the vector **v** with the coefficient of the highest power first.

polyval(v, s) is the value of the polynomial, with coefficients in **v**, at the scalar s. "polyval(v,A)" evaluates the polynomial at each entry of the matrix A.

"polyvalm(v, A)" is the value of the polynomial, with coefficients in **v**, at the matrix *A*.

"roots(v)" generates the roots of the polynomial whose coefficients are in the vector **v**. "roots(poly(A))" would generate the eigenvalues of *A*.

"eig(A)" = column vector of the eigenvalues of A. "[P, D] = eig(A)" generates a matrix *P* whose columns are the eigenvectors of *A* and a diagonal matrix *D* whose entries are the eigenvalues of *A*; $P^{-1}AP = D$. The results may be inaccurate if cond(P) is large.

"svd(A)" generates the singular value decomposition of *A*. "[U, S, V] = svd(A)" produces a diagonal matrix *S* and orthogonal (unitary) matrices *U* and *V* so that $A = USV^T$.

"qr(A)" generates the QR factorization of *A*. "[Q, R] = qr(A)" produces an upper-triangular matrix *R* and an orthogonal (unitary) matrix *Q* such that $A = QR$. Note that the diagonal entries of *R* may not be positive.

SCALAR FUNCTIONS

The usual scalar functions are available. Among them are: sqrt, abs, conj, sin, cos, tan asin, acos, atan, exp, log, log10, sinh, cosh, tanh, asinh, acosh, atanh, gamma, erf Note carefully, that if these functions are applied to a matrix or a vector, then the function is evaluated at every entry of the array. This can be very useful in preparing data for graphs.

GRAPHS

MATLAB has several ways of displaying information graphically. It will be helpful to run demos 2 and 9 to see the various possibilities. If you are running MATLAB on a mainframe, using a remote terminal, you should type "terminal" and identify your terminal type for MATLAB.

The basic Cartesian plot of a function $y = f(x)$ is obtained using the "plot" command. The first step is to construct two data vectors; the first contains the x coordinates of the points to be plotted and the second contains the y coordinates (the functional values).

As a simple example let us plot $y = e^{-3x} \sin 2x$ for $0 \leq x \leq 5$. The command "x = 0 : .1: 5;" constructs the vector **x** = [0 .1 .2 .3 ... 4.9 5], note that the semicolon suppresses the display of this long uninteresting vector. The command "y = (exp(-3*x)).*sin(2*x)" will construct and display a vector whose entries are the values of the function at the entries of the vector **x**; again the semicolon could be used to suppress the display. The command "plot(x,y)" will now cause the graph to be drawn and displayed. A title and axis labels can be added if desired using the commands, "title", "xlabel", and "ylabel". For various options regarding line type and point markings type "help plot". Polar plots, parametric plots, and various log plots are also available. It is also possible to plot several curves on the same graph using different line types for each graph.

MATLAB can also plot surfaces representing functions $z = f(x, y)$. The procedure is to define a rectangle in the xy plane that will be the domain of the function . As an example let us plot the function $z = \sin(x)\cos(y)$ for $-7 \leq x, y \leq 8$. The commands

```
x = -7: .5 : 8;  y = x;  [x1, y1] = meshdom(x, y);
```

will construct the mesh points for the domain and then the command "z = sin(x1).*cos(y1); " will generate a matrix of functional values for the function . Finally, the command "mesh(z)" will generate the surface. The graphics will not be part of the diary file. You can use the "print" command or do a "print screen" on the PC or use the MAC clipboard to get a hard copy.

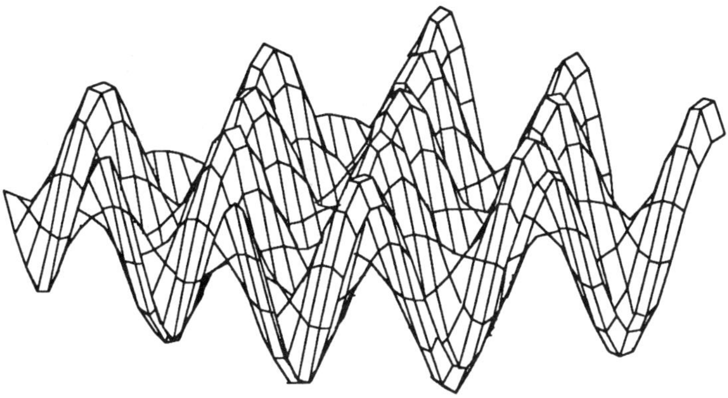

PROGRAMMING

MATLAB has significant programming capabilities and can be used instead of FORTAN or PASCAL to carry out complicated calculations, especially those which are iterative in nature. Looping is accomplished using "for", or "while" loops. Some simple examples will suffice to illustrate this capability. Avoid using *i* as an index.

1.

```
for j = 1:10
    w(j) = j^3 + 3*j
end
```

This code will construct the vector w = [4 14 36 76 140 234 364 536 756 1030]

2.

```
v = [ -1 2 3 5 7],
A = zeros(5);
for j = 1: 5

  for k = 1: 5

     A(j,k) = v(j)^(5-k);
  end

end
A
```

This code will generate a 5 × 5 Vandermonde matrix. Note that loops can be nested, but that there must be an "end" for each "for" statement.
The semicolon at the end of line 4 prevents the printing of 25 5 × 5 matrices. The final command prints the desired matrix.

3.

```
E = zeros(A);
F= eye(A);
k = 1;
while norm((E+F)-E, inf) > 0
    E = E + F;
    F = A*F/k;
    k = k+1;
end
k, E
```

This code will compute the exponential of the matrix A;

$$e^A = \sum_{k=0}^{\infty} \frac{A^k}{k!} = I + A + \frac{A^2}{2!} + \frac{A^3}{3!} + \cdots$$

The summation continues until the term added makes no difference.

M-FILES

MATLAB is normally used in a command-driven mode; when single-line commands are entered, MATLAB processes them immediately and displays the results. MATLAB is also capable of executing sequences of commands that are stored in files . These files are called m-files because they have a file type of ".m". These files can be constructed and edited using an editor or a word processor ; on a MAC they can be constructed without leaving MATLAB. Once they have been stored, they are available as a command. If, for example, the commands in the last example are stored in a file named exponA.m, then the command "exponA" will cause the program to compute the exponential of the matrix called A. Use of m-files makes it much easier to debug long programs and takes the burden off typing skill.

SAMPLE MATLAB DIARY

(User input in boldface, comments in parentheses or after %)

diary sample.dat (opens transcript or diary file named "sample.dat")

A=[17 -8 -12 14; 46 -22 -35 41; -2 1 4 -4; 4 -2 -2 3]

```
A =

   17   -8  -12   14
   46  -22  -35   41
   -2    1    4   -4
    4   -2   -2    3
```

det(A)

```
ans =

   -1
```

B = inv(A)

```
B =

  17.0000  -8.0000  -12.0000  14.0000
  42.0000 -20.0000  -31.0000  36.0000
  10.0000  -5.0000   -8.0000  11.0000
  12.0000  -6.0000  -10.0000  13.0000
```

K =[1 2 3 4]'

```
K =

   1
   2
   3
   4
```

X = A\K % computes the solution of the system AX = K

```
X =

  21.0000
  53.0000
  20.0000
  22.0000
```

rref([A K]) (row reduced echelon form of the augmented matrix; this will give better results if A is singular, or close to singular)

```
ans =

   1   0   0   0  21
   0   1   0   0  53
   0   0   1   0  20
   0   0   0   1  22
```

v = poly(A) %compute the characteristic polynomial of A and store coefficients in v.

v =

```
   1.0000  -2.0000  -0.0000   2.0000  -1.0000
```

($c(x) = \det(xI - A) = x^4 - 2x^3 + 2x - 1$)

polyvalm(v,A) % check the Cayley-Hamilton Theorem; $c(A) = 0$.

ans =

```
  1.0e-11 *

  -0.0442  -0.0711  -0.1080   0.1414
  -0.1421  -0.1891  -0.2728   0.3627
   0.0902  -0.0338  -0.0712   0.0739
   0.0469  -0.0462  -0.0831   0.0961
```

(this is the zero matrix!!! To see why this is not exactly *0*, change to long format and look at v more closely.)

A^3 -A^2 -A+eye(A)

ans =

```
   0   0   0   0
   0   0   0   0
   0   0   0   0
   0   0   0   0
```

(the minimum poly of A is $x^3 - x^2 - x + 1$)

[P, D] = eig(A)

P =

```
   0.3357   0.0000  -0.0000  -0.4586
   0.9231   0.2673   0.2673  -0.8762
   0.0839  -0.8018  -0.8018  -0.1234
   0.1678  -0.5345  -0.5345  -0.0823
```

D =

```
  -1.0000        0        0        0
        0   1.0000        0        0
        0        0   1.0000        0
        0        0        0   1.0000
```

inv(P)*A*P

```
ans =

  -1.0000   0.0000   0.0000  -0.0000
  -0.0000   1.0000  -0.0000   0.0000
   0.0000   0.0000   1.0000  -0.0000
  -0.0000  -0.0000  -0.0000   1.0000
```

cond(P)

```
ans =

  7.3724e+07
```

(The eigenvalue problem for this matrix is ill-conditioned)

A(3,3) = 5 %change the (3,3) entry of A to 5

```
A =

   17   -8  -12   14
   46  -22  -35   41
   -2    1    5   -4
    4   -2   -2    3
```

eig(A)

```
ans =

  -0.8160 + 1.1232i
  -0.8160 - 1.1232i
   3.6320
   1.0000
```

% note that the new *A* has two complex eigenvalues.

x=[1 1 1 1]'

```
x =

     1
     1
     1
     1
```

```
for j = 1:10
        z=A*x;                      %10 iterations of the power method
        x = z/norm(z,inf);
end
x

x =

   0.3286
   1.0000
  -0.3339
  -0.0287

A*x./x

ans =

   3.6241
   3.6242                          % x is an approximate eigenvector for the
   3.6295                            eigenvalue 3.62.
   3.6270

E = zeros(A); F = eye(A);k=1;
while norm((E+F)-E,inf)>0
                E = E+F;
                F = A*F/k;
                k = k+1;
end, k, E

k =

   31
E =

  41.2433  -19.2625  -58.3108   47.0301      (This is the sum of the first
 116.1248  -55.3441 -176.9186  143.5393        31 terms of the series
 -30.2550   15.1275   57.1689  -45.1076        defining e^A.)
  -1.0996    0.5498    3.9726   -2.1794
```

v = poly(A)

```
v =

   1.0000  -3.0000  -2.0000  -3.0000   7.0000
```

x=[-3:.1:6];
y=polyval(v,x);
plot(x,y,x,zeros(x)) (plots characteristic polynomial and the *x* axis)

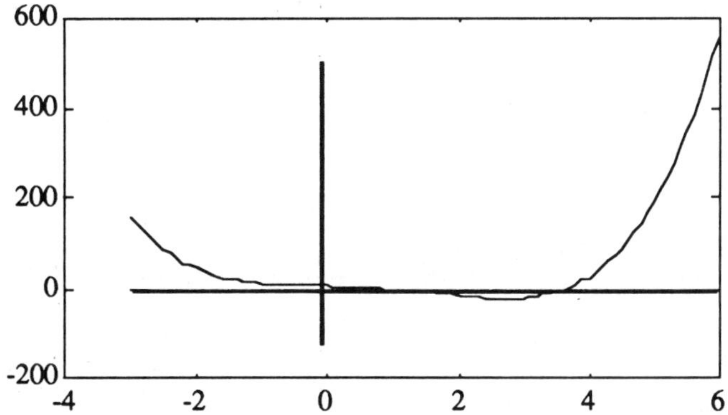

x=[-2: .1 : 4];
y=polyval(v,x);
plot(x,y,x,zeros(x))

(decrease the range of *x* so that the range of *y* is smaller)

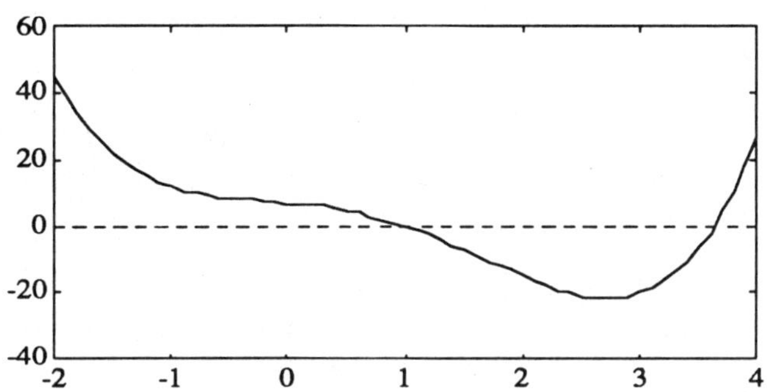

Appendix B

The MATALG Program

This help file describes MATALG, an interactive matrix algebra program. This program was developed by

John Burkardt and Charles G. Cullen
Department of Mathematics and Statistics,
University of Pittsburgh,
Pittsburgh, Pennsylvania, 15260
(E-mail to CULLEN@vms.cis.pitt.edu)

 This program is not copy protected. It is distributed free of charge with

INTRODUCTION TO NUMERICAL LINEAR ALGEBRA
Charles G. Cullen. PWS-Kent Publishing Company, Boston, Massachusetts, 1994.

A Macintosh version of this program is available from the author for the modest fee of $5.00 to cover postage and handling; please specify what type of MAC you are using.

Development of these programs was partially supported by a courseware development grant from the College of General Studies of the University of Pittsburgh.

This file contains information on the background and purpose of MATALG, how to run the program, what the commands are, what legal formulas and symbols are, and instructions for running the program on an IBM PC or compatible, or on a MACINTOSH.

This program can perform most common calculations with real scalars, vectors, and matrices. You name and dimension the variables, give them values, and define formulas involving them; the program interprets and evaluates those formulas. The program can be used as a simple programming language for carrying out matrix calculations. For example, the following commands declare A to be a 3 by 3 matrix, give it a value, and request the determinant. Then the value of A(3,3) is changed, the determinant is re-evaluated, and the matrix is inverted.

```
E A,3,3            ( Enter matrix A, 3 rows by 3 columns)
1,2,3              ( Entries are entered one row at a time,
4,5,6                separated by either commas or spaces.)
7 8 9
=DET(A)            ( Compute the determinant of A, which is 0!
                     Because there's nothing on the left hand side of
                     the = sign, we are implying that we want to
                     compute, but not save, the result.)
C A,3,3, 0.0       ( Change A(3,3) to 0.0)
=DET(A)            ( New value of determinant is 27)
B = INV(A)         ( Invert the new matrix and call it B)
```

The program will print out prompts, ask for missing input, help you decide what to do next, and show the intermediate results as you go along. These messages were not included in the example above.

Formulas can be more complicated, involving several functions. In order to compute the eigenvalues of A^TA in the above example, you could type the single line "= EVAL(TRANS(A)*A)"

RUNNING THE PROGRAM

IBM PC: The system must be booted and running already. Insert the MATALG disk in the disk drive and type 'MATALG'. When not running the program, the command 'DIR' will give a directory of all the files on the disk.

MACINTOSH: The system must be booted and running already. Insert the disk containing MATALG. Move the mouse to the MATALG disk and click it twice. All of the files in the disk will be displayed. Move the mouse to the 'MATALG APL' file and click twice.

COMMANDS

A command to MATALG is usually a single letter. The list of available commands follow. Most must be followed by one or more arguments. These arguments are listed following the command letter. You can give the full command on a single line. You do not have to give the full command on a single line, but can specify part of the command. In that case, MATALG will prompt you for the rest of the information it needs.

Command and Arguments		Short Definition
C	name, row, column, value	Change the value of a variable.
D	filename	Open /close a transcript file.
E	name, nrow, ncolumn	Enter a variable of the given size.
I		Initialize,
K	name	Clears specified variables
L	name, lo, hi, nstep	Tabulate a formula.
M	LU/QR/SVD, name	Factor matrix: LU, QR, or SVD
N		User notes. Terminated by a period in column 1.
P	name	Characteristic polynomial.
Q		Quit.
R	steps	Repeated evaluation of X = F(X).
S	name	Save formula value in a variable.
T	name/ALL/blank	Type out the given variables.
U	name	Partition a matrix.
V		Re-evaluate the last formula
W	filename, name	Write variable to file.
X	filename	Read variable from file.
variable = formula		Set variable to value of formula.
? Display extensive help.		

A command and its arguments may be entered on the same line, separated by commas or spaces, or each argument may be given on its own line. The program will prompt you for the next argument it needs if you do not supply it. Thus, "E A 3, 4" is the concise way to prepare to enter a matrix named A which has 3 rows and 4 columns. But if you just type "E" the program will ask you for the name of the variable, and then the number of rows and columns. Variable names may be up to 10 characters long, and may only consist of letters and numbers.

All variables are stored as real numbers inside the program, and the usual round-off problems can occur. When you are inputting a number to the program, you may use integer, rational, or real form. The following are all legal input values for a variable: 2, 3/4, 17.23

C (name, row, column, value) = CHANGE A VARIABLE

Changes the given entry of the variable to the new value. If the variable is a scalar or vector, you still must specify both row and column. For a scalar, the row and column should be 1.

"C Hilbert 3,4, 5.0" will change entry Hilbert(3,4) to 5. The program will echo print the variable after its value has been changed using the C command. The command "Hilbert(3,4) = 5" will do the same thing, but the new value will not be printed.

D (filename) = OPEN OR CLOSE TRANSCRIPT FILE

Requests that a disk file be opened with the given name, in which a transcript of the session will be stored. A second "D" command (with no argument) will close the file. File names may be up to 30 characters long. Transcript files can not be reopened from the program.

"D B:RECORD.DAT" would store the information in the file RECORD.DAT on the disk in drive B (or on the MAC disk named B). "D RECORD.DAT" would store the information in the file RECORD.DAT on the disk in the currently active drive.

E (variable, rows, columns) = ENTER A VARIABLE

Sets up a variable of the given name and size. The number of rows and columns must be no more than 12. You must use this command first, to set up variables, before entering their values and using them in a formula. The program will immediately request that you set the value of the variable. If the variable is a scalar (1 row and 1 column), then its value may be included on the same line. The program will echo print the variable when you have finished entering it. Complex numbers can not be entered.

"E SQUARE 5,4" would declare a matrix named SQUARE of five rows and four columns.
"E FRED 1,1, 17" or "FRED = 17" would declare a scalar named FRED whose value was 17.

"=" INDICATES FORMULA EVALUATION

An equal sign requests that the formula be interpreted and evaluated. Formulas such as "=1+2" need no variables, but a formula such as "=X+Y" requires that you have declared the variables X and Y earlier. "= A*B+A**2+INV(TRANS(A)*A)" is an example of formula input. Because nothing appears on the left hand side, the formula is evaluated, the value is printed, but nothing is done with the value. You can, if you like, assign the value of the formula to a variable using the S command. Thus you can say "=1+2" in which case the result will be printed, or "A=1+2" in which case the result will be printed and stored in A. The following are all legal commands: "A(1,2)=2", "X=4+SIN(Y)", "=X+Y". The program expects the formula to be in standard FORTRAN and will reject incorrect syntax as well as incompatible matrix operations.

H = HELP

Requests the display of the list of all the commands, with one line explanations, given at the beginning of this section.

I = INITIALIZE

This command initializes or reinitializes the program. If you issue this command you will be asked to affirm that you really mean it. It clears out a fixed amount of workspace to be used by the program. This is done automatically when the program begins, but it may be necessary to issue this command during program execution if all the workspace has been used up. WARNING: the "I" command erases the definitions and values of all user variables; to delete only part of the data, use the "K" command.

K (variable) = KILL (DELETE) THE LISTED VARIABLES

MATALG has a finite amount of space to store variables. To clear out all the space used so far, the "I" command will do. But the "I" command destroys all user variables and formulas. To delete just some variables, the "K" command can be used. It will delete the variables named, and return the space it was occupying to the free space available to the program. "K X,Y" will kill the variables named X and Y and return the memory allocated to X and Y to the free space.

L (Variable, start value, stop value, number of steps) = LIST A TABLE

This command allows you to list the successive values of a formula for equally spaced values of one of its arguments. The most recent formula you entered is the one that will be tabulated. For example, the two commands
"=DET(x-A)"

"L X, 0.0, 10, 100"
will list the value of the characteristic polynomial of A for every tenth of a unit from 0 to 10.

M (LU/QR/SVD, variable name) = MATRIX FACTORIZATION

This command allows you to carry out either the LU, QR, or SVD factorization of a rectangular matrix. Type either "LU", "QR", or "SVD" and then give the name of the matrix to be factored.

"M LU A" would carry out the LU factorization of the matrix A.

The LU factorization of an m by n matrix A returns an m by n unit lower triangular matrix L, an m by m permutation matrix P, and an m by n upper triangular matrix U, such that P*A=L*U. This factorization is useful in solving linear systems .

"M QR B" would carry out the QR factorization of the matrix B.

The QR factorization of an m by n matrix A returns an m by n orthogonal matrix Q, and an m by n upper triangular matrix R, such that A=Q*R. This factorization is useful in solving overdetermined systems in the least squares sense and is the essential ingredient in the best algorithms for finding the eigenvalues of a matrix.

"M SVD C" would carry out the singular value decomposition of the matrix C.

The SVD (singular value decomposition) of the m by n matrix returns an m by n orthogonal matrix U, an m by n diagonal matrix S, and an n by n orthogonal matrix V such that $U^TCV=S$. The singular value decomposition is useful in statistics and in solving least squares problems. It provides the best method of computing the rank of a matrix on a computer.

Once a factorization is carried out, you will be asked if you want to save the factors in variables. If so, you must provide names of the variables in which to store the factors

N = NOTE

Requests that the program pause while you write a note. This note will appear in the transcript file, if you have opened one. When you are done annotating your work, type a period in column 1 to resume running the program. The message will be preceded and followed by a line of stars in the transcript file.

P (Variable) = POLYNOMIAL

If the variable is a square matrix (of order n), this command computes the characteristic polynomial of the matrix. The result is a list of n+1 coefficients. The program will offer to store this information in a vector variable if you like. Then you can evaluate the polynomial using formulas involving the "POLY" function.

"P A", assuming A is a matrix, produces the coefficients of the characteristic polynomial of A. If we store these coefficients in a vector, say PCOEF, then the command "=POLY(PCOEF, A)" would evaluate the polynomial with the matrix A as its argument. In exact arithmetic, the result should be the zero matrix (Cayley-Hamilton Theorem).

If the variable is a vector, then it is assumed that the n roots of an nth. degree polynomial have been stored in the vector. The command computes the n+1 coefficients of the corresponding polynomial.

"P V" where V is the vector [1, 2, 3] would result in the computation of the coefficients of the polynomial (X-1)*(X-2)*(X-3), which equals $X^3 - 6X^2 + 11X - 6$.

Q = QUIT

This stops the program. Just in case you did not mean to type the letter Q, the program requires that you affirm your decision by asking you to enter "Y" or "YES" to quit.

R (steps) = REPEATED EVALUATION

This command only makes sense if you have typed in a formula of the form X = F(X). Here, we understand this equation to mean "Take the current value of X, evaluate F(X), and update X to have that value. The commands

"X=(A*X)/NORM2(A*X)"

"R 15"

where A is a matrix and X a vector, implements 15 steps of the power method for finding the dominant eigenvalue of A.

S (Variable) = SAVE A VARIABLE

If a formula is not of the form "variable=expression", the value returned by the formula is not going to be saved. If you decide you want to save the value, the S command will do it.

"=SQRT(16)"
"S X"

will save the value 4 in X.

T (Variable or 'All' or 'Debug' or blank) = TYPE A VARIABLE

Requests that the program print the value of one or more variables. If a variable name follows, as in "T X", its value will be typed out. 'All' requests that all the symbols used by the program be displayed. 'Debug' prints out all sorts of stuff, primarily for debugging.

U (Variable) = UNPACK (PARTITION) A MATRIX

Partition a matrix. This command lets you construct new matrices by partitioning an existing matrix. You will be asked:
The rows to keep, followed by a zero.
The columns to keep, followed by a zero.
The name of a matrix in which to store the result, or RETURN if you don't want to save it.

You actually define 4 new matrices by this process, and you can see and save the other three if you want. The second matrix is made by deleting the rows you listed, keeping the columns you deleted. The third and fourth matrices are made similarly. The program will let you know that you can examine them.

For example, if A is the matrix $\begin{bmatrix} 11 & 12 & 13 \\ 21 & 22 & 23 \\ 31 & 32 & 33 \end{bmatrix}$

Then if your row list is 1,0 and your column list is 1,3,0, the (first) submatrix will be

[11 13] and the others will be $\begin{bmatrix} 21 & 23 \\ 31 & 33 \end{bmatrix}$, $\begin{bmatrix} 22 \\ 32 \end{bmatrix}$, and [12] .

V = VALUE OF LAST FORMULA

Evaluate the current formula (presumably with a new value for at least one of the variables). For example, the sequence of commands

```
"E X, 1,1, 2"
"= X*3"
"C X, 1,1, 3"
"V"
```

will produce a value of 6 on the first evaluation, and a value of 9 on the second. This command can be used in place of the "R" command if you wish to monitor the calculation closely, perhaps to determine convergence of a sequence of approximations.

W (Filename, variable) =WRITE A VARIABLE TO A FILE

Writes the value of a variable into a file, which may later be read in by this program, or by MATMAN.

X (Filename) = READ A VARIABLE FROM A FILE

Requests that the given file be opened, and the names of the variables stored there be listed. Presumably, the file was made by this program with the "W" command, or byMATMAN. Once the program has listed the variables, you may pick one to read in.

The file "MATALG.DAT" contains some examples you can use.

? = DETAILED HELP

While running the program, you can always use the "H" command for quick help. Sometimes, though, that isn't enough. If you really want to see some detailed explanation of a particular command, you can use "?" instead of "H". Effectively, this causes the program to read the help file, MATALG.HLP, and to display the main topics. At this point your options are:

Type a topic, or enough of the topic to be unique. In this case you will get information on the given topic, plus possibly a list of subtopics.

Hit the RETURN key. This will back you up one 'level'. If you've just entered the help program, you will back up into the main program. If you're down one subtopic, you'll back up to the main topic, and so on.

Type CTRL-Z, "Control-Z", by holding down the "CONTROL" key on most terminals, or the "Clover-leaf" key on a Macintosh, and typing "Z", possibly followed by RETURN. This gets you out of the help program and back into the main program immediately, no matter how many levels down you are.

Type "?" to have information about the current topic repeated. You might give this command several times in a row, just to browse through the same information over and over again. But you might also issue it when backing up, because when backing up to a higher topic, that topic's information is not printed out again unless you specially request it with the "?".

FORMULAS

In order to use the formula option correctly, you must use the proper abbreviations for mathematical functions, and the right symbols for operations like multiplication and exponentiation.

Symbols and constants: The operators allowed are +, -, *, /, **, ^, =

* is standard scalar or matrix multiplication if both arguments are scalars or both are matrices. If one argument is a scalar, then each entry of the other argument will be multiplied by the scalar. If both arguments are row or column vectors, the dot product is taken. A row vector times a column vector yields a scalar. A column times a row vector yields a matrix.

/ is standard scalar division. You may 'divide' a matrix by a scalar, in which case each entry of the matrix is divided by the scalar, e.g. A/2 = (1/2)*A.

+ and - may be used for scalars, vectors, or matrices of the same order. Also, a NONSTANDARD shorthand allows you to add a scalar to a matrix. In this case, the scalar is added only to the diagonal. Thus, if B is a square matrix, B+2 is a legal formula, which really means B+2*I. This allows you to save storage space since the identity matrix need not be stored.

** (scalar exponentiation) is legal for scalar base and power, such as 2**3(= 2^3) or for a square matrix base and integer power, such as A**2 .

FUNCTIONS

S, S1, S2: Arguments may ONLY be scalar.
M : Arguments may be scalar or square matrix.
* : Arguments may be scalar, vector or matrix.
I : Arguments may only be a positive integer.

ABS(*)	- Absolute value of *.	
ACOS(S)	- The arc cosine of S.	$-1 < S < 1$
ALOG(S)	- Natural logarithm of S.	$S > 0$
ALOG10(S)	- Logarithm base 10 of S.	$S > 0$
ASIN(S)	- Arc sine of S.	$-1 < S < 1$
ATAN(S)	- Arc tangent of S.	
ATAN2(S1,S2)	- Arc tangent of (S1/S2). Correctly computes ATAN2(0,0)=PI/2.	
COS(S)	- Cosine of S.	
COSH(S)	- Hyperbolic cosine of S.	
DET(M)	- Determinant of matrix M.	
DIAG(M)	- Diagonal of matrix M, stored in a column vector.	
EPS	- Machine epsilon. This is the smallest power of two such that 1+EPS.GT.1.0	
EVAL(M)	- Real and imaginary parts of eigenvalues of matrix, stored in an array of m rows and 2 columns, real parts in first column, imaginary in second.	
EVEC(M)	- Eigenvectors of matrix, stored in square matrix of same size as M. (Eigenvectors corresponding to a complex pair: if eigenvalues j and j+1 are a complex pair, then the eigenvector for eigenvalue j is column j + i*column j+1, and the eigenvector for eigenvalue j+1 is column j - i*column j+1.)	
EXP(S)	- Exponential of S.	

ID(I)	- Matrix identity of order I.
INT(*)	- Replace real values by nearest integer.
INV(M)	- Inverse of matrix M.
IVAL(M)	- Imaginary parts of eigenvalues of matrix, stored in a column vector.
LN(S)	- Natural logarithm of S. 0 < S
LOG(S)	- Natural logarithm of S. 0 < S
LOG10(S)	- Logarithm base 10 of S. 0 < S
MAX(S1,S2)	- Maximum of S1, S2.
MIN(S1,S2)	- Minimum of S1, S2.
NEG(*)	- Changes sign of *.
NORM0(V)	- Infinity or Max-norm of a vector. NORM0(V) returns the maximum of the absolute values of the entries of V.
NORM1(V)	- 1-norm of a vector. NORM1(V) returns the sum of the absolute values of the entries of V.
NORM2(V)	- 2-norm, euclidean norm, or root-mean-square norm of the vector V. Returns the square root of the sum of the squares of the entries of V.
PI	- 3.14159265...
POLY(V,M)	- Polynomial evaluation. V contains the coefficients of the polynomial, with V(1) the coefficient of the highest order, and the last entry of V being the constant term. M is the square matrix argument of the polynomial.
RAN(*)	- Fills * with random numbers between 0 and 1.
RVAL(M)	- Real parts of eigenvalues of matrix, stored in a column vector.
SIN(S)	- Sine of S.
SINE(S)	- Sine of S.
SINH(S)	- Hyperbolic sine of S.
SQRT(S)	- Square root of S. 0 < S
STEP(S)	- Heavyside step function. STEP=0 if S<0, STEP=1 if S>0.
TAN(S)	- Tangent of S.
TANH(S)	- Hyperbolic tangent of S.
TRACE(M)	- Trace (sum of diagonal elements) of matrix.
TRANS(*)	- The transpose of *. A=TRANS(A) is a legal formula if A is square.
ZERO(I)	- The zero matrix of order I.
I!	- I factorial. Valid for I=0 to 25. I!=I*(I-1)*(I-2)*...*2*1.

SAMPLE MATALG TRANSCRIPT FILE (User input in boldface, comments in paranthesis.)

Enter command (H for menu or ? for full help)
E A 4 4
Added variable A
Enter A (1 1),A (1 2),A (1 3),A (1 4),...
17 -8 -12 14
Enter A (2 1),A (2 2),A (2 3),A (2 4),...
46,-22,-35,41
Enter A (3 1),A (3 2),A (3 3),A (3 4),...
-2 1 4 -4 4 -2 -2 3

A

17	-8	-12	14
46	-22	-35	41
-2	1	4	-4
4	-2	-2	3

Enter command (H for menu or ? for full help)
= DET(A)
FORMULA= -1.00000
Enter command (H for menu or ? for full help)
B=INV(A)
Implicit definition of B
B =

17.0000	-7.99999	-12.0000	14.0000	
41.9999	-20.0000	-31.0000	36.0000	($B=A^{-1}$)
9.99998	-4.99999	-8.00000	11.0000	
12.0000	-5.99999	-10.00000	13.0000	

Enter command (H for menu or ? for full help)
E K 4 1
Added variable K
Enter K(1),KX(2),K(3),K(4),
1 2 3 4
K TRANSPOSED

1 2 3 4

Enter command (H for menu or ? for full help)
=B*K
FORMULA= Transposed
21.0001 53.0002 20.0000 22.000 ($= A^{-1}K$)

```
Enter command (H for menu or ? for full help)
P A
X**4-2.00000*X**3+2.00000*X-1.00000  (CHARACTERISTIC
                                        POLYNOMIAL OF A )
Enter variable to save polynomial in, or RETURN.
COEF
Added variable COEF
Coefficients saved in COEF
Enter command (H for menu or ? for full help)
=POLY(COEF,A)

FORMULA=

    0       0       0       0
    0       0       0       0          (CAYLEY- HAMILTON
    0       0       0       0          THEOREM)
    0       0       0       0
Enter command (H for menu or ? for full help)
 =A**3-A**2-A+1
FORMULA=

    0       0       0       0     (MINIMUM  POLYNOMIAL
    0       0       0       0     IS  X3 - X2 - X + 1)
    0       0       0       0
    0       0       0       0
Enter command (H for menu or ? for full help)
 C A 3 3 5

A

   17      -8      -12      14
   46     -22      -35      41
   -2       1        5      -4
    4      -2       -2       3

= INV(A)
FORMULA=

 -.142856    .571428    1.71429    -4.85714
 -2.28571    2.14285    4.42857    -12.7143
 -1.42857    .714285    1.14286    -1.57143
 -2.28571    1.14286    1.42857    -2.71429
Enter command (H for menu or ? for full help)
= EVAL(A)
```

```
FORMULA=

  3.63196         0
 -.815968       1.12321        (TWO COMPLEX EIGENVALUES)
 -.815968      -1.12321
  .999966         0
Enter command (H for menu or ? for full help)
P = EVEC(A)
Implicit definition of P

FORMULA=

  0.297464         -0.213015      0.256110       0.447214
  0.905275         -0.571984      0.730475       0.894427
 -0.302205         -0.67449E-01   0.151782E-02   0.147742E-05
 -0.257684E-01     -0.134127      0.757088E-01   0.177456E-05
Enter command (H for menu or ? for full help)
= INV(P)*A*P
FORMULA=

 3.63198          0              0.26226E-05    -0.725591E-06
 0.148565E-04    -0.816006       1.12315         0.21901E-04
 0.483394E-04    -1.12317       -0.815976        0.485050E-04
-0.700168E-05     0.578165E-05  -0.579655E-05    1.00000
*************************************************************
 This matrix is block diagonal, except for round off error
*************************************************************

Enter command (H for menu or ? for full help)
E X 4 1
Added variable X
Enter X(1),X(2),X(3),X(4),
1 1 1 1

X  TRANSPOSED

       1  1  1  1

Enter command (H for menu or ? for full help)

X=A*X/NORM1(A*X)
FORMULA= Transposed
  .250000     .681818       0       0.681818E-01
```

```
Enter command (H for menu or ? for full help)
R 10                                          (POWER METHOD)

K= 1 Transposed
  -.207547   -.584907   -0.754718E-01  -.132076

K= 2 Transposed
  .239130    .630435    -0.217395E-01  .108694

K= 3 Transposed
  .222891    .650602    -.108436     0.180709E-01

K= 4 Transposed
  .119172    .409330    -.352329     -.119169

K= 5 Transposed
  .206362    .617455    -.175367     0.815608E-03

K= 6 Transposed
  .195753    .594960    -.193187     -0.160990E-01

K= 7 Transposed
  .191873    .585506    -.202642     -0.199797E-01

K= 8 Transposed
  .195242    .593481    -.195713     -0.155650E-01

K= 9 Transposed
  .194283    .591321    -.197438     -0.169573E-01

K= 10 Transposed
  .194221    .591143    -.197670     -0.169654E-0   (DOMINANT
                                                   EIGENVECTOR)
Enter command (H for menu or ? for full help)
S=1
Implicit definition of S
S = 1
Enter command (H for menu or ? for full help)
V = DET(S-A)
 Implicit definition of V
V= 0                          (1 IS AN EIGENVALUE OF A)
Enter command (H for menu or ? for full help)
```

L S, 0, 10, 11
Table of X, F(X)

X	F(X)	
.000000	7.00000	
.909091	1.04880	
1.81818	-12.1695	
2.72727	-21.5902	
3.63636	.243662	(THERE IS AN EIGENVALUE
4.54545	97.1816	BETWEEN 2.727 AND 3.636)
5.45455	329.466	
6.36364	773.730	
7.27273	1523.00	
8.18182	2686.70	
9.09091	4390.63	
10.0000	6777.00	

Appendix C

DECOMP and SOLVE

This appendix contains listings of the FORTRAN subroutines DECOMP and SOLVE from Forsyth, Malcolm, and Moler "Computer Methods for Mathematical Computation", pp. 52-53[FMM]. These two subroutines carry out the solution of the linear system AX = B; DECOMP carries out the LU factorization of A and estimates the condition number of A, SOLVE then solves the two triangular systems LY = B and UX = Y. In order to use these subroutines the user must construct a driver program, like the one in Appendix E, to dimension the arrays, input the data, call the subroutines, and output the solution.

```
C************************************************
C
   SUBROUTINE DECOMP(NDIM,N,A,COND,IPVT,WORK)
C
C************************************************
C
C SUBROUTINE DECOMP DECOMPOSES A MATRIX BY GAUSSIAN ELIMINATION
C AND ESTIMATES THE CONDITION OF THE MATRIX.
C
C ONCE THE SYSTEM HAS BEEN DECOMPOSED BY DECOMP, USE SUBROUTINE
C SOLVE TO COMPUTE SOLUTIONS TO THE LINEAR SYSTEM.
C
C INPUT
C
C NDIM   DECLARED ROW DIMENSION OF THE ARRAY CONTAINING A.
C        NDIM MUST BE GREATER THAN OR EQUAL TO N.
C        (NDIM IS THE FIRST DIMENSION IN DIMENSION STATEMENT FOR A).
C
C N      THE ORDER OF THE MATRIX (NUMBER OF EQUATIONS).
C
C A      A MATRIX OF DECLARED FIRST DIMENSION NDIM,
C        AND OF USED DIMENSIONS (N,N), WHICH CONTAINS THE COEFFICIENTS
C        OF THE LINEAR SYSTEM WHICH IS TO BE DECOMPOSED.
C
C OUTPUT
C
C A      CONTAINS AN UPPER TRIANGULAR MATRIX U AND A PERMUTED
C        VERSION OF A LOWER TRIANGULAR MATRIX I-L SO THAT
C        (PERMUTATION MATRIX)*A=L*U
C
C COND   AN ESTIMATE OF THE CONDITION OF NUMBER OF A .
C        CHANGES IN A AND B MAY CAUSE CHANGES COND TIMES AS LARGE IN X.
C        IF COND+1.0=COND, A IS SINGULAR TO WORKING PRECISION  COND
C        IS SET TO 1.0E+32 IF EXACT SINGULARITY IS DETECTED.
```

```
C
C IPVT  AN INTEGER VECTOR OF DIMENSION AT LEAST N, THE PIVOT VECTOR.
C       IPVT(K)=THE INDEX OF THE K-TH PIVOT ROW, FOR K=1,...,N-1
C       IPVT(N)=(-1)**(NUMBER OF ROW INTERCHANGES)
C
C WORK  A WORK VECTOR OF DIMENSION N WHOSE CONTENTS ARE IRRELEVANT
C       TO THE USER.
C
C IN THE CALLING PROGRAM, THE USER MUST DECLARE THE MATRIX A(*,*) AND
C THE VECTORS WORK(*) AND IPVT(*). THE FIRST DIMENSION OF A IS PASSED AS
C NDIM, AND THE SECOND DIMENSION OF A, AS WELL AS THE DIMENSIONS OF
C WORK AND IPVT, MUST BE AT LEAST N. NOTE - THE DETERMINANT OF A CAN BE
C COMPUTED ON RETURN FROM DECOMP BY
C  DET(A)=IPVT(N)*A(1,1)*A(2,2)*...*A(N,N)
C
C*****************************************************
C
      INTEGER NDIM,N
      DIMENSION A(NDIM,N),WORK(N)
      INTEGER IPVT(N),NM1,I,J,K,KP1,KB,KM1,M
      IPVT(N)=1
      IF(N.EQ.1)GOTO 150
      NM1=N-1
C
C GET COLUMN-SUM NORM OF MATRIX
C
      ANORM=0.0
      DO 20 J=1,N
       T=0.0
       DO 10 I=1,N
   10    T=T+ABS(A(I,J))
   20  IF(T.GT.ANORM)ANORM=T
C
C FIND PIVOT ENTRY AND ELIMINATE REMAINING ENTRIES
C
      DO 70 K=1,NM1
       KP1=K+1
C
C FIND PIVOT
C
       M=K
       DO 30 I=KP1,N
   30    IF(ABS(A(I,K)).GT.ABS(A(M,K)))M=I
       IPVT(K)=M
       IF(M.NE.K)IPVT(N)=-IPVT(N)
       CALL SWITCH(A(M,K),A(K,K))
C
C SKIP IF PIVOT IS ZERO
C
       IF(A(K,K).EQ.0.0)GOTO 70
C
C COMPUTE MULTIPLIERS
C
```

```
      DO 40 I=KP1,N
 40     A(I,K)=-A(I,K)/A(K,K)
C
C INTERCHANGE AND ELIMINATE BY COLUMNS
C
      DO 60 J=KP1,N
        CALL SWITCH(A(M,J),A(K,J))
        IF(A(K,J).EQ.0.0)GOTO 60
        DO 50 I=KP1,N
 50       A(I,J)=A(I,J)+A(I,K)*A(K,J)
 60     CONTINUE
 70   CONTINUE
C
C COND=(1-NORM OF A)*(ESTIMATE OF 1-NORM OF A-INVERSE) ESTIMATE
C OBTAINED BY ONE STEP OF INVERSE ITERATION FOR THE SMALL SINGULAR
C VECTOR. THIS INVOLVES SOLVING TWO SYSTEMS OF EQUATIONS
C (A-TRANSPOSE)*Y=E AND A*Z=Y WHERE E IS A VECTOR OF +1 OR -1 CHOSEN
C TO CAUSE GROWTH IN Y. ESTIMATE=(1-NORM OF Z)/(1-NORM OF Y)
C
C SOLVE (A-TRANSPOSE)*Y=E
C
      DO 100 K=1,N
        T=0.0
        IF(K.EQ.1)GOTO 90
        KM1=K-1
        DO 80 I=1,KM1
 80       T=T+A(I,K)*WORK(I)
 90     EK=1.0
        IF(T.LT.0.0)EK=-1.0
        IF(A(K,K).EQ.0.0)GOTO 160
        WORK(K)=-(EK+T)/A(K,K)
 100  CONTINUE
      DO 120 KB=1,NM1
        K=N-KB
        T=0.0
        KP1=K+1
        DO 110 I=KP1,N
 110      T=T+A(I,K)*WORK(K)
        WORK(K)=T
        M=IPVT(K)
        IF(M.EQ.K)GOTO 120
        CALL SWITCH(WORK(M),WORK(K))
 120  CONTINUE
      YNORM=0.0
      DO 130 I=1,N
 130  YNORM=YNORM+ABS(WORK(I))
C
C SOLVE A*Z=Y
C
      CALL SOLVE(NDIM,N,A,WORK,IPVT)
      ZNORM=0.0
      DO 140 I=1,N
 140  ZNORM=ZNORM+ABS(WORK(I))
```

```
C
C ESTIMATE CONDITION
C
    COND=ANORM*ZNORM/YNORM
    IF(COND.LT.1.0)COND=1.0
    RETURN
C
C 1 BY 1 SYSTEM
C
  150 COND=1.0
    IF(A(1,1).NE.0.0)RETURN
C
C EXACT SINGULARITY
C
  160 COND=1.0E+32
    RETURN
    END

C*****************************************************
C
    SUBROUTINE SOLVE(NDIM,N,A,B,IPVT)
C
C*****************************************************
C
C SUBROUTINE SOLVE TAKES AS INPUT THE MATRIX A WHICH HAS BEEN
C DECOMPOSED BY DECOMP, AND A RIGHT HAND SIDE B, AND RETURNS IN B
C THE SOLUTION X TO A*X=B.
C
C DO NOT ATTEMPT TO SOLVE A SYSTEM IF DECOMP HAS DETECTED SINGULARITY.
C
C INPUT
C
C NDIM   DECLARED ROW DIMENSION OF ARRAY CONTAINING A.
C        NDIM MUST BE GREATER THAN OR EQUAL TO N.
C        (NDIM IS THE FIRST DIMENSION IN DIMENSION STATEMENT FOR A.)
C
C N      THE ORDER OF THE MATRIX A, THE NUMBER OF EQUATIONS TO SOLVE.
C
C A      A MATRIX OF DECLARED FIRST DIMENSION NDIM, AND USED
C        DIMENSIONS (N,N) WHICH HAS ALREADY BEEN DECOMPOSED BY
C        CALLING DECOMP.
C
C B      A VECTOR OF DIMENSION AT LEAST N CONTAINING THE RIGHT
C        HAND SIDE OF THE LINEAR SYSTEM TO BE SOLVED.
C
C IPVT   AN INTEGER VECTOR OF DIMENSION AT LEAST N CONTAINING
C        THE PIVOT INFORMATION AS COMPUTED BY DECOMP.
C
C OUTPUT
C
C B      THE SOLUTION VECTOR FOR THE GIVEN INPUT SYSTEM.
C
C*****************************************************
```

```
C
      INTEGER NDIM,N,IPVT(N)
      DIMENSION A(NDIM,N),B(N)
C
C  SPECIAL CASE FOR 1 BY 1 SYSTEM
C
      IF(N.EQ.1)GOTO 50
C
C  FORWARD ELIMINATION
C
      NM1=N-1
      DO 20 K=1,NM1
        KP1=K+1
        M=IPVT(K)
        CALL SWITCH(B(M),B(K))
        DO 10 I=KP1,N
   10     B(I)=B(I)+A(I,K)*B(K)
   20   CONTINUE
C
C  BACK SUBSTITUTION
C
      DO 40 KB=1,NM1
        KM1=N-KB
        K=KM1+1
        B(K)=B(K)/A(K,K)
        T=-B(K)
        DO 30 I=1,KM1
   30     B(I)=B(I)+A(I,K)*T
   40   CONTINUE
   50 B(1)=B(1)/A(1,1)
      RETURN
      END
```

Appendix D

LINPACK

This appendix describes the subroutine naming convention used in LINPACK and gives descriptions of the LINPACK subroutine SGECO, SGESL, and SGEDI, from the LINPACK package[LUG]. A comparison is given of the subroutines DECOMP and SGECO and the subroutines SOLVE and SGESL.

LINPACK SUBROUTINE NAMING CONVENTION

Each LINPACK subroutine has a 5 letter name. The first letter indicates which kind of machine numbers are to be used:

S ≡ real single precision
D ≡ real double precision
C ≡ complex single precision.

The next two letters describe the form of the matrix or its decomposition:

GE ≡ general
GB ≡ general, banded
PO ≡ positive definite
PP ≡ positive definite, packed
PB ≡ Positive definite, banded
SI ≡ symmetric, indefinite
SP ≡ symmetric, indefinite, packed
HI ≡ Hermitian, indefinite
HP ≡ Hermitian, indefinite, packed
TR ≡ triangular
GT ≡ general tridiagonal
PT ≡ positive definite, tridiagonal
CH ≡ Cholesky decomposition
QR ≡ orthogonal - triangular decomposition
SV ≡ singular value decomposition.

The last two letters in the name of the subroutine indicate the computation done:

FA ≡ factor
CO ≡ factor and estimate the condition number
SL ≡ Solve
DI ≡ determinant, inverse, inertia

DC ≡ decompose
UD ≡ update
DD ≡ downdate
EX ≡ exchange.

Thus the subroutine SGECO factors (LU) the matrix A and estimates its condition number. The subroutine SGESL uses the output from SGECO to complete the solution of the system $AX = B$. The subroutine SCHDC computes the Cholesky factorization of the positive definite matrix A, while the subroutine CHPFA would factor a sparse complex matrix using packed storage.

Not all possible combinations are present, some because they are not needed and some because they make no sense. In the following tables a (√) indicates that the subroutine exists.

	SGE	SGB	SPO	SPP	SPB	SSI	SSP	CHI
CO	√	√	√	√	√	√	√	√
FA	√	√	√	√	√	√	√	√
SL	√	√	√	√	√	√	√	√
DI	√	√	√	√	√	√	√	√

	CHP	STR	SGT	SPT
CO	√	√		
FA	√			
SL	√	√	√	√
DI	√	√		

	DC	SL	UD	DD	EX
SCH	√		√	√	√
SQR	√	√			
SSV	√				

Complete listings of the code, calling sequences, and definitions of variables are given in the LINPACK Manual [LUG].

The following tables compare the variables in two of the basic LINPACK subroutines with the subroutines DECOMP and SOLVE.

SGECO	DECOMP	
A	A	Coefficient matrix A
LDA	NDIM	Size of the array in which A is stored
N	N	Size of A
IPVT	IPVT	Vector containing pivoting information
RCOND	COND	COND = 1/RCOND is an estimate of the condition number
Z	WORK	Work array

SGESL	SOLVE	
A	A	Packed LU factorization of A
LDA	NDIM	Size of array in which A is stored
N	N	Size of A
IPVT	IPVT	Pivot information from factorization subroutine
B	B	Right hand side
Job		Solve AX = B (job = 0) or $A^T X = B$ (job $\neq 0$)

```
C
      SUBROUTINE SGECO(A,LDA,N,IPVT,RCOND,Z)
      INTEGER LDA,N,IPVT(1)
      REAL A(LDA,1),Z(1)
      REAL RCOND
C
C     SGECO FACTORS A REAL MATRIX BY GAUSSIAN ELIMINATION
C     AND ESTIMATES THE CONDITION OF THE MATRIX.
C
C     IF  RCOND  IS NOT NEEDED, SGEFA IS SLIGHTLY FASTER.
C     TO SOLVE  A*X = B , FOLLOW SGECO BY SGESL.
C     TO COMPUTE  INVERSE(A)*C , FOLLOW SGECO BY SGESL.
C     TO COMPUTE  DETERMINANT(A) , FOLLOW SGECO BY SGEDI.
C     TO COMPUTE  INVERSE(A) , FOLLOW SGECO BY SGEDI.
C
C     ON ENTRY
C
C        A       REAL(LDA, N)
C                THE MATRIX TO BE FACTORED.
C
C        LDA     INTEGER
C                THE LEADING DIMENSION OF THE ARRAY  A .
C
C        N       INTEGER
C                THE ORDER OF THE MATRIX  A .
C
```

```
C   ON RETURN
C
C     A     AN UPPER TRIANGULAR MATRIX AND THE MULTIPLIERS
C           WHICH WERE USED TO OBTAIN IT.
C           THE FACTORIZATION CAN BE WRITTEN  A = L*U  WHERE
C           L  IS A PRODUCT OF PERMUTATION AND UNIT LOWER
C           TRIANGULAR MATRICES AND  U  IS UPPER TRIANGULAR.
C
C     IPVT    INTEGER(N)
C           AN INTEGER VECTOR OF PIVOT INDICES.
C
C     RCOND   REAL
C           AN ESTIMATE OF THE RECIPROCAL CONDITION OF  A .
C           FOR THE SYSTEM  A*X = B , RELATIVE PERTURBATIONS
C           IN  A  AND  B  OF SIZE  EPSILON  MAY CAUSE
C           RELATIVE PERTURBATIONS IN  X  OF SIZE  EPSILON/RCOND .
C           IF  RCOND  IS SO SMALL THAT THE LOGICAL EXPRESSION
C                  1.0 + RCOND .EQ. 1.0
C           IS TRUE, THEN  A  MAY BE SINGULAR TO WORKING
C           PRECISION.  IN PARTICULAR,  RCOND  IS ZERO  IF
C           EXACT SINGULARITY IS DETECTED OR THE ESTIMATE
C           UNDERFLOWS.
C
C     Z     REAL(N)
C           A WORK VECTOR WHOSE CONTENTS ARE USUALLY
C           UNIMPORTANT.  IF  A  IS CLOSE TO A SINGULAR MATRIX, THEN  C
C           Z IS AN APPROXIMATE NULL VECTOR IN THE SENSE THAT
C           NORM(A*Z) = RCOND*NORM(A)*NORM(Z) .
C
C   LINPACK. THIS VERSION DATED 08/14/78 .
C   CLEVE MOLER, UNIVERSITY OF NEW MEXICO, ARGONNE NATIONAL LAB.
C
C   SUBROUTINES AND FUNCTIONS
C
C   LINPACK SGEFA
C   BLAS SAXPY,SDOT,SSCAL,SASUM
C   FORTRAN ABS,AMAX1,SIGN
C

      SUBROUTINE SGESL(A,LDA,N,IPVT,B,JOB)
      INTEGER LDA,N,IPVT(1),JOB
      REAL A(LDA,1),B(1)
C
C   SGESL SOLVES THE REAL SYSTEM
C   A * X = B  OR  TRANS(A) * X = B
C   USING THE FACTORS COMPUTED BY SGECO OR SGEFA.
C
C   ON ENTRY
C
C     A     REAL(LDA, N)
C           THE OUTPUT FROM SGECO OR SGEFA.
C
C     LDA    INTEGER
```

```
C         THE LEADING DIMENSION OF THE ARRAY  A .
C
C     N      INTEGER
C         THE ORDER OF THE MATRIX  A .
C
C     IPVT   INTEGER(N)
C         THE PIVOT VECTOR FROM SGECO OR SGEFA.
C
C     B      REAL(N)
C         THE RIGHT HAND SIDE VECTOR.
C
C     JOB    INTEGER
C         = 0        TO SOLVE  A*X = B ,
C         = NONZERO  TO SOLVE  TRANS(A)*X = B  WHERE
C                  TRANS(A)  IS THE TRANSPOSE.
C
C   ON RETURN
C
C     B      THE SOLUTION VECTOR  X .
C
C   ERROR CONDITION
C
C     A DIVISION BY ZERO WILL OCCUR IF THE INPUT FACTOR CONTAINS A
C     ZERO ON THE DIAGONAL.  TECHNICALLY THIS INDICATES SINGULARITY
C     BUT IT IS OFTEN CAUSED BY IMPROPER ARGUMENTS OR IMPROPER
C     SETTING OF LDA .  IT WILL NOT OCCUR IF THE SUBROUTINES ARE
C     CALLED CORRECTLY AND IF SGECO HAS SET RCOND .GT. 0.0
C     OR SGEFA HAS SET INFO .EQ. 0 .
C
C   TO COMPUTE  INVERSE(A) * C  WHERE  C  IS A MATRIX
C   WITH  P  COLUMNS
C         CALL SGECO(A,LDA,N,IPVT,RCOND,Z)
C         IF (RCOND IS TOO SMALL) GO TO ...
C         DO 10 J = 1, P
C            CALL SGESL(A,LDA,N,IPVT,C(1,J),0)
C      10 CONTINUE
C
C   LINPACK. THIS VERSION DATED 08/14/78 .
C   CLEVE MOLER, UNIVERSITY OF NEW MEXICO, ARGONNE NATIONAL LAB.
C
C   SUBROUTINES AND FUNCTIONS
C
C   BLAS SAXPY,SDOT
C
      SUBROUTINE SGEDI(A,LDA,N,IPVT,DET,WORK,JOB)
      INTEGER LDA,N,IPVT(1),JOB
      REAL A(LDA,1),DET(2),WORK(1)
C
C   SGEDI COMPUTES THE DETERMINANT AND INVERSE OF A MATRIX
C   USING THE FACTORS COMPUTED BY SGECO OR SGEFA.
C
C   ON ENTRY
C
```

```
C     A       REAL(LDA, N)
C             THE OUTPUT FROM SGECO OR SGEFA.
C
C     LDA     INTEGER
C             THE LEADING DIMENSION OF THE ARRAY  A .
C
C     N       INTEGER
C             THE ORDER OF THE MATRIX  A .
C
C     IPVT    INTEGER(N)
C             THE PIVOT VECTOR FROM SGECO OR SGEFA.
C
C     WORK    REAL(N)
C             WORK VECTOR.  CONTENTS DESTROYED.
C
C     JOB     INTEGER
C             = 11   BOTH DETERMINANT AND INVERSE.
C             = 01   INVERSE ONLY.
C             = 10   DETERMINANT ONLY.
C
C   ON RETURN
C
C     A       INVERSE OF ORIGINAL MATRIX IF REQUESTED.
C             OTHERWISE UNCHANGED.
C
C     DET     REAL(2)
C             DETERMINANT OF ORIGINAL MATRIX IF REQUESTED.
C             OTHERWISE NOT REFERENCED.
C              DETERMINANT = DET(1) * 10.0**DET(2)
C              WITH  1.0 .LE. ABS(DET(1)) .LT. 10.0
C              OR  DET(1) .EQ. 0.0 .
C
C   ERROR CONDITION
C
C     A DIVISION BY ZERO WILL OCCUR IF THE INPUT FACTOR CONTAINS
C     A ZERO ON THE DIAGONAL AND THE INVERSE IS REQUESTED.
C     IT WILL NOT OCCUR IF THE SUBROUTINES ARE CALLED CORRECTLY
C     AND IF SGECO HAS SET RCOND .GT. 0.0 OR SGEFA HAS SET
C     INFO .EQ. 0 .
C
C   LINPACK. THIS VERSION DATED 08/14/78 .
C   CLEVE MOLER, UNIVERSITY OF NEW MEXICO, ARGONNE NATIONAL LAB.
C
C   SUBROUTINES AND FUNCTIONS
C
C   BLAS SAXPY,SSCAL,SSWAP
C   FORTRAN ABS,MOD
C
```

Appendix E

Interactive DRIVER Program

The following interactive FORTRAN driver solves a set of linear systems $AX = B_i$, $i = 1, 2, \ldots, M$ using the LINPACK subroutines SGECO and SGESL. The coefficient matrix A is assumed to be smaller than 50 by 50. The data (n, A, B) is entered from a terminal, echo printed at the terminal, the appropriate subroutines are called, and the answers are printed at the terminal and to a file on your directory. The condition number estimate and the factored matrix, in compact storage form, are also printed.

You may wish to modify the basic program so that it reads the data from a file. This subroutine can be easily modified to call either DECOMP and SOLVE or EISPACK subroutines . An appropriate LINK command must be used at compile time to tell the program where to find the subroutines called. At The University of Pittsburgh the three commands

```
FORTRAN DRIVER.FOR
LINK DRIVER, MTH:LINPKS/LIB
RUN DRIVER
```

will compile the program , link it to the MTH library , where the LINPACK subroutines are located, and run the program. These commands may vary from installation to installation. Check your local system documentation. The DRIVER program is constructed using a text editor or a word processor.

```
C       DRIVER.FOR
C
C       Dimension the needed arrays
C
        Real A(50,50), Z(50), B(50)
        Integer Ipvt(50)
        LDA = 50
C
C       Interactive data entry; the type statements are screen prompts
C
        Type*, 'ENTER THE SIZE OF THE SQUARE COEFFICIENT
          MATRIX '
```

```
      Read(5,*) N
      Do 50 I = 1, N
        Type*, ' ENTER THE ENTRIES OF A IN ROW ', I
        Read(5,*)(A(I,J), J = 1,N)
50    Continue
C
C     ECHO PRINT THE COEFFICIENT MATRIX A
C
      Write(6,*) ' THE COEFFICIENT MATRIX IS '
      Write(12,*) ' THE COEFFICIENT MATRIX IS '
      Do 55 I = 1,N
        Write(6,*) (A(I,J), J = 1,N)
        Write(12,*) (A(I,J), J = 1,N)
55    Continue
C
C     Do the LU factorization of the coefficient matrix A and output
C     the result
C
      CALL SGECO(A,LDA,N,IPVT,RCOND,Z)
      Write(6,*) ' THE DECOMPOSED MATRIX IS '
      Write(12,*) ' THE DECOMPOSED MATRIX IS '

      Do 60 I = 1,N
        Write(6,*) (A(I,J), J = 1,N)
        Write(12,*) (A(I,J), J = 1,N)
60    continue
      COND = 1/RCOND
      Write(6,*) ' THE CONDITION NUMBER ESTIMATE IS'
       COND
      Write(12,*) ' THE CONDITION NUMBER ESTIMATE IS'
       COND
C
C     Interactive data entry of the right hand sides
C
      Type*, ' ENTER THE NUMBER OF RHS '
      Read (5,*) M
      Do 70 K = 1,M
        Type* "ENTER THE NEXT RHS '
        Read(5,*)  (B(I), I = 1,N)
        Write(6,*) 'RHS', K,'IS'
        Write(12,*) 'RHS', K,'IS'
        Write(6,*) (B(I), I = 1,N)
        Write(12,*) (B(I), I = 1,N)
C
```

```
C       Solve the two triangular systems and output the solution
C
          CALL SGESL(A,LDA,N,IPVT,B,JOB)
          Write(6,*) 'THE SOLUTION OF SYSTEM  ', K
          Write(6,*) (B(I), I = 1,N)
          Write(12,*) (B(I), I = 1,N)

70      Continue
        Stop
        End
```

This code assumes that device 5 is the keyboard and that device 6 is the screen. The "write(12,*)" statements should generate an output file named FOR012.DAT containing the data and the solutions.

Appendix F

EISPACK Help File

The following help file and path chart for the EISPACK subroutines [EUG] was written by John Burkardt for the MTH library at the University of Pittsburgh. It is included here with his permission.

THIS IS MTH:EISPAK.HLP. A HELP FILE FOR USE WITH THE EIGENVALUE ANALYSIS PACKAGE EISPACK.

EISPACK CAN SOLVE

1) THE STANDARD EIGENVALUE PROBLEM $A*X = \lambda*X$

2) THE GENERALIZED PROBLEM $A*X = \lambda*B*X$

WHERE A (AND B) ARE GIVEN MATRICES, λ IS AN UNKNOWN EIGENVALUE (SCALAR), AND $X \neq 0$ IS AN UNKNOWN EIGENVECTOR

****

THIS DOCUMENT IS FOR USE WITH THE EISPACK PACKAGE OF SUBROUTINES FOR SOLVING THE EIGENVALUE PROBLEM $A*X=\lambda*X$. SINCE EISPACK OFFERS MANY OPTIONS FOR THE USER, THE PSEUDO-PROGRAM BELOW WILL ALLOW YOU TO DETERMINE, FOR YOUR PROBLEM, WHAT SEQUENCE OF CALLS TO EISPACK WILL GET THE DESIRED ANSWER.

TO USE THE PROGRAM, START WITH THE STATEMENT LABELED 0, AND FOLLOW THE DIRECTIONS. YOU SHOULD TRACE A PATH FROM 0 TO SOME GROUP OF STATEMENTS THAT END WITH A STOP. ALONG THE WAY YOU WILL HAVE SEEN THE CALLS TO THE EISPACK SUBROUTINES APPROPRIATE FOR YOUR PROBLEM. IN CASE YOU FORGOT THEM, YOU CAN FIGURE OUT WHERE YOU HAVE BEEN IN THE PROGRAM FROM WHERE YOU ARE, SINCE THE LABEL FOR THE FINAL STATEMENT CONTAINS THE LABELS OF ALL THE STATEMENTS IN THE PATH THAT LED YOU THERE. FOR EXAMPLE, IF THE STOP STATEMENT WAS IN THE GROUP OF STATEMENTS WITH LABEL 112321, THEN THE PATH WAS

1
11
112
1123
11232
112321.

THE SEQUENCE OF CALLS TO EISPACK MAY THUS BE RECONSTRUCTED.

****

EISPACK PARAMETER KEY

THE SUBROUTINES CALLED BELOW ARE LISTED WITH THEIR PARAMETERS. WE LIST HERE THE PARAMETERS, A BRIEF DEFINITION, THE TYPE AND THE DIMENSIONS. FULL EXPLANATIONS ARE AVAILABLE IN THE EISPACK DOCUMENT FILE [EUG].

NOTE THAT R OR I, SUFFIXED TO A PARAMETER NAME, DENOTES THE REAL OR IMAGINARY PART OF A COMPLEX VALUE. THUS A COMPLEX NUMBER A MIGHT BE STORED AS AR AND AI, WITH BOTH AR AND AI REAL NUMBERS, AND A = AR + AI i. FOR A REAL QUANTITY, ONLY AR WOULD BE USED.

DEFINITIONS OF THE EISPACK PARAMETERS

A THE MATRIX WHOSE EIGENPROBLEM IS OF INTEREST. AR CONTAINS THE REAL PART, AI THE IMAGINARY PART, IF ANY. AR AND AI ARE REAL VARIABLES OF DIMENSION (NM,N) OR (NM,3) OR (NM,2) WHERE THE SECOND DIMENSION DEPENDS ON WHETHER THE MATRIX IS FULL, TRIDIAGONAL, OR SYMMETRIC TRIDIAGONAL.

FOR SYMMETRIC TRIDIAGONAL STORAGE, STORE THE DIAGONAL ENTRIES IN AR(I,2), AND THE SUBDIAGONAL ELEMENTS IN AR(I,1), LEAVING AR(1,1) EMPTY. THAT IS, IF THE FULL MATRIX IS CALLED B, B(I,I)=AR(I,2), AND B(I,I-1)=AR(I,1).

FOR TRIDIAGONAL STORAGE, STORE THE SUBDIAGONAL

IN COLUMN 1, THE DIAGONAL IN COLUMN 2, AND THE SUPERDIAGONAL IN COLUMN 3. THUS B(I,I-1)=AR(I,1), B(I,I)=AR(I,2), AND B(I,I+1)=AR(I,3). AR(1,1) AND AR(N,3) ARE LEFT EMPTY.

ALFAI FOR THE GENERALIZED PROBLEM, A VECTOR OF LENGTH N CONTAINING THE IMAGINARY PARTS OF THE EIGENVALUES, WHICH MUST HOWEVER BE DIVIDED BY BETA.

ALFAR FOR THE GENERALIZED PROBLEM, A VECTOR OF LENGTH N CONTAINING THE REAL PARTS OF THE EIGENVALUES, WHICH MUST HOWEVER BE DIVIDED BY BETA.

B FOR THE GENERALIZED EIGENVALUE PROBLEM, THE RIGHT HAND SIDE MATRIX IN THE EQUATION A*X=λ*B*X. B SHOULD BE OF ORDER N, AND STORED WITH THE SAME FIRST ROW DIMENSION NM AS THE MATRIX A.

BETA FOR THE GENERALIZED EIGENVALUE PROBLEM, A VECTOR OF LENGTH N CONTAINING DIVISORS FOR THE EIGENVALUES. EIGENVALUE(J)=(ALFAR(J)+i*ALFAI(J))/BETA(J) WHERE i=SQRT(-1). NOTE THAT IF MATRIX B IS SINGULAR, SOME OF THE BETA'S MAY BE ZERO.

BD A VECTOR HOLDING BOUNDS FOR THE ERRORS IN THE EIGENVALUES COMPUTED BY RATQR. BD IS A REAL VECTOR OF DIMENSION N. D THE DIAGONAL ELEMENTS OF A TRIDIAGONAL MATRIX. D IS A REAL VECTOR OF DIMENSION N.

E THE SUBDIAGONAL ELEMENTS OF A TRIDIAGONAL MATRIX. E IS A REAL VECTOR OF DIMENSION N. E(1) DOES NOT REPRESENT AN ENTRY OF THE MATRIX, AND MAY BE CHANGED BY SOME SUBROUTINES. IF YOUR MATRIX IS ALREADY TRIDIAGONAL, THEN YOU MUST STORE YOUR SUBDIAGONAL ELEMENTS HERE. IF YOUR MATRIX IS NOT TRIDIAGONAL, THEN EISPACK WILL TRIDIAGONALIZE IT AND STORE THE ENTRIES IN E, SO IN THAT CASE YOU JUST PROVIDE STORAGE FOR E, BUT YOU DON'T HAVE TO PUT ANYTHING IN IT.

E2 E2(I)=E(I)**2. THE SQUARES OF THE SUBDIAGONAL ELEMENTS OF A TRIDIAGONAL MATRIX. E2 IS A REAL VECTOR OF DIMENSION N. E2(1) DOES NOT CORRESPOND TO AN ENTRY OF THE MATRIX, AND MAY BE CHANGED BY SOME SUBROUTINES. YOU MUST SET THIS VECTOR IN SOME CASES. IN OTHER CASES, IT WILL BE SET FOR YOU.

EPS1 A TOLERANCE PARAMETER FOR BISECT AND RATQR. EPS1 IS A REAL SCALAR.

H A COPY OF THE HESSENBERG MATRIX, WHOSE REAL PART WILL BE STORED IN HR, AND THE IMAGINARY PART IF ANY IN HI. HR AND HI ARE REAL ARRAYS OF DIMENSIONS (NM,N).

IDEF A PARAMETER TO RATQR. IDEF=+1 OR -1 ACCORDING AS THE INPUT MATRIX IS KNOWN TO BE POSITIVE DEFINITE OR NEGATIVE DEFINITE. IF ITS DEFINITENESS IS NOT KNOWN, IDEF=0. IDEF IS AN INTEGER SCALAR.

IERR AN ERROR INDICATING PARAMETER RETURNED BY THE ROUTINES. THE MEANING OF IERR DEPENDS ON THE ROUTINE. IERR IS AN INTEGER SCALAR.

IND A VECTOR THAT HOLDS SUBMATRIX TAGS FOR TINVIT. IND IS AN INTEGER VECTOR OF DIMENSION N.

INT A VECTOR DESCRIBING THE INTERCHANGES IN THE HESSENBERG PROCEDURE. INT IS AN INTEGER VECTOR OF DIMENSION N.

LB THE LOWER LIMIT OF AN INTERVAL IN WHICH EIGENVALUES ARE TO BE SOUGHT. NOTE THAT THOUGH THE NAME LB DEFAULTS TO INTEGER TYPE, LB IS TO BE DECLARED A REAL SCALAR.

M INDICATES THE NUMBER OF EIGENVALUES AND/OR EIGENVECTORS TO BE CALCULATED, OR RETURNS THE ACTUAL NUMBER CALCULATED. FOR A REAL GENERAL MATRIX, M INDICATES THE NUMBER OF COLUMNS OF ZR USED TO STORE EIGENVECTORS. M IS AN INTEGER SCALAR.

MATZ FOR THE GENERALIZED PROBLEM, A LOGICAL VARIABLE WHICH IS .FALSE. IF ONLY THE EIGENVALUES ARE WANTED, AND .TRUE. IF EIGENVECTORS ARE ALSO WANTED.

MM THE VALUE USED IN THE FORTRAN DIMENSION STATEMENTS OF THE CALLING PROGRAM FOR THE ARRAYS WHOSE NOMINAL EXTENT IS M. MM IS AN INTEGER SCALAR.

N THE ORDER OF THE MATRIX. N IS AN INTEGER SCALAR.

NM THE VALUE USED IN THE FORTRAN DIMENSION STATEMENTS OF THE CALLING PROGRAM FOR THE FIRST DIMENSION OF THE ARRAYS WHOSE NOMINAL EXTENT IS N. NM IS AN INTEGER SCALAR.

ORT A VECTOR WHICH CONTAINS INFORMATION ON THE ORTHOGONAL TRANSFORMATIONS CARRIED OUT BY ORTHES, ORTRAN AND ORTBAK. ORT IS A REAL VECTOR OF DIMENSION N.

RM1 A SCRATCH MATRIX. IT IS A REAL ARRAY OF DIMENSION (N,N).

RM2 A SCRATCH MATRIX. IT IS A REAL ARRAY OF DIMENSION (N,N).

RM3 A SCRATCH MATRIX. IT IS A REAL ARRAY OF DIMENSION (N,N+2).

RV1 A SCRATCH VECTOR. IT IS A REAL VECTOR OF DIMENSION N.

RV2 A SCRATCH VECTOR. IT IS A REAL VECTOR OF DIMENSION N.

RV3 A SCRATCH VECTOR. IT IS A REAL VECTOR OF DIMENSION N.

RV4 A SCRATCH VECTOR. IT IS A REAL VECTOR OF DIMENSION N.

RV5 A SCRATCH VECTOR. IT IS A REAL VECTOR OF DIMENSION N.

RV6 A SCRATCH VECTOR. IT IS A REAL VECTOR OF DIMENSION N.

SCALE A VECTOR HOLDING SCALING FACTORS AND PERMUTATIONS FOR BALANC AND CBAL. IT IS A REAL VECTOR OF DIMENSION N.

SELECT A LOGICAL VECTOR SUCH THAT IF SELECT(I) IS TRUE, THEN THE EIGENVECTOR CORRESPONDING TO THE I-TH EIGENVALUE IS TO BE COMPUTED. NO MORE THAN M ELEMENTS OF SELECT SHOULD BE SET TRUE. SELECT IS A LOGICAL VECTOR OF DIMENSION N.

TYPE A LOGICAL PARAMETER TO RATQR WHICH IS .TRUE. IF THE SMALLEST EIGENVALUES ARE TO BE FOUND, AND .FALSE. IF THE LARGEST EIGENVALUES ARE TO BE FOUND.

TAU A VECTOR CONTAINING INFORMATION TO BE PASSED FROM HTRIDI TO HTRIBK. TAU IS A REAL VECTOR OF DIMENSION 2*N.

UB THE UPPER BOUND OF THE INTERVAL IN WHICH EIGENVALUES ARE TO BE COMPUTED. UB IS A REAL SCALAR.

W CONTAINS THE EIGENVALUES OF THE MATRIX. WR CONTAINS THE REAL PART AND WI CONTAINS THE IMAGINARY PART IF ANY. WR AND WI ARE REAL VECTORS OF DIMENSION N OR M.

Z CONTAINS THE EIGENVECTORS OF THE MATRIX STORED AS COLUMNS. ZR CONTAINS THE REAL PART OF THE EIGENVECTOR, ZI THE IMAGINARY PART IF ANY. ZR AND ZI ARE REAL ARRAYS OF DIMENSION (NM,N) OR (NM,M) DEPENDING ON THE ROUTINE WHICH IS CALLED. IF THE J-TH EIGENVALUE IS COMPLEX WITH POSITIVE COMPLEX PART, THEN THE J-TH AND THE J+1-ST COLUMN OF Z MUST BE USED TO COMPUTE THE CORRESPONDING EIGENVECTOR. THE J-TH COLUMN CONTAINS THE REAL, THE J+1-ST COLUMN THE

IMAGINARY PART OF THE EIGENVECTOR. FOR THE J+1-ST EIGENVALUE, WHICH IS THE COMPLEX CONJUGATE OF THE J-TH EIGENVALUE, JUST USE THE COMPLEX CONJUGATE OF THE EIGENVECTOR CONSTRUCTED FOR THE J-TH EIGENVALUE.

****

```
 0     FOR THE STANDARD PROBLEM, WITH REAL MATRIX, GO
         TO 1.
       FOR THE STANDARD PROBLEM, WITH COMPLEX
         MATRIX, GO TO 2.
       FOR THE GENERALIZED PROBLEM, GO TO 3.

 1     IF MATRIX IS SYMMETRIC, GO TO 11
       IF MATRIX IS NOT SYMMETRIC, GO TO 12

 11    IF MATRIX IS ALREADY TRIDIAGONAL, GO TO 111
       IF MATRIX IS NOT ALREADY TRIDIAGONAL, GO TO 112

 111   IF ALL EIGENVALUES AND EIGENVECTORS DESIRED,
         GO TO 1111
       IF ALL EIGENVALUES AND NO EIGENVECTORS DESIRED,
         GO TO 1112
       IF SOME EIGENVALUES AND EIGENVECTORS DESIRED,
         GO TO 1113

 1111  DO I=2,N
         DO J=1,N
          ZR(I,J)=0.0
           IF(I.EQ.J)ZR(I,J)=1.0
         CONTINUE
       CONTINUE
       CALL IMTQL2(NM,N,AR(1,2),AR(1,1),ZR,IERR)
       (THE I-TH EIGENVALUE IS RETURNED IN AR(I,2)).
       STOP

 1112  CALL IMTQL1(N,AR(1,2),AR(1,1),IERR)
       (THE I-TH EIGENVALUE IS RETURNED IN AR(I,2)).
       STOP

 1113  IF THE RATIONAL QR METHOD IS TO BE USED, GO TO
                11131
       IF BISECTION IS TO BE USED, GO TO 11132
```

```
11131 IF WANT CORRESPONDING EIGENVECTORS, GO TO
            111311.
      DO I=2,N
       E2(I)=AR(I,1)**2
       CONTINUE
       CALL RATQR(N,EPS1,AR(1,2),AR(1,1),E2,M,WR,IND,BD,
            TYPE, IDEF,IERR)
      STOP

111311 DO I=2,N
        E2(I)=AR(I,1)**2
        CONTINUE
        CALL RATQR(N,EPS1,AR(1,2),AR(1,1),E2,M,WR,IND,BD,
            TYPE,  IDEF,IERR)
        CALL TINVIT(NM,N,AR(1,2),AR(1,1),E2,M,WR,IND,ZR,
            IERR,RV1, RV2,RV3,RV4,RV6)
      STOP

11132  IF WANT CORRESPONDING EIGENVECTORS, GO TO
            111321
      DO I=2,N
       E2(I)=AR(I,1)**2
       CONTINUE
       CALL BISECT(N,EPS1,AR(1,2),AR(1,1),E2,LB,UB,MM,M,
        WR,IND,IERR,RV4,RV5)
      STOP

111321  DO I=2,N
       E2(I)=AR(I,1)**2
       CONTINUE
       CALL BISECT(N,EPS1,AR(1,2),AR(1,1),E2,LB,UB,MM,M,
            WR,IND,IERR,RV4,RV5)
       CALL TINVIT(NM,N,AR(1,2),AR(1,1),E2,M,WR,IND,ZR,
            IERR,RV1, RV2,RV3,RV4,RV6)
      (THE CALLS TO BISECT AND TINVIT MAY BE REPLACED
            BY
       CALL TSTURM(NM,N,EPS1,D,E,E2,LB,UB,MM,M,WR,
            ZR,IERR,RV1, RV2,RV3,RV4,RV5,RV6) )
      STOP

112    IF ALL EIGENVALUES AND EIGENVECTORS DESIRED,
            GO TO 1121
      IF ALL EIGENVALUES AND NO EIGENVECTORS DESIRED,
```

```
          GO TO 1122
    IF SOME EIGENVALUES AND EIGENVECTORS DESIRED,
          GO TO 1123

1121   CALL TRED2(NM,N,AR,WR,E,ZR)
       CALL TQL2(NM,N,WR,E,ZR,IERR)
    (IMTQL2 MAY REPLACE TQL2, WITH IDENTICAL
          CALLING SEQUENCE)
    STOP

1122  CALL TRED1(NM,N,AR,WR,E,E2)
    CALL TQL1(N,WR,E,IERR)
    (IMTQL1 MAY REPLACE TQL1 WITH IDENTICAL CALLING
          SEQUENCE)
    STOP

1123   IF THE RATIONAL QR METHOD IS TO BE USED, GO TO
          11231
       IF BISECTION IS TO BE USED, GO TO 11232

11231 IF CORRESPONDING EIGENVECTORS ARE DESIRED, GO
          TO 112311
    CALL TRED1(NM,N,AR,D,E,E2)
     CALL RATQR(N,EPS1,D,E,E2,M,WR,IND,BD,TYPE,

          IDEF,IERR)
    STOP

112311  CALL TRED1(NM,N,AR,D,E,E2)
     CALL RATQR(N,EPS1,D,E,E2,M,WR,IND,BD,TYPE,
          IDEF,IERR)
     CALL TINVIT(NM,N,D,E,E2,M,WR,IND,ZR,IERR,RV1,
          RV2,RV3,RV4,RV6)
    CALL TRBAK1(NM,N,AR,E,M,ZR)
    STOP

11232  IF CORRESPONDING EIGENVECTORS ARE DESIRED, GO
          TO 112321
     CALL TRED1(NM,N,AR,D,E,E2)
     CALL BISECT(N,EPS1,D,E,E2,LB,UB,MM,M,WR,IND,
          IERR,RV4,RV5)
    STOP

112321  CALL TRED1(NM,N,AR,D,E,E2)
```

```
 CALL BISECT(N,EPS1,D,E,E2,LB,UB,MM,M,WR,IND,
         IERR,RV4,RV5)
 CALL TINVIT(NM,N,D,E,E2,M,WR,IND,ZR,IERR,RV1,
         RV2,RV3, RV4,RV6)
 CALL TRBAK(NM,N,AR,E,M,ZR)
 STOP
 (THE CONSECUTIVE CALLS TO BISECT AND TINVIT MAY
BE REPLACED
 BY A SINGLE CALL TO
TSTURM(NM,N,EPS1,D,E,E2,LB,UB,MM,
         M,WR,ZR,IERR,RV1,RV2,RV3,RV4,RV5,RV6))

12     IF MATRIX IS ALREADY TRIDIAGONAL, GO TO 121
       IF MATRIX IS NOT ALREADY TRIDIAGONAL, GO TO 122

121    IF ALL EIGENVALUES AND EIGENVECTORS DESIRED, GO
         TO 1211
       IF ALL EIGENVALUES AND NO EIGENVECTORS DESIRED,
         GO TO 1212
       IF SOME EIGENVALUES AND EIGENVECTORS DESIRED,
         GO TO 1213

1211   CALL FIGI2(NM,N,AR,WR,E,ZR,IERR)
       CALL IMTQL2(NM,N,WR,E,ZR,IERR)
       STOP

1212   CALL FIGI(NM,N,AR,WR,E,E2,IERR)
       CALL IMTQL1(N,WR,E,IERR)
       STOP

1213   CALL FIGI(NM,N,AR,D,E,E2,IERR)
       IF RATIONAL QR METHOD TO BE USED, GO TO 12131
       IF BISECTION TO BE USED, GO TO 12132

12131  CALL RATQR(N,EPS1,D,E,E2,M,WR,IND,BD,TYPE,
         IDEF,IERR)
       IF CORRESPONDING EIGENVECTORS DESIRED, GO TO
         121311
       STOP

121311 CALL TINVIT(NM,N,D,E,E2,M,WR,IND,ZR,IERR,RV1,
         RV2,RV3,RV4,RV6)
       CALL BAKVEC(NM,N,AR,E,M,ZR,IERR)
       STOP
```

```
12132    CALL BISECT(N,EPS1,D,E,E2,LB,UB,MM,M,WR,IND,
              IERR,RV4,RV5)
       IF CORRESPONDING EIGENVECTORS DESIRED, GO TO
              121321
       STOP

121321   CALL TINVIT(NM,N,D,E,E2,M,WR,IND,ZR,IERR,RV1,
              RV2,RV3, RV4,RV6)
       CALL BAKVEC(NM,N,AR,E,M,ZR,IERR)
       (THE CONSECUTIVE CALLS TO BISECT AND TINVIT MAY
              BE REPLACED
       BY A SINGLE CALL TO TSTURM(NM,N,EPS1,D,E,E2,LB,
          UB,MM, M,WR,ZR,IERR,RV1,RV2,RV3,RV4,RV5,RV6)

122     IF MATRIX TO BE BALANCED, GO TO 1221
       IF  NOT, SET LOW = 1,  IGH = N AND   GO TO 1222

1221    CALL BALANC(NM,N,AR,LOW,IGH,SCALE)
       IF ELEMENTARY SIMILARITIES ARE TO BE USED, GO TO
              12211
       IF ORTHOGONAL SIMILARITIES ARE TO BE USED, GO TO
              12212

12211   CALL ELMHES(NM,N,LOW,IGH,AR,INT)

       IF ALL EIGENVALUES AND EIGENVECTORS DESIRED, GO
              TO 122111
       IF ALL EIGENVALUES AND NO EIGENVECTORS DESIRED
              GO TO 122112
       IF SOME EIGENVALUES AND EIGENVECTORS DESIRED,
              GO TO 122113

122111  CALL ELTRAN(NM,N,LOW,IGH,AR,INT,ZR)
         CALL HQR2(NM,N,LOW,IGH,AR,WR,WI,ZR,IERR)
        M=N
        CALL BALBAK(NM,N,LOW,IGH,SCALE,M,ZR)
        STOP

122112  CALL HQR(NM,N,LOW,IGH,AR,WR,WI,IERR)
        STOP

122113  DO I=1,N
        DO J=1,N
```

```
HR(I,J)=AR(I,J)
CONTINUE
CONTINUE
CALL HQR(NM,N,LOW,IGH,HR,WR,WI,IERR)
CALL INVIT(NM,N,AR,WR,WI,SELECT,MM,M,ZR,
        IERR,RM3,RV1,RV2)
CALL ELMBAK(NM,LOW,IGH,AR,INT,M,ZR)
CALL BALBAK(NM,N,LOW,IGH,SCALE,M,ZR)
STOP

12212 CALL ORTHES(NM,N,LOW,IGH,AR,ORT)
  IF ALL EIGENVALUES AND EIGENVECTORS DESIRED, GO
        TO 122121
  IF ALL EIGENVALUES AND NO EIGENVECTORS DESIRED
        GO TO 122122
  IF SOME EIGENVALUES AND EIGENVECTORS DESIRED,
        GO TO 122123


122121 CALL ORTRAN(NM,N,LOW,IGH,AR,ORT,ZR)
  CALL HQR2(NM,N,LOW,IGH,AR,WR,WI,ZR,IERR)
  M=N
  CALL BALBAK(NM,N,LOW,IGH,SCALE,M,ZR)
  STOP

122122 CALL HQR(NM,N,LOW,IGH,AR,WR,WI,IERR)
  STOP

122123 DO I=1,N
  DO J=1,N
  HR(I,J)=AR(I,J)
  CONTINUE
  CONTINUE
  CALL HQR(NM,N,LOW,IGH,HR,WR,WI,IERR)
  CALL INVIT(NM,N,AR,WR,WI,SELECT,MM,M,ZR,
        IERR,RM3,RV1,RV2)
  CALL ORTBAK(NM,LOW,IGH,AR,ORT,M,ZR)
  CALL BALBAK(NM,N,LOW,IGH,SCALE,M,ZR)
  STOP

1222 IF ELEMENTARY SIMILARITIES ARE TO BE USED, GO TO
        12221
  IF ORTHOGONAL SIMILARITIES ARE TO BE USED, GO TO
        12222
```

```
12221  CALL ELMHES(NM,N,LOW,IGH,AR,INT)
       IF ALL EIGENVALUES AND EIGENVECTORS DESIRED, GO
              TO 122211
       IF ALL EIGENVALUES AND NO EIGENVECTORS DESIRED
              GO TO 122212
       IF SOME EIGENVALUES AND EIGENVECTORS DESIRED,
              GO TO 122213

122211  CALL ELTRAN(NM,N,LOW,IGH,AR,INT,ZR)
         CALL HQR2(NM,N,LOW,IGH,AR,WR,WI,ZR,IERR)
        STOP

122212  CALL HQR(NM,N,LOW,IGH,AR,WR,WI,IERR)
        STOP

122213   DO I=1,N
           DO J=1,N
            HR(I,J)=AR(I,J)
           CONTINUE
         CONTINUE
          CALL HQR(NM,N,LOW,IGH,HR,WR,WI,IERR)
          CALL INVIT(NM,N,AR,WR,WI,SELECT,MM,M,ZR,
               IERR,RM3,RV1,RV2)
          CALL ELMBAK(NM,LOW,IGH,AR,INT,M,ZR)
          STOP

12222  CALL ORTHES(NM,N,LOW,IGH,AR,ORT)
        IF ALL EIGENVALUES AND EIGENVECTORS DESIRED,
              GO TO 122221
        IF ALL EIGENVALUES AND NO EIGENVECTORS
              DESIRED GO TO 122222
        IF SOME EIGENVALUES AND EIGENVECTORS DESIRED,
              GO TO 122223

122221  CALL ORTRAN(NM,N,LOW,IGH,AR,ORT,ZR)
         CALL HQR2(NM,N,LOW,IGH,AR,WR,WI,ZR,IERR)
        STOP

122222  CALL HQR(NM,N,LOW,IGH,AR,WR,WI,IERR)
        STOP

122223  DO I=1,N
         DO J=1,N
```

```
      HR(I,J)=AR(I,J)
     CONTINUE
    CONTINUE
    CALL HQR(NM,N,LOW,IGH,HR,WR,WI,IERR)
    CALL INVIT(NM,N,AR,WR,WI,SELECT,MM,M,ZR,
           IERR,RM3,RV1,RV2)
    CALL ORTBAK(NM,LOW,IGH,AR,ORT,M,ZR)
    STOP

2     IF MATRIX IS HERMITIAN, GO TO 22
      IF MATRIX IS NOT HERMITIAN, GO TO 21

21    IF MATRIX IS TO BE BALANCED GO TO 211
      IF MATRIX IS NOT TO BE BALANCED, GO TO 212

211   CALL CBAL(NM,N,AR,AI,LOW,IGH,SCALE)
      CALL COMHES(NM,N,LOW,IGH,AR,AI,INT)
      IF ALL EIGENVALUES AND EIGENVECTORS DESIRED, GO
           TO 2111
      IF ALL EIGENVALUES AND NO EIGENVECTORS DESIRED,
           GO TO 2112
      IF SOME EIGENVALUES AND EIGENVECTORS DESIRED,
           GO TO 2113

2111  CALL COMLR2(NM,N,LOW,IGH,INT,AR,AI,WR,WI,
           ZR,ZI,IERR)
      M=N
      CALL CBABK2(NM,N,LOW,IGH,SCALE,M,ZR,ZI)
      STOP

2112  CALL COMLR(NM,N,LOW,IGH,AR,AI,WR,WI,IERR)
      STOP

2113   DO I=1,N
       DO J=1,N
         HR(I,J)=AR(I,J)
         HI(I,J)=AI(I,J)
        CONTINUE
       CONTINUE
        CALL COMLR(NM,N,LOW,IGH,HR,HI,WR,WI,IERR)
        CALL CINVIT(NM,N,AR,AI,WR,WI,SELECT,MM,M,
           ZR,ZI,IERR,RM1,RM2,RV1,RV2)
        CALL COMBAK(NM,LOW,IGH,AR,AI,INT,M,ZR,ZI)
        CALL CBABK2(NM,N,LOW,IGH,SCALE,M,ZR,ZI)
```

```
        STOP

212    LOW=1
       IGH=N
       CALL COMHES(NM,N,LOW,IGH,AR,AI,INT)
       IF ALL EIGENVALUES AND EIGENVECTORS DESIRED, GO
              TO 2121
       IF ALL EIGENVALUES AND NO EIGENVECTORS DESIRED,
              GO TO 2122
       IF SOME EIGENVALUES AND EIGENVECTORS DESIRED,
              GO TO 2123

2121   CALL COMLR2(NM,N,LOW,IGH,INT,AR,AI,WR,
              WI,ZR,ZI,IERR)
       STOP

2122   CALL COMLR(NM,N,LOW,IGH,AR,AI,WR,WI,IERR)
       STOP

2123    DO I=1,N
        DO J=1,N
          HR(I,J)=AR(I,J)
          HI(I,J)=AI(I,J)
         CONTINUE
        CONTINUE
         CALL COMLR(NM,N,LOW,IGH,HR,HI,WR,WI,IERR)
        CALL CINVIT(NM,N,AR,AI,WR,WI,SELECT,MM,M,
              ZR,ZI,IERR,RM1,RM2,RV1,RV2)
        CALL COMBAK(NM,LOW,IGH,AR,AI,INT,M,ZR,ZI)
        STOP

22     IF ALL EIGENVALUES AND EIGENVECTORS DESIRED,
              GO TO 221
       IF ALL EIGENVALUES AND NO EIGENVECTORS DESIRED,
              GO TO 222
       IF SOME EIGENVALUES AND EIGENVECTORS DESIRED,
              GO TO 223

221     DO I=1,N
        DO J=1,N
          ZR(I,J)=0.0
          IF(I.EQ.J)AR(I,J)=1.0
         CONTINUE
        CONTINUE
```

```
      CALL HTRIDI(NM,N,AR,AI,WR,E,E2,TAU)
      CALL TQL2(NM,N,WR,E,ZR,IERR)
      CALL HTRIBK(NM,N,AR,AI,TAU,M,ZR,ZI)
      STOP
      (IMTQL2 MAY REPLACE TQL2 WITH SAME CALLING
            SEQUENCE)

222   CALL HTRIDI(NM,N,AR,AI,WR,E,E2,TAU)
      CALL TQL1(N,WR,E,IERR)
      STOP
      (IMTQL1 MAY REPLACE TQL1 WITH SAME CALLING
            SEQUENCE)

223   IF RATIONAL QR IS TO BE USED, GO TO 2231
      IF BISECTION IS TO BE USED, GO TO 2232

2231  CALL RATQR(N,EPS1,D,E,E2,M,WR,IND,BD,
            TYPE,IDEF,IERR)
      IF CORRESPONDING VECTORS DESIRED, GO TO 22311
      STOP

22311 CALL TINVIT(NM,N,D,E,E2,M,WR,IND,ZR,IERR,
            RV1,RV2,RV3,RV4,RV6)
      CALL HTRIBK(NM,N,AR,AI,TAU,M,ZR,ZI)
      STOP

2232  CALL BISECT(N,EPS1,D,E,E2,LB,UB,MM,M,WR,
            IND,IERR,RV4,RV5)
      IF CORRESPONDING EIGENVECTORS DESIRED, GO TO
            22321
      STOP

22321 CALL TINVIT(NM,N,D,E,E2,M,WR,IND,ZR,IERR,
            RV1,RV2,RV3,RV4,RV6)
      CALL HTRIBK(NM,N,AR,AI,TAU,M,ZR,ZI)
      STOP
      (THE CONSECUTIVE CALLS TO BISECT AND TINVIT
       MAY BE REPLACED BY A SINGLE CALL TO
       TSTURM(NM,N,EPS1,D,E,E2,LB,UB,MM,
            M,WR,ZR,IERR,RV1,RV2,RV3,RV4,RV5,RV6)

3     IF ONLY EIGENVALUES DESIRED, GO TO 31.
      IF EIGENVALUES AND EIGENVECTORS DESIRED, GO TO
```

```
              32.

31     MATZ=.FALSE.
       CALL QZHES(NM,N,A,B,MATZ,Z)
       CALL QZIT(NM,N,A,B,EPS1,MATZ,Z,IERR)
       CALL QZVAL(NM,N,A,B,ALFR,ALFI,BETA,MATZ,Z)
       STOP

32     MATZ=.TRUE.
       CALL QZHES(NM,N,A,B,MATZ,Z)
       CALL QZIT(NM,N,A,B,EPS1,MATZ,Z,IERR)
       CALL QZVAL(NM,N,A,B,ALFR,ALFI,BETA,MATZ,Z)
       CALL QZVEC(NM,N,A,B,ALFR,ALFI,BETA,Z)
       STOP
```

****

LIST OF SUBROUTINES IN EISPAK AND THEIR CALLING SEQUENCES

```
BAKVEC(NM,N,AR,E,M,ZR,IERR)

BALANC(NM,N,AR,LOW,IGH,SCALE)

BALBAK(NM,N,LOW,IGH,SCALE,M,ZR)

BISECT(N,EPS1,D,E,E2,LB,UB,MM,M,WR,IND,IERR,RV4,RV5)

 OR

 BISECT(N,EPS1,AR(1,2),AR(1,1),E2,LB,UB,MM,M,WR,
          IND,IERR,RV4,RV5)

CBABK2(NM,N,LOW,IGH,SCALE,M,ZR,ZI)

CBAL(NM,N,AR,AI,LOW,IGH,SCALE)

 CINVIT(NM,N,AR,AI,WR,WI,SELECT,MM,M,ZR,ZI,
          IERR,RM1,RM2,RV1,RV2)

COMBAK(NM,LOW,IGH,AR,AI,INT,M,ZR,ZI)

COMHES(NM,N,LOW,IGH,AR,AI,INT)
```

COMLR(NM,N,LOW,IGH,AR,AI,WR,WI,IERR)

OR

COMLR(NM,N,LOW,IGH,HR,HI,WR,WI,IERR)

COMLR2(NM,N,LOW,IGH,INT,AR,AI,WR,WI,ZR,ZI,IERR)

ELMBAK(NM,LOW,IGH,AR,INT,M,ZR)

ELMHES(NM,N,LOW,IGH,AR,INT)

ELTRAN(NM,N,LOW,IGH,AR,INT,ZR)

FIGI(NM,N,AR,D,E,E2,IERR)

OR

FIGI(NM,N,AR,WR,E,E2,IERR)

FIGI2(NM,N,AR,WR,E,ZR,IERR)

HQR(NM,N,LOW,IGH,HR,WR,WI,IERR)

OR

HQR(NM,N,LOW,IGH,AR,WR,WI,IERR)

HQR2(NM,N,LOW,IGH,AR,WR,WI,ZR,IERR)

HTRIBK(NM,N,AR,AI,TAU,M,ZR,ZI)

HTRIDI(NM,N,AR,AI,WR,E,E2,TAU)

IMTQL1(N,AR(1,2),AR(1,1),IERR)

OR

IMTQL1(N,WR,E,IERR)

IMTQL2(NM,N,AR(1,2),AR(1,1),ZR,IERR)

OR

```
IMTQL2(NM,N,WR,E,ZR,IERR)

INVIT(NM,N,AR,WR,WI,SELECT,MM,M,ZR,IERR,RM3,RV1,RV2)

ORTBAK(NM,LOW,IGH,AR,ORT,M,ZR)

ORTHES(NM,N,LOW,IGH,AR,ORT)

ORTRAN(NM,N,LOW,IGH,AR,ORT,ZR)

QZHES(NM,N,A,B,MATZ,Z)

QZIT(NM,N,A,B,EPS1,MATZ,Z,IERR)

QZVAL(NM,N,A,B,ALFR,ALFI,BETA,MATZ,Z)

QZVEC(NM,N,A,B,ALFR,ALFI,BETA,Z)

RATQR(N,EPS1,D,E,E2,M,WR,IND,BD,TYPE,IDEF,IERR)

 OR

RATQR(N,EPS1,AR(1,2),AR(1,1),E2,M,WR,IND,BD,TYPE,
            IDEF,IERR)

 TINVIT(NM,N,D,E,E2,M,WR,IND,ZR,IERR,RV1,
            RV2,RV3,RV4,RV6)

 OR

TINVIT(NM,N,AR(1,2),AR(1,1),E2,M,WR,IND,ZR,IERR,
            RV1,RV2,RV3,RV4,RV6)

TQL1(N,WR,E,IERR)

TQL2(NM,N,WR,E,ZR,IERR)

TRBAK1(NM,N,AR,E,M,ZR)

TRED1(NM,N,AR,D,E,E2)

 OR

TRED1(NM,N,AR,WR,E,E2)
```

TRED2(NM,N,AR,WR,E,ZR)

TSTURM(NM,N,EPS1,D,E,E2,LB,UB,MM,M,W,ZR,IERR,
RV1,RV2,RV3,RV4,RV5,RV6)

****

Answers and Hints for Selected Exercises

CHAPTER 1

1.1 ■ Exercise Set

1. $[A \mid I] \rightarrow \begin{bmatrix} 1 & 0 & 0 & 1 & -1 & 1 \\ 0 & 1 & 0 & -1 & 2 & 1 \\ 0 & 0 & 1 & -1 & 1 & 1 \\ 0 & 0 & 0 & 0 & 0 & 0 \end{bmatrix} \Rightarrow X = \begin{bmatrix} 1 & -1 & 1 \\ -1 & 2 & 1 \\ -1 & 1 & 1 \end{bmatrix}$

2. $A^{-1} = \begin{bmatrix} 0 & 1 & 2 \\ -1 & 3 & 0 \\ 1 & -2 & 1 \end{bmatrix}$

3. $X = \begin{bmatrix} \frac{-3834}{871} & \frac{228}{871} & \frac{5841}{1742} & \frac{-47}{13} & \frac{3459}{1742} \end{bmatrix}^T$

6. a. $AB = \begin{bmatrix} 12 & 0 & 5 \\ 17 & 2 & 8 \\ 9 & 11 & 4 \\ 28 & -23 & 18 \end{bmatrix}$; BA is not defined.

b. $\det(C) = 0$, $\det(C^{-1}) = -2.28 \times 10^{15}$, with a warning that C is likely to be singular.

c. B^{-1} is not defined.

d. $(B^TB)^{-1} = \begin{bmatrix} 0.1818 & 0 & -0.0909 \\ 0 & 0.2 & 0 \\ -0.0909 & 0 & 0.5455 \end{bmatrix}$

e. $\begin{bmatrix} 348 & 676 & 662 & 295 \\ 434 & 858 & 799 & 320 \\ -62 & -73 & -234 & -230 \\ 1491 & 2834 & 3030 & 1506 \end{bmatrix}$

f. $\begin{bmatrix} 17 & -3 & 84 & 11 \\ -6 & -6 & 25 & 11 \\ -89 & -195 & -168 & 46 \\ 247 & 570 & 297 & -137 \end{bmatrix}$

g. 70

7. $H^{-1} = \begin{bmatrix} 25 & -300 & 1050 & -1400 & 630 \\ -300 & 4800 & -18900 & 26880 & -12600 \\ 1050 & -18900 & 79380 & -117600 & 56700 \\ -1400 & 26880 & -117600 & 179200 & -88200 \\ 630 & -12600 & 56700 & -88200 & 44100 \end{bmatrix}$ exactly.

MATLAB computes

H ^{1}=1.0e+05*

$$\left[\begin{array}{lll} 0.00025000000000 & -0.00300000000000 & 0.01049999999999 \\ -0.00300000000000 & 0.04799999999996 & -0.18899999999982 \\ 0.01049999999999 & -0.18899999999982 & 0.79379999999930 \\ -0.01399999999998 & 0.26879999999975 & -1.17599999999901 \\ 0.00629999999999 & -0.12599999999988 & 0.56699999999954 \end{array}\right.$$

$$\left.\begin{array}{ll} -0.01399999999998 & 0.00629999999999 \\ 0.26879999999975 & -0.12599999999988 \\ -1.17599999999902 & 0.56699999999954 \\ 1.79199999999862 & -0.88199999999936 \\ -0.88199999999936 & 0.44099999999971 \end{array}\right]$$

H^*H ^{1}=

$$\left[\begin{array}{lll} 1.00000000000033 & 0.00000000000041 & -0.00000000000010 \\ 0.00000000000030 & 0.99999999999981 & 0.00000000000008 \\ 0.0000000000030 & 0.00000000000016 & 1.0000000000228 \\ 0.0000000000024 & -0.00000000000023 & 0.0000000000091 \\ 0.0000000000022 & 0.00000000000003 & 0.0000000000066 \end{array}\right.$$

$$\left.\begin{array}{ll} -0.00000000000680 & 0.0000000000522 \\ -0.00000000000282 & 0.0000000000293 \\ -0.00000000000424 & 0.0000000000316 \\ 0.99999999999636 & 0.0000000000182 \\ -0.00000000000249 & 1.0000000000165 \end{array}\right]$$

and

H ^{1}H=

$$\begin{bmatrix} 1.00000000000002 & -0.00000000000001 & -0.00000000000000 & 0.00000000000001 & -0.00000000000000 \\ 0.00000000000022 & 1.00000000000027 & -0.00000000000016 & 0.00000000000023 & 0.00000000000025 \\ 0.00000000000063 & 0.00000000000069 & 1.00000000000020 & 0 & -0.00000000000075 \\ 0.00000000000120 & 0.00000000000082 & -0.00000000000060 & 1.00000000000000 & -0.00000000000006 \\ -0.00000000000024 & -0.00000000000071 & -0.00000000000048 & -0.00000000000091 & 0.99999999999983 \end{bmatrix}$$

12. $A\boldsymbol{\varepsilon}_i = \text{Col}_i(A) = 0$ for $i = 1, 2, \ldots, n$ implies $A = 0$.

1.2 ■ Exercise Set

1. For $\sqrt{3}: 0.1732 \times 10^1$; for $\pi: 0.3142 \times 10^1$; for $e: 0.2718 \times 10^1$; for $1/(57!): 0.2467 \times 10^{-76}$; and for $68!: 0.2480 \times 10^{97}$.

2. Absolute errors are 1.264×10^{-3} and 2.667×10^{-7}. Relative errors are 4.025×10^{-4} and 8.489×10^{-8}.

4. **a.** $e^{-6} \approx 8.774 \times 10^{-3}$; $e^6 \approx 403.4$.

b. Relative error for e^{-6} is 2.540; for e^6 is 7.137×10^{-5}.

5. **a.** $0.0012335 \le x < 0.0012345$

b. $356.65 \le x < 356.75$.

6. **a.** -14.99

b. -15.01

c. 1.023×10^{-3}, 3.10×10^{-4}

d. Seven for the direct method and five using part b.

7. If x, y, and z are large enough that their squares overflow, the second method still gives an accurate result.

9. The smallest machine number ε such that $\varepsilon + 1.0 = 1.0$. Machine epsilon.

CHAPTER 2

2.1 ■ Exercise Set

1. **a.** Step 2: Change $m - 1$ to m

c. Step 3: Change to $k = 1, 2, \ldots, m, \neq i$.

2. $A \to \begin{bmatrix} 5 & 7 & 6 & 5 \\ 0 & \frac{1}{5} & -\frac{2}{5} & 0 \\ 0 & 0 & 2 & 3 \\ 0 & 0 & 0 & \frac{1}{2} \end{bmatrix}$, $B \to \begin{bmatrix} 4 & -1 & -1 & 0 & 1 \\ 0 & \frac{15}{4} & -\frac{5}{4} & -1 & \frac{9}{4} \\ 0 & 0 & \frac{10}{3} & -\frac{4}{3} & 4 \\ 0 & 0 & 0 & \frac{16}{5} & \frac{31}{5} \end{bmatrix}$

3. $A \to \begin{bmatrix} 7 & 10 & 8 & 7 \\ 0 & -\frac{4}{7} & \frac{22}{7} & 3 \\ 0 & 0 & \frac{5}{2} & \frac{17}{4} \\ 0 & 0 & 0 & \frac{1}{10} \end{bmatrix}$, $B \to \begin{bmatrix} 4 & -1 & -1 & 0 & 1 \\ 0 & \frac{15}{4} & -\frac{5}{4} & -1 & \frac{9}{4} \\ 0 & 0 & \frac{10}{3} & -\frac{4}{3} & 4 \\ 0 & 0 & 0 & \frac{16}{5} & \frac{31}{5} \end{bmatrix}$

5. a. $x_1 = 0.00$, $x_2 = 1.00$
b. $x_1 = x_2 = 1.00$
c. $x_1 = 10000/9999$, $x_2 = 9998/9999$

7. $\displaystyle\int_{i=1}^{m-1} \int_{k=i+1}^{m} \left(1 + \int_{j=i}^{n} dj\right) dk\, di$

8. b. $\displaystyle\sum_{i=1}^{n} i^3 = \frac{n^2(n+1)^2}{4}$

10. a. $n(n-1)(n/3 + 5/6) = n^33 + \cdots$
b. $(n/2)(n+1) = n^2/2 + \cdots$

11. $m(m-1)(n - m/3 + 11/6)$; approximately twice the number of multiplications.

12. The reduced matrix has at least one zero row.

2.2 ■ Exercise Set

1. For $i > j$, $\text{ent}_{ij}(AB) = \displaystyle\sum_{k=1}^{n} a_{ik}b_{kj} = \sum_{k=1}^{i-1} \underbrace{a_{ik}}_{=0} b_{kj} + \sum_{k=i}^{n} a_{ik} \underbrace{b_{kj}}_{=0} = 0$

3. $A \to L = \begin{bmatrix} 1 & 0 & 0 \\ 3 & 1 & 0 \\ 3 & \frac{1}{5} & 1 \end{bmatrix}$, $U = \begin{bmatrix} 1 & 1 & 1 \\ 0 & -5 & -20 \\ 0 & 0 & 0 \end{bmatrix}$;

$$B \to L = \begin{bmatrix} 1 & 0 & 0 & 0 \\ 1 & 1 & 0 & 0 \\ 2 & 2 & 1 & 0 \\ -2 & 0 & 0 & 1 \end{bmatrix}, \ U = \begin{bmatrix} 2 & 3 & -1 & 2 \\ 0 & 1 & 5 & -3 \\ 0 & 0 & -6 & 9 \\ 0 & 0 & 0 & 5 \end{bmatrix}$$

4. For A the second pivot is zero. $B = \begin{bmatrix} 1 & 0 & 0 & 0 \\ 1 & 1 & 0 & 0 \\ \frac{3}{2} & 0 & 1 & 0 \\ 2 & -3 & 10 & 1 \end{bmatrix}\begin{bmatrix} 2 & 4 & 3 & 2 \\ 0 & 1 & -1 & -5 \\ 0 & 0 & \frac{1}{2} & -1 \\ 0 & 0 & 0 & 5 \end{bmatrix}.$

5. Do the reduction of A to U using MATMAN; build L by hand.

6. $AX = B \Leftrightarrow QRX = B \Leftrightarrow RX = Q^TX$; solve by back substitution.

8. $A \to L = \begin{bmatrix} 1 & 0 & 0 \\ 1 & 1 & 0 \\ \frac{1}{3} & \frac{5}{12} & 1 \end{bmatrix}, \ U = \begin{bmatrix} 3 & -2 & -17 \\ 0 & 4 & 16 \\ 0 & 0 & 0 \end{bmatrix}, \ P = \begin{bmatrix} 0 & 1 & 0 \\ 0 & 0 & 1 \\ 1 & 0 & 0 \end{bmatrix};$

$$B \to L = \begin{bmatrix} 1 & 0 & 0 & 0 \\ 0.25 & 1 & 0 & 0 \\ 0.50 & -0.857143 & 1 & 0 \\ 0.25 & -0.714286 & 0.837945 & 1 \end{bmatrix},$$

$$U = \begin{bmatrix} 4 & 1 & -11 & -1 \\ 0 & 1.75 & 35.75 & 83.25 \\ 0 & 0 & 36.1429 & 73.8571 \\ 0 & 0 & 0 & 0.826083 \end{bmatrix}, \ P = \begin{bmatrix} 0 & 0 & 1 & 0 \\ 0 & 0 & 0 & 1 \\ 0 & 1 & 0 & 0 \\ 1 & 0 & 0 & 0 \end{bmatrix}$$

9. $X_1 = \begin{bmatrix} 0 \\ -0.285714 \\ 0.428571 \end{bmatrix}, \ X_2 = \begin{bmatrix} 0.214286 \\ 0.183673 \\ -0.132653 \end{bmatrix}, \ X_3 = \begin{bmatrix} -0.173469 \\ -0.097676 \\ 0.161808 \end{bmatrix}$

12. a. $A = L(DR)$, $A = A^T = R^T(D^TL^T) \Rightarrow L = R^T$ by Theorem 2.1.
$A = LDL^T = LD^{1/2}D^{1/2}L^T = (LD^{1/2})(LD^{1/2})^T = BB^T$.

b. Show that $X^TAX = X^T(LDL^T)X = (L^TX)^TD(L^TX) > 0$ for all X if and only if all $d_{ii} > 0$.

15. $\begin{bmatrix} 0 & 0 & 1 & 0 \\ 0 & 1 & 0 & 0 \\ 0 & 0 & 0 & 1 \\ 1 & 0 & 0 & 0 \end{bmatrix} A = \begin{bmatrix} 1 & 0 & 0 & 0 \\ -\frac{2}{3} & 1 & 0 & 0 \\ \frac{1}{3} & x & 1 & 0 \\ \frac{1}{3} & x & \frac{1}{5} & 1 \end{bmatrix}\begin{bmatrix} 6 & 9 & 22 & 14 \\ 0 & 0 & \frac{65}{3} & \frac{19}{3} \\ 0 & 0 & \frac{25}{2} & -\frac{7}{2} \\ 0 & 0 & 0 & \frac{6}{5} \end{bmatrix}$

$x = 0$ with hand calculations and $x = -\frac{1}{2}$ with machine calculations. Give a reason for the difference.

16. b. $U = \begin{bmatrix} 6 & 9 & 22 & 14 \\ 0 & 0 & \frac{65}{3} & \frac{19}{3} \\ 0 & 0 & 0 & -\frac{93}{13} \\ 0 & 0 & 0 & 0 \end{bmatrix}$

2.3 ■ Exercise Set

1.
$$\left.\begin{aligned} v(X) &= v(Y + X - Y) \\ &\leq v(Y) + v(X - Y) \\ &\Rightarrow v(X) - v(Y) \leq v(X - Y) \\ v(Y) &= v(X + Y - X) \\ &\leq v(X) + v(Y - X) \\ &\Rightarrow v(Y) - v(X) \leq v(X - Y) \end{aligned}\right\} \Rightarrow |v(Y) - v(X)| \leq v(X - Y)$$

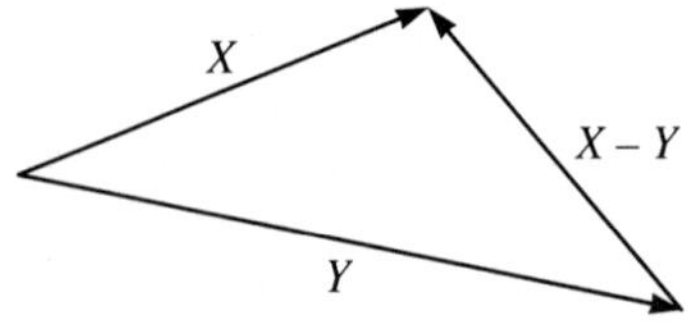

4. The function m is a norm if and only if A is nonsingular.

6. (1) For $X \neq 0$, $X^TAX > 0 \Rightarrow v(X) = \sqrt{X^TAX} > 0$.
(2) $v(kX) = \sqrt{(kX)^TA(kX)} = \sqrt{k^2X^TAX} = |k|\, v(X)$.
(3) By Exercise 12 in Section 2.2, $A = BB^T$.
$v(X + Y) = \sqrt{(X + Y)^TBB^T(X + Y)} = \sqrt{(B^T(X + Y))^T(B^T(X + Y))}$
$= \|(B^T(X + Y)\|_2 \leq \|B^TX\|_2 + \|B^TY\|_2 = \sqrt{(B^TX)^T(B^TX)}$
$+ \sqrt{Y^TBB^TY} = \sqrt{X^TAX} + \sqrt{Y^TAY} = v(X) + v(Y)$

8. b. Let U be a vector whose entries are ± 1 such that $u_ix_i = |x_i|$. By Cauchy-Schwarz, $\|X\|_1 = \sum_{i=1}^n |x_i| = U^TX \leq \|U\|_2\, \|X\|_2 = \sqrt{n}\, \|X\|_2$.

$$\begin{aligned} \|X\|_1^2 &= (|x_1| + |x_2| + \cdots + |x_n|)^2 \\ &= \sum_{i=1}^{n} |x_i|^2 + 2\sum_{i \neq j} |x_ix_j| \geq \|X\|_2^2. \quad \text{Thus, } \|X\|_1 \geq \|X\|_2. \end{aligned}$$

9. Using MATLAB:

```
A=hilb(5), T=rand(5,10), for i=1:10, X=T(:,i),
v(i)=norm(A*X,1)/norm(X,1), end, M=max(v), m=min(v).
```

11. By Exercise 8a, $\lim_{n\to\infty} \|X^{(n)} - X\|_\infty = 0 \Leftrightarrow \lim_{m\to\infty} \|X^{(n)} - X\|_2 = 0$. The result then follows from Theorem 2.6 for the ∞-norm.

12. $\|A\|_1 = 2.2833,\ \|A^{-1}\|_1 = 413280.$

14. $$\|A^{-1}\| = \max\{\|A^{-1}X\| : \|X\| = 1\} = \max\left\{\frac{\|A^{-1}X\|}{\|X\|} : X \neq 0\right\}$$

$$= (\text{let } X = AY) \max\left\{\frac{\|Y\|}{\|AY\|} : Y \neq 0\right\} = \max\left\{\frac{1}{\|AY\|} : \|Y\| = 1\right\}$$

$$= \frac{1}{\min\{\|AY\| : \|Y\| = 1\}}.$$

15. b. Let Y be a unit vector such that $\|AY\|_2 = \|A\|_2$. Let X be a unit vector in the direction of AY. Show that $|X^TAY| = \|A\|_2$.

16. The first assertion follows directly from Exercise 15 of this section.

$$\|A^TA\|_2 \leq \|A^T\|_2\,\|A\|_2 = \|A\|_2^2,$$

so

$$\|A^TA\| = \max\{|Y^TA^TAX| : \|X\|_2 = \|Y\|_2 = 1\} \leq \|A\|_2^2.$$

Now let X be a unit vector such that $\|AX\|_2 = \|A\|_2$. Then $|X^TA^TAX| = \|A\|_2^2$.

17. a. From Theorem 2.5, $\dfrac{\|AX\|_1}{\|X\|_1} \leq \dfrac{n\,\|AX\|_\infty}{\|X\|_\infty}$. Thus $\|A\|_1$

$$=\max\left\{\frac{\|AX\|_1}{\|X\|_1} : X \neq 0\right\} \leq \max\left\{n\frac{\|AX\|_\infty}{\|X\|_\infty} : X \neq 0\right\} = n\,\|A\|_\infty.$$

2.4 ■ Exercise Set

2. Since $\|AX\|_2^2 = X^TA^TAX = X^TX = \|X\|_2^2$ for all X, $\|A\|_2 = 1$.
By Exercise 16 in Section 2.3, $\text{cond}_2(A) = \|A\|_2\,\|A^{-1}\|_2 = \|A\|_2\,\|A^T\|_2 = \|A\|_2^2 = 1$.

$$\text{cond}_1(H) = \|A\|_1\,\|A^T\|_1 = \|A\|_1\,\|A\|_\infty = \|A^T\|_\infty\,\|A\|_\infty = \text{cond}_\infty(A).$$

3. 39601, 51.6667, 2310

4. $\text{cond}_\infty(H) = 943655.86$

7. $X = \begin{bmatrix} 2.0000 & 1.8367 & 2.1663 \\ 1.5000 & 1.6683 & 1.3332 \\ -2.5000 & -2.3417 & -2.6658 \\ 2.5000 & 2.1883 & 2.8302 \end{bmatrix}$

$$\frac{\|\Delta X_1\|_1}{\|X\|_1} = 0.0943, \quad \frac{\|\Delta B_1\|_1}{\|B\|_1} = 0.0530 \Rightarrow \text{cond}_1(A) \geq 1.78$$

$$\frac{\|\Delta X_2\|_1}{\|X\|_1} = 0.0975, \quad \frac{\|\Delta B_2\|_1}{\|B\|_1} = 0.0505 \Rightarrow \text{cond}_1(A) \geq 1.931$$

9. $H^{-1} = \begin{bmatrix} 9 & -36 & 30 \\ -36 & 192 & -180 \\ 30 & -180 & 180 \end{bmatrix},$

$$(H + \Delta H)^{-1} = \begin{bmatrix} 5.087 & -12.5217 & 6.5217 \\ -12.5217 & 51.1304 & -39.1304 \\ 6.5217 & -39.1304 & 39.1304 \end{bmatrix}, \ \text{cond}_\infty(H) = 748.$$

10. $A^{-1} = \frac{1}{6}\begin{bmatrix} -2 & 2 & 6 & -2 \\ -1 & 1 & 0 & 2 \\ 5 & -5 & -6 & 2 \\ -1 & 7 & 6 & -4 \end{bmatrix},$

$$(A + \Delta M)^{-1} = \begin{bmatrix} -0.3329 & 0.3329 & 0.9997 & -0.3332 \\ -0.1673 & 0.1673 & 0.0008 & 0.3331 \\ 0.8331 & -0.8331 & -1.0002 & 0.3334 \\ -0.1656 & 1.1656 & 0.9998 & -0.6663 \end{bmatrix}.$$

Using $B = \begin{bmatrix} 1 \\ 0 \\ 0 \\ 0 \end{bmatrix}, \ \dfrac{\dfrac{\|\Delta X\|_1}{\|X + \Delta X\|_1}}{\dfrac{\|\Delta A\|_1}{\|A\|_1}} = 4.78 \leq \text{cond}_1(A).$

11. $\text{cond}_1(A) = 18$, $\text{cond}_1(B) = 1.9673 \times 10^3$, $\text{cond}_1(C) = 2.8375 \times 10^4$

14. $\text{cond}_1(J) = 8.4444$, $\text{cond}_2(J) = 5.0755$, $\text{cond}_\infty(J) = 7.3333$

15. Use Exercise 17 from Section 2.3.

2.5 ■ Exercise Set

2. $X^TA^TAX = (AX)^T(AX) = \|AX\|_2^2 \geq 0$
$\text{ent}_{ii}(A^TA) = \text{Row}_i(A^T)\,\text{Col}_i(A) = \text{Col}_i(A)^T\,\text{Col}_i(A) = \|\text{Col}_i(A)\|_2^2 \geq 0$
Recall Theorem 1.1.

3. $A^TA = \begin{bmatrix} 10 & 14 \\ 14 & 20 \end{bmatrix}$ is not diagonally dominant. For $X \neq 0$, $X^TA^TAX = 10x_1^2 + 28x_1x_2 + 20x_2^2 = 10(x_1 + 1.4x_2)^2 + 0.4x_2^2 > 0$.

4. $AX = \lambda X, X \neq 0 \Rightarrow X^TAX = \lambda X^TX \Rightarrow \lambda = \dfrac{X^TAX}{X^TX} > 0$

5. $X^TAX = ax_1^2 + 2bx_1x_2 + cx_2^2 = a\left(x_1 + \dfrac{b}{a}x_2\right)^2 + \left(\dfrac{ac - b^2}{a}\right)x_2^2$

7. For $X = x_i\mathbf{e_i} + x_j\mathbf{e_j}$, $X^TAX = a_{ii}x_i^2 + 2a_{ij}x_ix_j + a_{jj}x_j^2 = [x_i \;\; x_j]\begin{bmatrix} a_{ii} & a_{ij} \\ a_{ji} & a_{jj} \end{bmatrix}\begin{bmatrix} x_i \\ x_j \end{bmatrix} > 0$.
Thus $\begin{bmatrix} a_{ii} & a_{ij} \\ a_{ji} & a_{jj} \end{bmatrix}$ is positive definite. The result now follows from Exercise 5 in this section.

8. $A = LL^T$ where $L^T = \begin{bmatrix} 2.0000 & -0.5000 & 0 & 0 & 0 \\ 0 & 1.9365 & -0.5164 & 0 & 0 \\ 0 & 0 & 1.9322 & -0.5175 & 0 \\ 0 & 0 & 0 & 1.9319 & -0.5176 \\ 0 & 0 & 0 & 0 & 1.9319 \end{bmatrix}$

10. $AX = L(L^TX) = B \Rightarrow Y = L^TX = \begin{bmatrix} 0.5000 \\ 1.1619 \\ 1.8632 \\ 2.5697 \\ 2.2414 \end{bmatrix}$; $L^TX = Y \Rightarrow X = \begin{bmatrix} 0.4936 \\ 0.9744 \\ 1.4038 \\ 1.6410 \\ 1.1603 \end{bmatrix}$.

12. $B = LL^T$, where $L^T = \begin{bmatrix} 2.0000 & -0.5000 & -0.5000 & 0 & 0 \\ 0 & 1.9365 & -0.6455 & -0.5164 & 0 \\ 0 & 0 & 1.8257 & -0.7303 & -0.5477 \\ 0 & 0 & 0 & 1.7889 & -0.7826 \\ 0 & 0 & 0 & 0 & 1.7571 \end{bmatrix}$

CHAPTER 3

3.1 ■ Exercise Set

1. a. $\lambda^3 - 12\lambda^2 + 21\lambda - 10$; 1, 1, 10; $P = \begin{bmatrix} 1 & 0 & 2 \\ 0 & 1 & 2 \\ -2 & -2 & 1 \end{bmatrix}$.

c. $\lambda^4 - 13\lambda^2 + 36$; 2, −2, 3, −3; $P = \begin{bmatrix} 1 & 1 & 1 & 1 \\ 2 & -2 & 3 & -3 \\ 4 & 4 & 9 & 9 \\ 8 & -8 & 27 & -27 \end{bmatrix}$.

2. a. $\lambda^2 - 8\lambda + 15$; 3, 5; $X_1 = c_1\begin{bmatrix} 1 \\ 1 \end{bmatrix}$, $X_2 = \begin{bmatrix} 1 \\ -1 \end{bmatrix}$; $P = \frac{1}{\sqrt{2}}\begin{bmatrix} 1 & 1 \\ 1 & -1 \end{bmatrix}$.

c. $\lambda^3 - 6\lambda^2 + 9\lambda - 4$; 1, 1, 4; $X_{1,2} = c_1\begin{bmatrix} 0 \\ 1 \\ 1 \end{bmatrix} + c_2\begin{bmatrix} 1 \\ 1 \\ 0 \end{bmatrix}$, $X_3 = c_3\begin{bmatrix} 1 \\ -1 \\ 1 \end{bmatrix}$;

$$P = \begin{bmatrix} 0 & \frac{2}{\sqrt{6}} & \frac{1}{\sqrt{3}} \\ \frac{1}{\sqrt{2}} & \frac{1}{\sqrt{6}} & \frac{-1}{\sqrt{3}} \\ \frac{1}{\sqrt{2}} & \frac{-1}{\sqrt{6}} & \frac{1}{\sqrt{3}} \end{bmatrix}.$$

3. b. $Y = c_1\begin{bmatrix} -1 \\ 1 \\ 1 \end{bmatrix}e^{4t} + c_2\begin{bmatrix} 0 \\ 1 \\ -1 \end{bmatrix}e^{3t} + c_3\begin{bmatrix} 2 \\ 1 \\ 1 \end{bmatrix}e^{-5t}$.

c. $c_1 = \frac{1}{3}$, $c_2 = 0$, $c_3 = \frac{2}{3}$.

5. Suppose that λ_1 is an eigenvalue of A of algebraic multiplicity e and geometric multiplicity f. Let $\{X_1, X_2, \ldots, X_f\}$ be a basis for $\mathscr{E}\mathscr{S}(\lambda_1)$. Complete this independent set to a basis $\{X_1, X_2, \ldots, X_f, X_{f+1}, \ldots, X_n\}$ for $\mathscr{R}^n$, so that $P = [X_1, X_2, \ldots, X_f, X_{f+1}, \ldots, X_n]$ is nonsingular. Then

$$AP = [\lambda_1 X_1 \quad \lambda_1 X_2 \cdots \lambda_1 X_f \quad PX_{f+1} \cdots PX_n] = P\left[\begin{array}{ccc|c} \lambda_1 & & & \\ & \ddots & & B_{12} \\ & & \lambda_1 & \\ \hline & 0 & & B_{22} \end{array}\right],$$

so that

$$P^{-1}AP = \begin{bmatrix} \lambda_1 I_f & B_{12} \\ 0 & B_{22} \end{bmatrix} = B.$$

The characteristic polynomial of B, which is the same as the characteristic polynomial of A, has $(\lambda - \lambda_1)^f$ as a factor. Since e is the algebraic multiplicity of λ_1, it follows that $f \leq e$.

7. (3) The trace is the sum of the eigenvalues by Exercise 4b.

(4) $\det(A)$ is the product of the eigenvalues by Exercise 4b.

(5) Similar matrices are equivalent and hence have the same rank.

8. $$\begin{bmatrix} a_{11} & a_{12} + ib_{12} & a_{13} + ib_{13} \\ a_{12} - ib_{12} & a_{22} & a_{23} + ib_{23} \\ a_{13} - ib_{13} & a_{23} - ib_{23} & a_{33} \end{bmatrix}$$

9. $\lambda^4 - 7\lambda^3 - 122\lambda^2 - 279\lambda - 57$

10. The polynomial has four distinct real zeros, the largest of which is greater than 6.

11. $X_{20} = [0.5935 \ \ 0.9397 \ \ 0.9744 \ \ 1]^T$. $\quad AX_{20} \approx 15.8328X_{20}$

14. **a.** $\det(\lambda I - A) = \lambda^3 - 2\lambda^2 - 5\lambda + 6$,
$\det(\lambda I - B) = \lambda^4 - 10\lambda^3 + 15\lambda^2 + 34\lambda - 36$

3.2 ■ Exercise Set

2. characteristic polynomial: $c(x) = (x - 2)^3(x - 3)^2$, $(x - 5)^3(x - 6)(x - 7)$, $(x - 3)(x - 4)^5$
minimum polynomial: $m(x) = (x - 2)^3(x - 3)^2$, $(x - 5)^2(x - 6)(x - 7)$, $(x - 3)(x - 4)^2$

3. $H^H = (I - 2XX^H)^H = I^H - 2X^{H^H}X^H = I - 2XX^H = H$; therefore, H is Hermitian. $H^HH = H^2 = I - 4XX^H + 4X(X^HX)X^H = I - 4XX^H + 4XX^H = I$; therefore H is unitary.

5. **c.** If U is upper triangular, then $\text{ent}_{11}(UU^H) = \sum_{j=1}^n u_{1j}\overline{u_{1j}} = \sum_{j=1}^n |u_{1j}|^2$ and $\text{ent}_{11}(U^HU) = |u_{11}|^2$. Since U is normal, $\sum_{j=2}^n |u_{1j}|^2 = 0$ so that $u_{12} = u_{13} = \cdots = u_{1n} = 0$. Similarly, $\text{ent}_{22}(UU^H) = \text{ent}_{22}(U^HU) \Rightarrow u_{23} = u_{24} = \cdots = u_{2n} = 0$. Continue this argument or use induction.

7. Suppose that A is positive (semi) definite and $AX = \lambda X$ with $X \neq 0$.

Then $X^TAX = X^T\lambda X = \lambda X^TX$, so $\lambda = \dfrac{X^TAX}{X^TX} > (\geq)0$.

Conversely, if A is symmetric, then, by Theorem 3.5 there exists an orthonormal basis for $\mathcal{R}^n$ consisting of eigenvectors of A. Suppose that $AX_i = \lambda_iX_i$ with $X_i^TX_j = 0$ for $i \neq j$. For any X in $\mathcal{R}^n$,

$X = a_1X_1 + a_2X_2 + \cdots + a_nX_n$ and $X^TAX = (\sum_{i=1}^n a_iX_i^T)(\sum_{j=1}^n a_j\lambda_jX_j) = \sum_{i=1}^n \sum_{j=1}^n a_ia_j\lambda_jX_i^TX_j = \sum_{i=1}^n a_i^2\lambda_iX_i^TX_i = \sum_{i=1}^n a_i^2\lambda_i > 0$, therefore, A is positive definite.

9. a. If $a_0 \neq 0$, then $A(A^{n-1} + a_{n-1}A^{n-2} + \cdots + a_1I) = -a_0I$, which implies that $A^{-1} = -\dfrac{1}{a_0}(A^{n-1} + a_{n-1}A^{n-2} + \cdots + a_1I)$.

b. $AX = \lambda X \Rightarrow X = \lambda A^{-1}X \Rightarrow A^{-1}X = \dfrac{1}{\lambda}X$.

11. For the unit vector $Y = \dfrac{1}{\|X\|_2}X$, $Y^TAY = \dfrac{X^TAX}{\|X\|_2^2} = \dfrac{X^TAX}{X^TX}$. Thus $\left\{\dfrac{X^TAX}{X^TX}: X \neq 0\right\} = \{Y^TAY: \|Y\|_2 = 1\}$.

12. $P = \begin{bmatrix} 1 & 1 \\ 0 & \varepsilon \end{bmatrix}$. It is true that $\begin{bmatrix} \lambda & 1 & 0 \\ 0 & \lambda & 1 \\ 0 & 0 & \lambda \end{bmatrix}$ is similar to $\begin{bmatrix} \lambda & \varepsilon & 0 \\ 0 & \lambda & \varepsilon \\ 0 & 0 & \lambda \end{bmatrix}$, and so on.

13. $X = \begin{bmatrix} 13 & -87 \\ 3 & -8 \end{bmatrix}$

14. a. $k = \dfrac{a_{23}}{\lambda_1 - \lambda_2}$

3.3 ■ Exercise Set

1. (T)

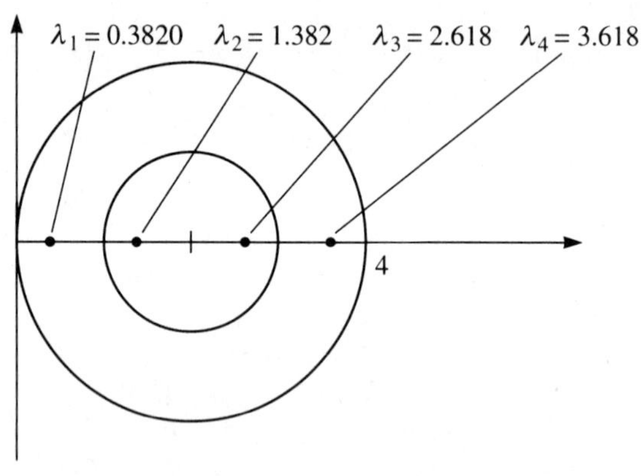

2. $0.997 \leq \lambda \leq 3.002$

3. $P(\varepsilon) = \begin{bmatrix} 1 & 1 \\ 0 & \varepsilon \end{bmatrix}$. $\lim_{\varepsilon \to 0} P(\varepsilon) = \begin{bmatrix} 1 & 1 \\ 0 & 0 \end{bmatrix}$, which is singular.

5. This follows from the fact that for a real polynomial, complex zeros occur in conjugate pairs.

6. $|\lambda_i - \tilde{\lambda}_i| \le \|P^{-1}\,\Delta A P\|_1 \le \|P\|_1\,\|P^{-1}\|_1\,\|\Delta A\|_1$.

7. For $P = \begin{bmatrix} -10 & -11 \\ 9 & 10 \end{bmatrix}$, $P^{-1}AP = \mathrm{Dg}\{2, 1\}$. $\mathrm{cond}_2(P) = 402$. New eigenvalues are $1 \pm \sqrt{99}\,i$.

9.

Eigenvalues of B	Eigenvalues of $B + \Delta B$	Relative Change
1	1.000049996	0.000050
3	1.999950004	0.000017
2	2	0
3.732050808	3.731964206	0.000023
0.267949192	0.268035794	0.000323

11. If $AY = \lambda Y$, then $\dfrac{\|AY\|}{\|Y\|} = \dfrac{\|\lambda Y\|}{\|Y\|} = |\lambda|$. Since $\|A\| = \max\left\{\dfrac{\|AX\|}{\|X\|}; X \neq 0\right\}$, $\|A\| \ge |\lambda| \Rightarrow \|A\| \ge \rho(A)$.

14. For $X^{-1}BX = \begin{bmatrix} 1 & \frac{5}{6} & \frac{15}{14} \\ \frac{3}{10} & 4 & \frac{9}{14} \\ \frac{21}{60} & \frac{7}{18} & 6 \end{bmatrix}$, the Gerschgorin circles do not overlap.

3.4 ■ Exercise Set

1. a. The dominant eigenvalue is 12 with eigenvector $[1\ \ -1\ \ 1\ \ 1]^T$.

b. The dominant eigenvalue of $A - 10I$ is likely to be complex.

c. The dominant eigenvalue of $(A - 3I)^{-1}$ is -1 with eigenvector $[1\ \ 1\ \ -1\ \ 1]^T$. Thus $\lambda = 2$ is an eigenvalue of A for the same eigenvector.

3. (A) The dominant eigenvalue is 3 with eigenvector $[1\ \ 1\ \ -\frac{1}{2}\ \ -\frac{1}{2}]^T$.

(C) The sequence oscillates wildly, so the dominant eigenvalue is most likely complex.

5. The dominant eigenvalue of $A - \mu_1 I$ is the eigenvalue of A farthest from μ_1.

7. Compute $(A - \lambda_1 X_1 X_1^T)X_i$ using the orthogonality of the eigenvectors of A.

8. a.

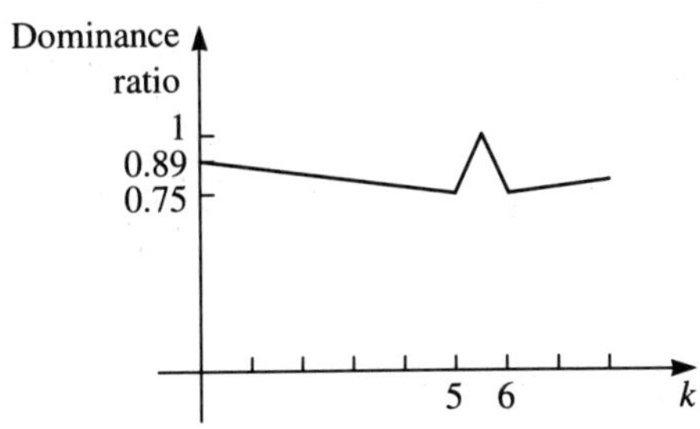

10. Use Exercise 12 in Section 3.3 and the fact that $\text{cond}_2(P) = 1$ if A is symmetric.

12. With $X_0 = [1, i, 1, -i]^T$, $X_{20} = [1, 1, 0.25, -0.25i, -0.5]^T$ is an eigenvector of $C - kI$ for $\lambda = -4i$. The corresponding eigenvalue of C is $-4i + k = 2 - 3i$.

13. **a.** $\lambda^4 - 7\lambda^3 - 122\lambda^2 - 279\lambda - 57$

b. About $4n^3/3$.

3.5 ■ Exercise Set

1. $\lambda = 2$, $X = [-1 \ -1 \ 1 \ 1]^T$

2. The eigenvalues of A are 10.5920, −2.3663, −0.4170, 0.1913.

5. **a.** The dominant eigenvalue is real and positive.

b. The dominant eigenvalue of $(A - 2I)^{-1}$ is negative, so there is an eigenvalue of A, near 2 but less than 2.

c. The dominant eigenvalue of $(A + 2I)^{-1}$ is complex, so the eigenvalues of A nearest −2 are complex.

6. The eigenvalues of A are 13.6241, −1.5414, 4.5414, −4.6241.

7. **a.** $2.9347 \pm 2.7369i$, -0.8694.

c. 1.9691, $-0.7482 \pm 1.0804i$, -0.6864, 0.9638

CHAPTER 4

4.1 ■ Exercise Set

3. $w = \dfrac{1}{\sqrt{26}}\begin{bmatrix} -1 \\ 3 \\ 4 \end{bmatrix}$, $H = \dfrac{1}{26}\begin{bmatrix} 24 & 6 & 8 \\ 6 & 8 & -24 \\ 8 & -24 & -6 \end{bmatrix}$,

$$HAH^{-1} = \begin{bmatrix} 14.59 & -2.468 & 0.3254 \\ -1.391 & 2.249 & 2.485 \\ -1.982 & 3.716 & 3.160 \end{bmatrix}$$

4. $w = \dfrac{1}{\sqrt{20}}\begin{bmatrix} 0 \\ -2 \\ 4 \end{bmatrix}, H = \dfrac{1}{5}\begin{bmatrix} 5 & 0 & 0 \\ 0 & 3 & 4 \\ 0 & 4 & -3 \end{bmatrix}, HAH^{-1} = \begin{bmatrix} 12 & 3.8 & -1.6 \\ 5 & 8.4 & -1.8 \\ 0 & 0.2 & -0.4 \end{bmatrix}$

5. $L = \begin{bmatrix} 1 & 0 & 0 & 0 \\ 2 & 1 & 0 & 0 \\ -3 & 0 & 1 & 0 \\ -\frac{1}{2} & 0 & 0 & 1 \end{bmatrix}, LAL^{-1} = \begin{bmatrix} 20 & 1 & 6 & 4 \\ 32 & 7 & 14 & 8 \\ -48 & -1 & -15 & -11 \\ -2 & 1.5 & 0 & 2 \end{bmatrix}$

6. $L = \begin{bmatrix} 1 & 0 & 0 & 0 \\ 0 & 1 & 0 & 0 \\ 0 & \frac{3}{2} & 1 & 0 \\ 0 & \frac{1}{4} & 0 & 1 \end{bmatrix}, LAL^{-1} = \begin{bmatrix} 0 & -9 & 6 & 4 \\ -4 & 2 & 2 & 0 \\ 0 & 0.25 & 6 & 1 \\ 0 & -3 & 3.5 & 4 \end{bmatrix}$

9. a. $R_1 = \begin{bmatrix} 0.7415 & -0.6710 & 0 \\ 0.6710 & 0.7415 & 0 \\ 0 & 0 & 1 \end{bmatrix}, A^{(1)} = \begin{bmatrix} -2.5249 & 0 & 4.3078 \\ 0 & 7.5249 & -1.2011 \\ 4.3078 & -1.2011 & -3 \end{bmatrix}$

c. $A^{(3)} = \begin{bmatrix} 1.5519 & -0.8241 & -0.0491 \\ -0.8241 & 7.5769 & 0 \\ -0.0491 & 0 & -7.1288 \end{bmatrix}$

4.2 ■ Exercise Set

5. b. $A_1 = \begin{bmatrix} 14.3000 & -2.9850 & -2.6579 & 3.5827 \\ -2.9850 & -2.6636 & 1.2899 & 2.1604 \\ -2.6579 & 1.2899 & -2.1030 & -0.3819 \\ 3.5827 & 2.1604 & -0.3819 & -2.5333 \end{bmatrix}$

$$A = QR \Rightarrow Q^TA = R \Rightarrow A_1 = RQ = Q^TAQ$$

d. $A_2 = Q_1^TA_1Q_1 = Q_1^T(Q^TAQ)Q_1 = (QQ_1)^TA(QQ_1)$

$A_3 = Q_2^TA_2Q_2 = (QQ_1Q_2)^TA(QQ_1Q_2)$, and so on.

e. $A_6 = \begin{bmatrix} 15.8328 & 0.0331 & 0.0006 & 0.0000 \\ 0.0331 & -5.9209 & -0.0082 & -0.0000 \\ 0.0006 & -0.0082 & -2.6854 & -0.0000 \\ 0.0000 & -0.0000 & -0.0000 & -0.2264 \end{bmatrix}$ is almost upper triangular.

6. $A = \begin{bmatrix} 1 & 1 & 3 \\ 0 & 1 & 0 \\ 1 & 1 & 4 \\ 0 & 1 & 12 \end{bmatrix} = QR$

$$= \begin{bmatrix} -0.7071 & 0 & 0.0587 & 0.7047 \\ 0 & -0.7071 & 0.7047 & -0.0587 \\ -0.7071 & 0 & -0.0587 & -0.7047 \\ 0 & -0.7071 & -0.7047 & 0.0587 \end{bmatrix} \begin{bmatrix} -1.4142 & -1.4142 & -4.9497 \\ 0 & -1.4142 & -0.84853 \\ 0 & 0 & -8.5147 \\ \hdashline 0 & 0 & 0 \end{bmatrix}$$

$$= \begin{bmatrix} -0.7071 & 0 & 0.0587 \\ 0 & -0.7071 & 0.7047 \\ -0.7071 & 0 & -0.0587 \\ 0 & -0.7071 & -0.7047 \end{bmatrix} \begin{bmatrix} -1.4142 & -1.4142 & -4.9497 \\ 0 & -1.4142 & -0.84853 \\ 0 & 0 & -8.5147 \end{bmatrix} = Q_1 R_1.$$

The columns of Q_1 are the desired orthonormal basis.

8. The columns of the matrix $\begin{bmatrix} -0.2308 & -0.6697 & 0.0415 \\ 0 & 0 & 0.9983 \\ -0.3077 & -0.6380 & -0.0415 \\ -0.9231 & 0.3801 & 0.0035 \end{bmatrix}$ are the desired basis.

9. $X = [1.9123 \ \ -1.4035 \ \ 0.0877 \ \ 0.1579]^T$

10. a. $0 = X_1^T X_2 = aX_1^T X_1 + X_1^T Y_2 \Rightarrow a = -\dfrac{X_1^T Y_2}{X_1^T X_1}$.

b. $0 = X_i^T X_{k+1} = a_i X_i^T X_i + X_i^T Y_{k+1} \Rightarrow a_i = -\dfrac{X_i^T Y_{k+1}}{X_i^T X_i}$.

4.3 ■ Exercise Set

1. $RQ = \begin{bmatrix} 15.0000 & 0.0769 & -5.1534 & -4.2938 \\ 10.4000 & 9.1065 & -3.2471 & -25.3741 \\ 0 & 16.5701 & 7.3019 & -6.9350 \\ 0 & 0 & -0.0888 & 1.5916 \end{bmatrix}$

4. $H = \begin{bmatrix} 0 & 25 & 0 & 0 \\ 25 & 0 & 625 & 0 \\ 0 & 625 & 0 & 125 \\ 0 & 0 & 125 & 0 \end{bmatrix}$

5. $H = \begin{bmatrix} 0 & 25 & 0 & 0 \\ 25 & -3.8400 & 630.1095 & -3.6462 \\ 0 & 622.1304 & 5.2587 & 124.9696 \\ 0 & 0 & 120.1696 & -1.4187 \end{bmatrix}$

7. About $2n^3/3$ multiplications.

8. By looking at the subdiagonal of $H - \lambda I$, it is clear that $\text{rank}(H - \lambda I) = n - 1$. Thus the dimension of the associated eigenspace is 1. In the Jordan form there is only one block for each eigenvalue.

9. The result follows from Exercise 8 in this section.

4.4 ■ Exercise Set

2. $A_5 = \begin{bmatrix} 3.5243 & -0.2922 & 0.0000 & 0.0000 \\ -0.2922 & 2.7009 & -0.1180 & -0.0000 \\ 0 & -0.1180 & 1.3928 & 0.0026 \\ 0 & 0 & 0.0026 & 0.3820 \end{bmatrix}$

3. $A_7 = \left[\begin{array}{cc|cc|cc} 10.0001 & 5.0001 & -3.4641 & -1.7321 & -14.6969 & 7.3484 \\ -5.0001 & 10.0001 & 1.7321 & -3.4641 & 7.3484 & 14.6969 \\ \hline 0.0001 & 0.0000 & 4.0000 & 2.0000 & -2.8285 & 1.4143 \\ -0.0000 & 0.0001 & -2.0000 & 4.0000 & 1.4143 & 2.8285 \\ \hline 0.0000 & 0.0000 & -0.0000 & -0.0000 & 1.9999 & -1.0000 \\ 0.0000 & -0.0000 & -0.0000 & 0.0000 & 1.0000 & 1.9999 \end{array}\right]$

The eigenvalues are approximately $10 \pm 5i$, $4 \pm 2i$, $2 \pm i$.

4. $A_8 = \text{Dg}\{-6.8776, -2.1782, 8.7667, 5.4329, 4.8562\}$

6. c. Using the first shift strategy, the convergence is quite slow. After nine iterations, $H(12, 11) \approx 0$, and we find the eigenvalue $\lambda = -5$. After 30 iterations, $H(10, 9) \approx 0$, and we find the eigenvalues $4 \pm 3i$. After 40 iterations, $H(8, 7) \approx 0$, and we find the eigenvalues $3 \pm 4i$.

8. b. After 20 iterations of Algorithm (4.4) on $B = C_1 + 2I$, $B(3, 2) \approx 0$

and the eigenvalues $2.7071 \pm 0.7071i$ and $1.2929 \pm 0.7071i$ are found for B. The corresponding eigenvalues of C_1 are $0.7071 \pm 0.7071i$ and $-0.7071 \pm 0.7071i$, which are the fourth roots of -1.

4.5 ■ Exercise Set

1. $$\begin{bmatrix} 1.0000 & 0 & 0 & 0 \\ 0 & 0.6667 & -0.6800 & -0.3053 \\ 0 & 0.6667 & 0.3608 & 0.6522 \\ 0 & 0.3333 & 0.6383 & -0.6938 \end{bmatrix} = Q$$

CHAPTER 5

5.1 ■ Exercise Set

1. a. $\sigma_i = |\lambda_i| = \sqrt{\lambda_i^2}$.

b. The symmetric matrix $Q^TBQ = Q^{-1}BQ = (PR^{-1})^{-1}B(PR^{-1}) = R(P^{-1}BP)R^{-1} = RDR^{-1}$ is also upper triangular and hence diagonal.

2. $$U^TAV = \begin{bmatrix} -0.214837 & -0.887231 & 0.408248 \\ -0.520587 & -0.249644 & -0.816497 \\ -0.826338 & 0.387943 & 0.408248 \end{bmatrix}^T \begin{bmatrix} 1 & 2 & 3 \\ 4 & 5 & 6 \\ 7 & 8 & 9 \end{bmatrix}$$

$$\times \begin{bmatrix} -0.479671 & 0.776691 & -0.408248 \\ -0.572368 & 0.0756866 & 0.816496 \\ -0.665065 & -0.625318 & -0.408248 \end{bmatrix}$$

$$= \text{Dg}\{16.8481, 1.06837, 0.432418 \times 10^{-7}\}$$

3. $$U = \begin{bmatrix} -0.347140 & -0.566937 & 0.556322 & 0.486495 \\ -0.372713 & -0.495456 & -0.557736 & -0.0831902 \\ -0.437183 & -0.155326 & -0.262657 & -0.363899 \\ -0.514619 & 0.274353 & 0.502134 & -0.548285 \\ -0.533495 & 0.577677 & -0.241473 & 0.568650 \end{bmatrix},$$

$$V = \begin{bmatrix} -0.109009 & 0.626652 & -0.571063 & -0.518952 \\ -0.277577 & 0.042648 & -0.588902 & 0.757843 \\ -0.437989 & -0.725867 & -0.354159 & -0.394784 \\ -0.848076 & 0.280367 & 0.449057 & 0.0225467 \end{bmatrix}$$

$U^TAV = \text{Dg}\{65.4082, 3.13087, 0.956027, 0.234068\}$

4. $\|A\|_2 = \|U\Sigma V^T\| = \|\Sigma\|_2 = \sigma_1 = \sqrt{\rho(AA^T)}$.

5. $\text{cond}_2(A) = \|A\|_2 \, \|A^{-1}\|_2 = \sigma_1 \dfrac{1}{\sigma_n}$.

7. $\mathcal{NS}(A) = \text{Basis}\{\text{Col}_3(V)\}$, $\mathcal{NS}(A)^\perp = \text{Basis}\{\text{Col}_1(V), \text{Col}_2(V)\}$
$\mathcal{CS}(A) = \text{Basis}\{\text{Col}_1(U), \text{Col}_2(U)\}$, $\mathcal{CS}(A)^\perp = \text{Basis}\{\text{Col}_3(U)\}$.

9. $(PAQ)(PAQ)^T = PA(QQ^T)A^TP^T = P(AA^T)P^T$ are similar and hence have the same eigenvalues. It follows that A and PAQ have the same singular values.

5.2 ■ Exercise Set

2. $y = 3.1714x + 1.400$

3. $z = 7.4 + 3x - 5.5y$

6. $X = [0.8333 \;\; 0.3333 \;\; -0.1667]^T$

8. $X = [5.5069 \;\; -6.7927 \;\; -2.0439 \;\; 3.6066]^T$

5.3 ■ Exercise Set

1. Roots of $x^2 - 4.002x + 12.9989$, which are $2.001 \pm 2.999i$

4. $y = 10^9(-3.44958 + 0.00185190x)$; for $x = 1990$, $y = 235{,}697{,}000$.

7. Using the last set of coefficients obtained in the text, we have pop(1991) $\approx$ 257,959,000 and pop(2000) $\approx$ 285,167,000.

Using the additional data and solving by the **SVD** method (MATLAB) yields pop(1991) $\approx$ 253,879,000 and pop(2000) $\approx$ 279,468,000.

CHAPTER 6

6.1 Exercise Set

1. a. $M = \begin{bmatrix} -1 & 1 & 0 \\ 1 & -2 & 1 \\ 0 & 1 & -1 \end{bmatrix}$, $\rho(M) = 3$

b. $M = \begin{bmatrix} 0 & \frac{1}{2} & 0 \\ \frac{1}{3} & 0 & \frac{1}{3} \\ 0 & \frac{1}{2} & 0 \end{bmatrix}$, $\rho(M) = 0.5774$

c. $\begin{bmatrix} 0 & \frac{1}{2} & 0 \\ 0 & \frac{1}{6} & \frac{1}{3} \\ 0 & \frac{1}{12} & \frac{1}{6} \end{bmatrix}$, $\rho(M) = 0.3333$

d. $M = \begin{bmatrix} 0.5 & 0.25 & 0 \\ 0.0833 & 0.5417 & 0.1667 \\ 0.0208 & 0.1354 & 0.5417 \end{bmatrix}$, $\rho(M) = 0.75$

2. c. $M = \begin{bmatrix} 0 & \frac{1}{4} & \frac{1}{4} & 0 \\ 0 & \frac{1}{16} & \frac{1}{16} & \frac{1}{4} \\ 0 & \frac{1}{16} & \frac{1}{16} & \frac{1}{4} \\ 0 & \frac{1}{32} & \frac{1}{32} & \frac{1}{8} \end{bmatrix}$, $\rho(M) = 0.25$

d. $M = \begin{bmatrix} -0.5 & 1.5 & 1.5 & 0 \\ -7.5 & -2.75 & 2.25 & 1.5 \\ -7.5 & 2.25 & -2.75 & 1.5 \\ -22.5 & -0.75 & -0.75 & -0.5 \end{bmatrix}$, $\rho(M) = 5$

3. a. By Exercise 5 in Section 3.2, B is unitarily similar to a diagonal matrix. If $P^H BP = D$, then $\|B\|_2 = \|PDP^H\|_2 = \|D\|_2 = \rho(D) = \rho(B)$.

5. $\lim_{n\to\infty} T^n = \lim_{n\to\infty} \begin{bmatrix} \frac{1}{2^n} & \frac{n}{2^{n-1}} \\ 0 & \frac{1}{2^n} \end{bmatrix} = 0.$

7. $X_n = X_k$ for all $n > k$.

9. $\rho(A) \geq 0$ for all A, but $\rho(A) = 0$ does not imply $A = 0$. $\rho(kA) = |k|\,\rho(A)$, but in general it is not true that $\rho(A + B) \leq \rho(A) + \rho(B)$ and $\rho(AB) \leq \rho(A)\rho(B)$.

6.2 ■ Exercise Set

1. Gerschgorin says that $\rho(J) \leq 1$ and $\rho(\mathscr{L}) \leq 0.8125$,
$\Lambda(J) = \{0, 0, 0, \pm 0.3536, \pm 0.3536, \pm 0.7071\}$,
$\Lambda(\mathscr{L}) = \{0, 0, 0, 0, 0, 0.5, 0.125, 0.125, +8.62 \times 10^{-10}\}$.

2. **a.** The exact solution, correct to 16 places, is $U = \begin{bmatrix} 7.142857142857144 \\ 9.821428571428573 \\ 7.142857142857143 \\ 18.75000000000000 \\ 25.00000000000000 \\ 18.75000000000000 \\ 42.85714285714286 \\ 52.67857142857143 \\ 42.85714285714286 \end{bmatrix}$

b. Twenty iterations of Gauss-Seidel yield $X_{20} = \begin{bmatrix} 7.142799326351711 \\ 9.821370754923139 \\ 7.142828234604426 \\ 18.74994218349457 \\ 24.99994218349457 \\ 18.74997109174728 \\ 42.85711394889015 \\ 52.67854252031871 \\ 42.85712840301650 \end{bmatrix}$

3. $J = \dfrac{1}{2}\begin{bmatrix} 0 & 1 & 0 & 0 & 0 \\ 1 & 0 & 1 & 0 & 0 \\ 0 & 1 & 0 & 1 & 0 \\ 0 & 0 & 1 & 0 & 1 \\ 0 & 0 & 0 & 1 & 0 \end{bmatrix}$, $\rho(J) = 0.8660$

$$\mathcal{L} = \frac{1}{32}\begin{bmatrix} 0 & 16 & 0 & 0 & 0 \\ 0 & 8 & 16 & 0 & 0 \\ 0 & 4 & 8 & 16 & 0 \\ 0 & 2 & 4 & 8 & 16 \\ 0 & 1 & 2 & 4 & 8 \end{bmatrix}, \ \rho(\mathcal{L}) = 0.75$$

4. The exact solution is $X = (1/6)[23\ 29\ 31\ 37]^T$.

6. $$J = \frac{1}{4}\left[\begin{array}{ccccc:cccc} 0 & 0 & 0 & 0 & 0 & 1 & 1 & 0 & 0 \\ 0 & 0 & 0 & 0 & 0 & 1 & 0 & 1 & 0 \\ 0 & 0 & 0 & 0 & 0 & 1 & 1 & 1 & 1 \\ 0 & 0 & 0 & 0 & 0 & 0 & 1 & 0 & 1 \\ 0 & 0 & 0 & 0 & 0 & 0 & 0 & 1 & 1 \\ \hdashline 1 & 1 & 1 & 0 & 0 & 0 & 0 & 0 & 0 \\ 1 & 0 & 1 & 1 & 0 & 0 & 0 & 0 & 0 \\ 0 & 1 & 1 & 0 & 1 & 0 & 0 & 0 & 0 \\ 0 & 0 & 1 & 1 & 1 & 0 & 0 & 0 & 0 \end{array}\right], \ \rho(J) = 0.7071$$

$$\mathcal{L} = \begin{bmatrix} 0 & 0 & 0 & 0 & 0 & 0.2500 & 0.2500 & 0 & 0 \\ 0 & 0 & 0 & 0 & 0 & 0.2500 & 0 & 0.2500 & 0 \\ 0 & 0 & 0 & 0 & 0 & 0.2500 & 0.2500 & 0.2500 & 0.2500 \\ 0 & 0 & 0 & 0 & 0 & 0 & 0.2500 & 0 & 0.2500 \\ 0 & 0 & 0 & 0 & 0 & 0 & 0 & 0.2500 & 0.2500 \\ 0 & 0 & 0 & 0 & 0 & 0.1875 & 0.1250 & 0.1250 & 0.0625 \\ 0 & 0 & 0 & 0 & 0 & 0.1250 & 0.1875 & 0.0625 & 0.1250 \\ 0 & 0 & 0 & 0 & 0 & 0.1250 & 0.0625 & 0.1875 & 0.1250 \\ 0 & 0 & 0 & 0 & 0 & 0.0625 & 0.1250 & 0.1250 & 0.1875 \end{bmatrix}, \ \rho(\mathcal{L}) = 0.5.$$

8. With $X_0 = [1\ 1\ 1\ 1\ 1\ 1]^T$, $X_{20} = [1.8590\ 1.4646\ 1.9716\ 2.0280\ 2.5351\ 2.1408]^T$.

6.3 ■ Exercise Set

1. Matrix A:

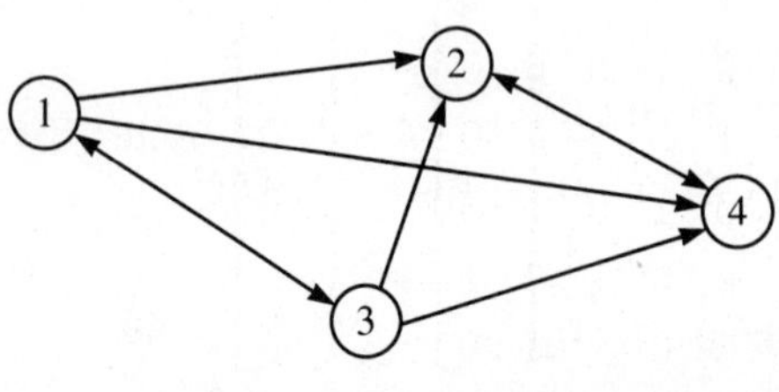

The matrix A is reducible, there is no path from node 4 to node 1 or node 3.

Matrix B:

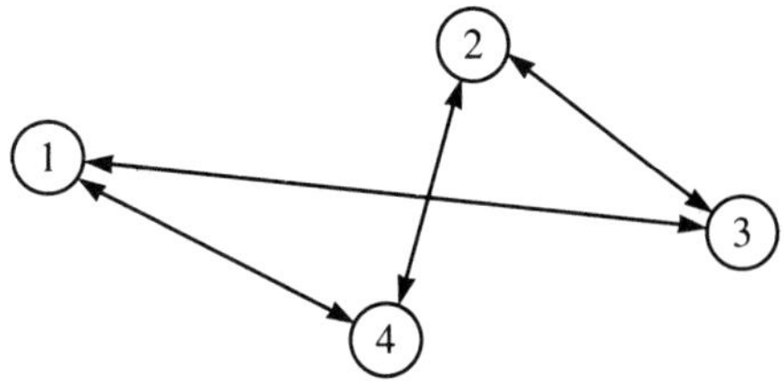

The matrix B is irreducible.

2. For $X^{-1}BX$, the first circle is larger, but the second and third are smaller and do not overlap as they do for B.

3. $\|\mathscr{L}\|_\infty = 0.8125$, $\|\mathscr{L}\|_1 = 0.8672$, $\|\mathscr{L}\|_{A^{1/2}} = 0.5976$

4. $$B^{1/2} = \begin{bmatrix} 1.3585 & -0.3871 & -0.0632 & -0.0211 & -0.0075 \\ -0.3871 & 1.2954 & -0.4082 & -0.0706 & -0.0211 \\ -0.0632 & -0.4082 & 1.2879 & -0.4082 & -0.0632 \\ -0.0211 & -0.0706 & -0.4082 & 1.2954 & -0.3871 \\ -0.0075 & -0.0211 & -0.0632 & -0.3871 & 1.3585 \end{bmatrix}$$

6. $$J = \begin{bmatrix} 0 & 0.5000 & -0.1667 & 0 \\ -0.2857 & 0 & 0.8571 & -0.7143 \\ 0.5000 & 0.5000 & 0 & 2.0000 \\ -0.1250 & -0.1250 & 0.7500 & 0 \end{bmatrix},$$

$$\mathscr{L} = \begin{bmatrix} 0 & 0.5000 & -0.1667 & 0 \\ 0 & -0.1429 & 0.9048 & -0.7143 \\ 0 & 0.1786 & 0.3690 & 1.6429 \\ 0 & 0.0893 & 0.1845 & 1.3214 \end{bmatrix}$$

$\rho(J) = 1.3733$, $\rho(\mathscr{L}) = 1.6025$

7. $$T = \frac{1}{2}\begin{bmatrix} 0 & 1 & 0 & 0 & 0 & 0 & 0 \\ 1 & 0 & 1 & 0 & 0 & 0 & 0 \\ 0 & 1 & 0 & 1 & 0 & 0 & 0 \\ 0 & 0 & 1 & 0 & 1 & 0 & 0 \\ 0 & 0 & 0 & 1 & 0 & 1 & 0 \\ 0 & 0 & 0 & 0 & 1 & 0 & 1 \\ 0 & 0 & 0 & 0 & 0 & 1 & 0 \end{bmatrix} T + \begin{bmatrix} 0 \\ 0 \\ 0 \\ 0 \\ 0 \\ 0 \\ 50 \end{bmatrix}.$$ After 60 iterations of Gauss-Seidel, we obtain $T = [12.4970\ 24.9950\ 37.4939\ 49.9939\ 62.4948\ 74.9963\ 87.4982]^T$.

6.4 ■ Exercise Set

1. If $A = I - L - R$, where L is strictly lower triangular, and R is strictly upper triangular, then $Q = (\omega^{-1}I - L)$ is the SOR splitting matrix.

2. $\omega_b = 1.0557$

3. $\omega_b = 1.0775$

4.
$$\mathcal{L}_\omega = \frac{1}{2}\begin{bmatrix} 2(1-\omega) & \omega & 0 & 0 & 0 \\ \omega(1-\omega) & \frac{\omega^2}{2} + 2(1-\omega) & \omega & 0 & 0 \\ \frac{\omega^2}{2}(1-\omega) & \frac{\omega^3}{4} + \omega(1-\omega) & \frac{\omega^2}{2} + 2(1-\omega) & \omega & 0 \\ \frac{\omega^3}{4}(1-\omega) & \frac{\omega^4}{8} + \frac{\omega^2}{2}(1-\omega) & \frac{\omega^3}{4} + \omega(1-\omega) & \frac{\omega^2}{2} + 2(1-\omega) & \omega \\ \frac{\omega^4}{8}(1-\omega) & \frac{\omega^5}{16} + \frac{\omega^3}{4}(1-\omega) & \frac{\omega^4}{8} + \frac{\omega^2}{2}(1-\omega) & \frac{\omega^3}{4} + \omega(1-\omega) & \frac{\omega^2}{2} + 2(1-\omega) \end{bmatrix}$$

5. $X_5 = [6.9675\ 9.6221\ 7.0737\ 18.5700\ 24.8688\ 18.7085\ 42.7986\ 52.6405\ 42.8460]^T$

6. $\omega_b \approx 1.1714$

6.5 ■ Exercise Set

1. $A = LL^T \Rightarrow L^{-1}A(L^T)^{-1} = I \Rightarrow \text{Row}_i(L^{-1})\,\text{Col}_j(A(L^T)^{-1})$

$$= \langle \text{Col}_i((L^T)^{-1}), A\,\text{Col}_j((L^T)^{-1})\rangle = \begin{cases} 0 & \text{if } i \neq j \\ 1 & \text{if } i = j \end{cases}$$

4. After entering A(m,m), B, X, m, the following MATLAB commands will implement Algorithm 6.10.

```
R=B-A*X, V=R, C=R'*R
for k=0:m-1
     Z=A*V, t=c/(V'*Z), X=X+t*V
     R=R-t*Z,d=R'*R,N(k+1)=sqrt(d),V=R+(d/c)*V,c=d
end
```

6. $X = [1.8592\ \ 1.4648\ \ 1.9718\ \ 2.0282\ \ 2.5352\ \ 2.1408]^T$

7. $n^2 + 6n + 2$

Bibliography

Coleman, T. F., and van Loan, C. F. 1988. *Handbook for Matrix Computations.* Philadelphia: SIAM.

Cullen, C. G. 1972. *Matrices and Linear Transformations.* 2nd ed. New York: Dover Publications Inc.

Dongarra, J. J., Moler, C. B., Bunch, J. R., and Stewart, G. W. 1979. *LINPACK Users Guide.* Philadelphia: SIAM.

Forsythe, G. E., Malcolm, M. A., and Moler, C. B. 1977. *Computer Methods for Mathematical Computations.* Englewood Cliffs, NJ: Prentice Hall.

Golub, G. H., and van Loan, C. F. 1989. *Matrix Computations.* 2nd ed. Baltimore: The Johns Hopkins University Press.

Halmos, P. R. 1974. *Finite-Dimensional Vector Spaces.* New York: Springer-Verlag New York Inc.

Kincaid, D., and Cheney, W. 1991. *Numerical Analysis.* Pacific Grove, CA: Brooks/Cole Publishing Co.

Kolman, B. and Hill, D. 1993. *Introductory Linear Algebra with Applications. 5th ed.* New York: Macmillan Inc.

The Math Works. 1992. *The Student Edition of MATLAB.* Englewood Cliffs, NJ: Prentice Hall.

Parlett, B. N. 1967. Global convergence of the basic QR algorithm on Hessenberg matrices. *Mathematics of Computation* 21: 223–29.

Smith, B. T., Boyle, J. M., Garbow, B. S., Ikebe, Y., Klema, V. C., and Moler, C. B. 1976. *Matrix Eigensystems Routines—EISPACK Guide.* 2nd ed. New York: Springer-Verlag Inc.

Stewart, G. W. 1973. *Introduction to Matrix Computations.* New York: Academic Press Inc.

Varga, R. S. 1962. *Matrix Iterative Analysis.* Englewood Cliffs, NJ: Prentice Hall.

Wilkinson, J. H. 1965. *The Algebraic Eigenvalue Problem.* Oxford: Clarendon Press.

Wolfram, S. 1988. *Mathematica, a System for Doing Mathematics by Computer.* Reading, MA: Addison-Wesley Publishing Co Inc.

Young, D. M. 1971. *Iterative Solutions of Large Linear Systems.* New York: Academic Press Inc.

Young, D. M., and Hageman, L. 1981. *Applied Iterative Methods.* New York: Academic Press Inc.

Index

This package cannot be returned once this disk packaging has been opened or tampered with.